Aircraft Performance

Theory and practice

Martin E Eshelby
BSc, PhD, CEng, FRAeS, AFAIAA, MSFTE
College of Aeronautics
Cranfield University, UK

ELSEVIER

Elsevier Ltd
Linacre House, Jordan Hill, Oxford OX2 8DP
200 Wheeler Road, Burlington, MA 01803

Co-published in North America by
American Institute of Aeronautics and Astronautics, Inc.,
1801 Alexander Bell Drive,
Reston, VA 20191-4344
Transferred to digital printing 2004
© 2000 Martin E Eshelby

British Library Cataloguing in Publication Data
A catalogue record for this book is available from the British Library

Library of Congress Cataloging-in-Publication Data
A catalog record for this title is available from the Library of Congress

ISBN 0 340 75897 X

1 2 3 4 5 6 7 8 9 10

Commissioning Editor: Matthew Flynn
Production Editor: James Rabson
Production Controller: Priya Gohil
Cover Design: Terry Griffiths

Typeset in 10/12 Times by Academic & Technical Typesetting, Bristol

Contents

Preface   vii
Notation   xi
Abbreviations   xv

1. An introduction to the performance of fixed-wing aircraft   1
   1.1 Introduction   1
   1.2 The mission profile   2
   1.3 Performance estimation   5
   1.4 Performance measurement   6
   1.5 Operational performance   7
   1.6 Conclusions   8
   Bibliography   9

2. The atmosphere and air data measurement   10
   2.1 Introduction   10
   2.2 The characteristics of the atmosphere   10
   2.3 Vertical development of the atmosphere   12
   2.4 The standard atmosphere model   14
   2.5 Measurement of air data   24
   2.6 Practical considerations of air data measurement   34
   2.7 Air data computers   35
   2.8 Conclusions   36
   Bibliography   37

3. The force system of the aircraft and the equations of motion   39
   3.1 The equations of motion for performance   39
   3.2 The aircraft force system   40
   3.3 The aerodynamic force characteristics   42
   3.4 The propulsive forces   53
   3.5 Aerodynamic relationships   60
   3.6 Conclusions   61
   Bibliography   62

4. Cruising performance   63
   4.1 Introduction   63

4.2 Specific air range and specific endurance   64
4.3 Range and endurance for aircraft with thrust-producing engines   65
4.4 Range and endurance for aircraft with power-producing engines   79
4.5 Aircraft with mixed powerplants   81
4.6 Conclusions   82
Bibliography   83

5. Climb and descent performance   85
5.1 Introduction   85
5.2 Climb and descent performance analysis   88
5.3 Measurement of best climb performance   95
5.4 Climb performance in aircraft operations   96
5.5 Descent performance in aircraft operations   98
5.6 The effect of wind on climb and descent performance   102
5.7 High-performance climb   104
5.8 Conclusions   105
Bibliography   106

6. Take-off and landing performance   107
6.1 Introduction   107
6.2 Take-off performance   110
6.3 Landing performance   116
6.4 STOL and VTOL considerations   122
6.5 Conclusions   124
Bibliography   124

7. Aircraft manoeuvre performance   125
7.1 Introduction   125
7.2 The manoeuvre envelope   127
7.3 Aircraft manoeuvres   130
7.4 Transport aircraft manoeuvre performance   137
7.5 Military aircraft manoeuvre performance   137
7.6 Conclusions   138
Bibliography   139

8. Aircraft performance measurement and data handling   140
8.1 Introduction   140
8.2 Parametric performance data analysis   143
8.3 The equivalent-weight method   155
8.4 Performance data reduction   159
8.5 Conclusions   165
Bibliography   166

9. Scheduled performance   167
9.1 Introduction   167
9.2 Flight planning   169
9.3 Take-off performance   171

9.4 Take-off net flight path   181
9.5 En-route performance   187
9.6 Landing performance   189
9.7 Summary of performance planning   192
9.8 Fuel planning   192
9.9 Conclusions   196
Bibliography   197

10. The application of performance   198
     10.1 Introduction   198
     10.2 The performance summary   201
     10.3 Operational analysis   207
     10.4 Flight planning   213
     10.5 Conclusions   227
     Bibliography   227

11. Performance examples   228
     11.1 Introduction   228
     11.2 Aircraft characteristics   228
     11.3 Basic operating points determined by aerodynamic characteristics 231
     11.4 Estimation of flight path performance   234
     11.5 Payload–range diagram   254
     11.6 Fuel required for a specified mission   258
     11.7 Extension of the performance analysis   260

Appendix A: Reference axis systems   261

Appendix B: The performance equations of motion   268

Appendix C: The International Standard Atmosphere model   274

Index   282

Preface

An aircraft is only a viable vehicle if it has the performance necessary to carry out its mission safely; it is its performance that sells an aircraft.

Many textbooks have addressed the subject of aircraft performance; of these, a few have concentrated on performance only but most have been textbooks on aircraft aerodynamics that have included a section on flight-path related performance. Almost all these texts have been concerned with the estimation of the aerodynamic characteristics of the aircraft and using these characteristics in the calculation of the principal elements of performance; for example, cruising performance in terms of the distance travelled on a given load of fuel, the rate of climb and the take-off and landing distances required. In many cases the expressions developed are based on the optimum case only and do not consider performance under non-optimum conditions. Few texts address performance measurement or the implications of performance on the airworthiness of the aircraft.

In practice, the design process needs to consider the aircraft under practical operational conditions, which may not be the optimum, and to be able to assess the effect of design changes or modifications on the performance of the aircraft. This requires a much more flexible approach to the development of the performance expressions for design. The performance implications on the safety of the aircraft also need to be considered. The aircraft must have a certificate of airworthiness before it can be operated legally and part of the certification process is to demonstrate that the aircraft has sufficient performance to guarantee safe operation throughout its mission. The operational performance, which is based on the requirements of the airworthiness certification, limits the operation of the aircraft to ensure that it complies with the necessary safety regulations, this is known as *performance scheduling*.

The objective of this treatment of performance is to widen the topic beyond the estimation of the performance of the aircraft in its design stage. It extends to include the measurement of performance needed to validate the design estimates and to produce data for the certification of the aircraft and for the construction of the aircraft performance manual. It also covers the principles of performance scheduling and the practical considerations of operational performance.

This study of performance follows the aircraft through from its design to its operation as a practical vehicle.

The treatment starts with a study of the properties of the atmosphere, which can be regarded as the 'working fluid' of the aircraft. This extends to the definition and measurement of the air data relationships that are fundamental to performance. These are airspeed, Mach number, altitude and air temperature (Chapter 2 and Appendix C).

The aerodynamic characteristics of the aircraft and the propulsive characteristics of the powerplant need to be determined and to be expressed in forms that can readily be used to form the expressions for performance in the most adaptable manner (Chapter 3 and Appendices A and B). The performance expressions for the basic elements of the flight path, take-off, climb, cruise, descent, landing and manoeuvre, need to be developed in a form that will allow any operational mission to be analysed and evaluated and the effects of practical restrictions to be taken into account (Chapters 4–7).

Performance measurement is an essential part of the development of the aircraft, and this is carried out through the flight testing process. Since the performance estimation process is not exact, the achieved performance needs to be measured to verify the design estimates and to provide validated data for the airworthiness certification process and for the performance manual (Chapter 8).

The practical application of performance to the aircraft in operation is concerned with performance scheduling, which covers performance planning and fuel planning, and which together form part of the flight plan; this is a direct application of the airworthiness criteria to the operation in order to ensure flight safety (Chapter 9).

Performance data are required for many purposes beyond the design of the aircraft. These purposes include the marketing of the aircraft, the construction of the performance manual and certification of the aircraft, each of which requires a different approach to the presentation of the performance data (Chapter 10).

To illustrate the performance estimation process in the design of the aircraft, and to show how performance data are developed and applied to the operation of the aircraft, a series of examples is given (Chapter 11).

In any practical treatment of performance it is necessary to refer to various sets of regulations and requirements where the airworthiness of the aircraft is the main issue. In this study, several such documents will be referred to and used in the development of the topic. These documents are lengthy, complex and contain a great amount of detail that would be impossible to absorb into a textbook intended for an educational purpose. Extracts from the regulations and requirements will be used, often in a paraphrased form for clarity, as examples and illustrations of, or as criteria for, the topic under consideration. It must be remembered that, in all such cases, rules, regulations and requirements change from time to time in the light of improved knowledge and experience and that this study cannot guarantee to refer to the latest version. Therefore, this text should be regarded as a guide to the principles and must not be used as being definitive to any form of rule, regulation or requirement; the reader must refer to the latest version of the approved document if definitive information is required.

Only subsonic aircraft are considered in this study, although reference will be made to Mach number effects in the transonic region and incidental reference will be made to supersonic operation where it is necessary for completeness. The reason for this is that, with the exception of Concorde, most supersonic aircraft are military aircraft that rarely cruise supersonically and their performance consists mainly of unsteady, manoeuvring flight. Supersonic operation is a special case that requires a completely different approach to its analysis and does not follow logically along the performance path of conventional, subsonic aircraft.

In preparing the material for this book, I have accumulated a large amount of information from many sources, some of which is difficult to attribute or reference. It is characteristic of people who are involved with aircraft to talk about aircraft at almost any opportunity and to share their experiences with others in the same field. Over many years I have taken part in many conversations, discussions and arguments about aircraft from which I have gleaned gems of knowledge and snippets of information that would never have appeared in print, but which have made it possible to put this book together. My thanks are due to all those who have contributed to this book, wittingly or unwittingly (but always most generously), their knowledge, experience and opinions.

MEE

Notation

a	Speed of sound, m/s; acceleration, m/s^2
C	Specific fuel consumption, kg/N hr, kg/kW hr
C_D	Coefficient of drag
C_L	Coefficient of lift
C_p, C_v	Specific heat of air at constant pressure, constant volume
C_p	Coefficient of pressure
C_P	Engine power coefficient
C_X	General coefficient form of aerodynamic force or moment, X
D	Drag force, N
D_M	Standard momentum drag, N
D_z	Zero lift drag, N
E	Lift-drag ratio; endurance, hr; energy, N m
E_s	Specific energy, m, ft
F	General force, N
F_N	Standard net thrust, N
F_G	Gross thrust, N
g	Gravitational acceleration, m/s^2
g_0	Standard gravitational acceleration, 9.80665 m/s^2
h	Take-off, landing screen height
H	Geopotential height, m, ft
H_p	Pressure height, m, ft
J	Propeller advance ratio
K	Lift dependent drag factor
K_0	Volume dependent wave drag factor
L	Temperature lapse rate, K/m; Lift force, N
m	Mass, kg
M	Mach number
n	Load factor, L/W; constant relating to specific fuel consumption (sfc)
N	Engine rotational speed, rpm, rps
p	pressure N/m^2
p_0	Datum pressure, $=1.225$ N/m^2
p_p	Pitot or total pressure, N/m^2

P	Power, kW
q	Dynamic pressure, N/m^2
Q_f	Fuel mass flow, kg/hr
Q_a	Air mass flow, kg/s
r	Air thermometer recovery factor
R	Gas constant = 287.05287 Nm/kg K; range, km, n miles; runway reaction force, N; radius of turn, m
R_e	Reynolds number
$\Re$	Landing retardation force, N
S	Gross wing area, m^2
S_A, S_a	Take-off, landing airborne distance, m
S_{fr}, S_b	Landing free-roll and braking distances, m
S_G, S_g	Take-off, landing ground run distance, m
S_{TO}, S_{ldg}	Take-off, landing distances respectively, m
t	Time, s, hr
T	Static air temperature, K; thrust, N
T_0	Datum temperature = 288.15 K
T_t	Total temperature, K
u	Relative speed, V/V_{md}
v	Dimensionless rate of climb, $\mathrm{d}H/\mathrm{d}t/V_{md}$
v_c	Rate of climb, $\mathrm{d}H/\mathrm{d}t$
V	True airspeed [*N.B. all speeds are in knots or m/s*]
V_A	Design manoeuvre speed
V_c	Calibrated airspeed
V_C, M_C	Design cruising speed or Mach number
V_e	Equivalent airspeed
V_{EF}	Engine failure recognition speed
V_D	Design diving speed
V_H	Maximum level flight speed
V_i	Indicated airspeed
V_{LOF}	Lift-off speed
V_{md}	Minimum drag speed
V_{mP}	Minimum power speed
V_{mc}	Minimum control speed
$V_{max\,ref}$	Maximum refusal speed
$V_{min\,con}$	Minimum continue speed
V_{MU}	Minimum unstick speed
V_R	Rotation speed
V_{REF}	Reference landing speed
V_S	Stalling speed
V_{td}, V_{nd}	Landing touchdown and nose wheel down speeds
V_w	Wind speed
V_1	Decision speed
V_2, V_3	Take-off safety speeds
W	Weight, N
x	Length or distance, m, km, nautical miles (nm)
Y	Sideforce, N; lift independent drag factor, $\frac{1}{2}\rho_0 S C_{Dz}$

z	Geometric height, m, ft
Z	Lift dependent drag factor, $2K/\rho_0 S$

Greek Symbols

α	Angle of attack, deg, rad
β	Angle of sideslip, deg, rad; Prandtl–Glauert factor
γ	Ratio of specific heats of air, 1.4
$\gamma_1, \gamma_2, \gamma_3$	Aircraft velocity axis angles, bank angle, climb gradient, track respectively
δ	Relative pressure, p/p_0
ϕ	Bank attitude
η	Propeller efficiency
λ	Dimensionless power, $\eta P/V_{md}D_{min}$
μ_R	Runway rolling coefficient of friction
θ	Relative temperature, T/T_0, pitch attitude,
ρ	Air density, kg/m^3
ρ_0	Datum density, 1.225 kg/m^3
σ	Relative density, ρ/ρ_0
τ	Dimensionless net thrust, F_N/D_{min}
τ_1, τ_2	Engine gross thrust axis angles
ω	Fuel ratio, W_i/W_f
ϖ	Speed ratio V/V_{emP}
ψ	Yaw attitude or heading

Abbreviations

Airspeeds

ASIR	Airspeed indicator reading
CAS	Calibrated airspeed
EAS	Equivalent airspeed
IAS	Indicated airspeed
TAS	True airspeed

General

ADC	Air data computer
ADD	Airflow direction detector
CG	Centre of gravity
FMS	Flight management system
ISA	International Standard Atmosphere

Performance

aeo	All engines operating
Alt	Altitude
DCA	Diversion cruise altitude
ESHP	Equivalent shaft horsepower
MCP	Maximum continuous power
nm	Nautical miles
oei	One engine inoperative
SAR	Specific air range
SC	Specific climb
SE	Specific endurance, Specific energy
SEP	Specific excess power
WAT	Weight, altitude, temperature

Regulatory and documentary

AN(G)R	Air Navigation (General) Regulations
CA	Certificate of Airworthiness
CAA	Civil Airworthiness Authority, (UK)
FAA	Federal Airworthiness Authority, (USA)
FAR	Federal Airworthiness Regulations, (USA)
ICAO	International Civil Aviation Organization
JAR	Joint Airworthiness Requirements (European)

Take-off and landing distances

ASDA	Accelerate-stop distance available
EMDA	Emergency distance available
LDA	Landing distance available
LDR	Landing distance required
TODA	Take-off distance available
TODR	Take-off distance required
TORA	Take-off run available
TORR	Take-off run required

Weights

APS	Aircraft prepared for service weight
MFL	Maximum fuel load
MLW	Maximum landing weight
MSP	Maximum structural payload
MTOW	Maximum take-off weight
MTWA	Maximum take-off weight authorized
MZFW	Maximum zero fuel weight
OEW	Operating empty weight
TOW	Take-off weight

1

An introduction to the performance of fixed-wing aircraft

This study of aircraft performance will be based on material related to the flight path of fixed-wing aircraft operating in the atmosphere; the performance of rotary-wing aircraft, or helicopters, is a separate topic and needs a different approach. However, the material can be applied to aircraft with powered lift; for example, vectored thrust, boundary layer control or slipstream-induced lift.

1.1 Introduction

In terms of an aircraft, performance can be defined as a measure of the ability of the aircraft to carry out a specified task. In this study the expression 'performance' will be taken to refer to tasks relating to the flight path of the aircraft rather than to those involving its stability, control or handling qualities.

Performance can be used as a measure of the capability of the aircraft in many ways. In the case of a civil transport aircraft it determines an element of the cost of the operation of the aircraft and hence it contributes to its economic viability as a transport vehicle. In military combat operations, time, manoeuvre and radius of action are some of the more critical performance parameters in the overall evaluation of the effectiveness and air superiority of the aircraft. Performance can also be regarded as a measure of safety. Whilst an aircraft has an excess of thrust over drag it can increase its energy by either climbing or accelerating; if the drag exceeds the thrust then it will be losing energy as it either decelerates or descends. In safe flight, the aircraft must not be committed to a decrease of energy that would endanger it so that, at all critical points in the mission, the thrust available must exceed the drag; this is a consideration of the performance aspect of the airworthiness of the aircraft. Airworthiness and performance are intimately associated. However, in any conflict between efficiency and flight safety the airworthiness criterion relating to the safety of the aircraft must be considered to be dominant. Airworthiness, however, must not cloud the issue; it is only a code of practice and does not prevent the aircraft from having performance outside the limits set by that code. All that the airworthiness code of practice does is to determine practical, safe limitations on the operation of the aircraft so that the risk of unsafe operation is reduced to an acceptably low level. The airworthiness codes of practice vary with the size of the aircraft, number of engines and operational purpose. They are nationally based, although generally they all contain very similar recommendations, and are subject to amendment from time to time in the light of experience and new knowledge. Obviously, to consider all the airworthiness regulations in this study of performance would be impractical and so reference to them will be made only in a general sense where it is necessary for the development of the topic under consideration. Where reference is made, it will be to the European code, known as the Joint Airworthiness Requirements, JAR 25, which is almost identical to the American code of Federal Airworthiness Regulations, FAR 25, both of which relate to large civil transport aircraft. Codes of practice covering the performance of military aircraft in non-combat operations are sufficiently similar in their concept and application to the civil codes of practice that they need not be considered separately.

The design of an aircraft starts from a statement of the flight-path-related performance that the aircraft is expected to achieve. The basic statement of performance will be concerned with the payload the aircraft will be required to carry and the mission profile it will be required to fly. The payload of a civil transport aircraft may be defined in terms of numbers of passengers, tonnage of freight, volume of freight, or as combinations of freight and passengers. The definition of military aircraft mission payloads may cover a wide range of possibilities including personnel, troops, support equipment and supplies in transport aircraft, and internally carried stores, externally carried stores and sensor pods on combat aircraft. The mission will be defined, in the first instance, in terms of the range or the radius of action over which the payload is required to be carried or the endurance of the aircraft engaged on a patrol mission. There may be secondary considerations relating to the route, which may influence the design, and these might refer, for example, to the state of the atmosphere or to the size of the airfields into and out of which the aircraft will be expected to operate.

Before taking the discussion of the performance of the aircraft, and its influence on the design process, any further, the flight path of the aircraft in typical missions should be considered.

1.2 The mission profile

In broad terms, aircraft operations can be classified into Civil Operations, which are commercial flights transporting passengers or cargo from one geographical location to another, or Military Operations, which are concerned with defensive, or offensive, flight operations or their associated support operations. Whilst civil operations are, fundamentally, intended to be profit making, military operations are more concerned with the need to achieve their objective than with the cost of doing so; this distinction points to the different performance criteria that will need to be applied to the mission.

A typical mission profile of a civil transport aircraft is shown in Fig. 1.1. The primary mission is to fly a payload from the departure point to the destination. This requires the aircraft to take off from the departure point, climb to the cruising height and cruise to the destination, where the aircraft descends and lands. However,

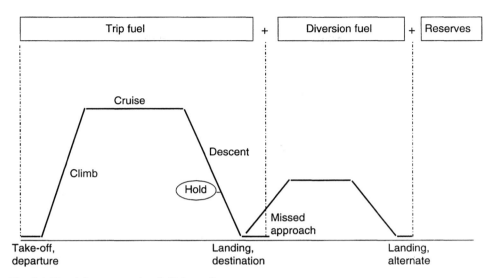

Fig. 1.1 The civil transport aircraft flight profile.

should the aircraft be unable to land at the destination when it arrives, it will have to divert to an alternate airfield and the flight plan will need to include provision for the diversion. The fuel planning for the flight will have to include reserves of fuel for the diversion and for safety; this will be considered in Chapters 9 and 10.

The civil transport mission profile shows that the flight path consists of a number of elements, or manoeuvres, which make up the total mission but which can be analysed separately, these are; take-off, climb, cruise, descent and landing; with additional manoeuvres such as turning or flying a holding pattern. In the design process, each element of the mission can be analysed individually and the performance of the aircraft estimated to show that it can achieve the necessary performance to carry out each individual manoeuvre within the limitations imposed by the specification or by airworthiness safety criteria. The overall, or block, performance can then be deduced from the performance in each of the individual elements integrated over the mission.

Military aircraft missions follow a much wider spectrum of profiles than civil operations, but generally they can be broken down into the same basic units for the performance estimation process. Figure 1.2 shows some typical military aircraft mission profiles. One of the significant differences between civil and military aircraft missions is that the military aircraft usually operates from a support base to which it returns after the mission.

The military Transport/Supply mission may be similar to the civil transport mission if the aircraft is able to land and refuel at its destination. However, when supplying forward battlefield areas it may not be possible to land or refuel and supplies may have to be air dropped with the aircraft then returning to its support base. This means that the aircraft may need to carry fuel for the outward and return legs and it may return without payload, although it may be possible to use in-flight refuelling to extend the range of the aircraft or to allow take-off at reduced weight.

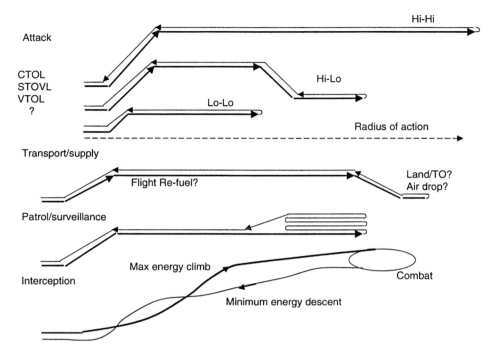

Fig. 1.2 Military aircraft mission profiles.

Patrol or surveillance missions often require the aircraft to remain on station for as long as possible. In this case, it is not the range that is important but the endurance, and the mission is usually flown at the speed that will maximize the time the aircraft can remain airborne on its fuel load.

Missions involving attacks on ground targets can take many forms. In some cases they can be made at high altitude, taking advantage of the benefits in performance to increase their radius of action. If it is necessary to fly all or part of the mission at low altitude to avoid detection the radius of action will be reduced. The aircraft will need the agility to enable it to manoeuvre evasively or to position itself for its final attack and the operational effectiveness of the aircraft can be increased if it can take off from shorter airfields or unprepared airstrips. Short take-off, assisted take-off or vertical take-off capability has been a feature of the development of ground attack aircraft for many years.

Airborne interception by a fighter on a hostile aircraft must be made as quickly as possible and on the terms most favourable to the intercepting aircraft if the attack, and subsequent combat manoeuvre, is to be successful. This requires a climb and acceleration, in the minimum time, to a height and Mach number at which there is a performance advantage to the fighter over its adversary. This calls for a large excess of thrust over drag to enable the aircraft to perform those combat manoeuvres involving high normal accelerations, or 'g', as the aircraft turns. Aircraft designed for the airborne interception role need to have a very high thrust-to-weight ratio to achieve the climb rates, acceleration and manoeuvres called for in their missions.

The very high rates of climb achieved by the fighter mean that the flight path is never steady, since acceleration is inevitably associated with the climb. Therefore, the analysis of the interception mission needs a different approach from that used for lower performance aircraft since the flight path cannot be separated into basic elements that can be analysed individually.

Once the mission profile and the payload of the aircraft have been specified the design process can commence. From the performance standpoint the total design process extends from the initial project design estimations right through to the delivery of the aircraft into service. In the first phase of the procedure the aerodynamic characteristics of the aircraft required to provide the specified performance are determined, the structural layout and design phase then progress into hardware. The complete aircraft then needs to be tested to verify the design, and developed, where necessary, to meet the design specification. In the final phase the aircraft is prepared for its operational role. The overall procedure can be divided into three broad areas: *Performance estimation*, in the initial design phase, *Performance measurement* in the testing phase and *Operational performance*, which covers the certification of the aircraft and the provision of validated operating data.

1.3 Performance estimation

Performance estimation involves the prediction of the capabilities of the aircraft from the consideration of its aerodynamic design, powerplant and operating environment. It can be applied to:

- the design of a new type of aircraft;
- the modification of an existing aircraft type in respect to design changes affecting its aerodynamic characteristics or powerplant; or
- to supplement, or extend, the full-scale measured performance of an aircraft type for conditions outside those already established.

The performance estimation process begins with the proposal of some performance target, for example, the requirement to carry a specified payload over a given route that forms the mission profile, possibly within time and cost limits. Implicit in the specification is the requirement that the aircraft can carry out that task with the required level of performance needed to comply with airworthiness regulations relating to its safe operation. The process involves the estimation of the aircraft aerodynamic characteristics and the powerplant output necessary to meet the critical points of the specification, bearing in mind the airworthiness requirements that will have to be met. (In the case of a modification exercise, or the extension of the performance of an existing aircraft, some of the aerodynamic characteristics and powerplant data may be known already and will provide a base for the estimation process.) Initially, the estimation will centre on individual elements of the flight path, for example the take-off, the climb or the cruise, and later it will progress to the integrated performance and the estimation of the time and fuel required for the mission.

The estimation process will include an assumption of the environmental conditions of the atmosphere in which the aircraft will operate. Since the atmosphere is a very variable medium in that its pressure and temperature vary with height, and also with geographical location and time, performance estimation is usually based on the assumption of a simplified atmosphere model, the *International Standard Atmosphere, ISA*. This is a linearized model of the temperate atmosphere, which represents the mean global atmosphere state with respect to seasonal changes and latitude, and is used as the basis for aircraft design. Alternative models, *design atmospheres*, cater for the need to estimate performance in hotter or colder climates, representing the tropical and arctic regions. The atmosphere models are discussed in Chapter 2.

The performance estimation process is iterative; each cycle bringing the estimation towards a design that will fulfil the requirements of the proposal and satisfy the airworthiness requirements. When the estimations show that the performance targets can be met, the aircraft can proceed to manufacture, or modification, and the next phase is to verify that the aircraft achieves its estimated performance by flight measurement.

1.4 Performance measurement

Performance measurement is required for three main purposes:

- to verify that the aircraft achieves the estimated design performance targets;
- to demonstrate that the aircraft can satisfy the safety criteria set down in the airworthiness requirements; and
- to provide validated performance data for the performance section of the flight manual.

Since the performance estimation process was based on mathematical models of the aircraft and its powerplant, which contained simplifications and assumptions, the performance estimations derived from those models will not be exact. Differences between the estimated performance of the aircraft and the actual performance achieved by the aircraft in flight will exist and will need to be evaluated so that the true performance can be established and the estimation processes and design techniques can be improved. The performance of the aircraft is measured in development trials and compared with the estimated performance; where there is a difference, the characteristics of the aircraft and of the powerplant can be measured and compared with those used in the models. By improving the model in this way a better database can be developed for the aircraft, and the design methods improved for the future. The aircraft design can then be modified and the aircraft re-tested until it meets its performance targets.

As the design of the aircraft is developed, and the flight trials show that it is meeting its performance targets, data are measured for submission to the airworthiness authority for the *Certification* of the aircraft. These data must demonstrate that the aircraft is capable of operating with sufficient performance margin to guarantee that it will be safe. This must be shown either in the case of a flight in which no incident occurs or in the case of a flight in which an incident of predictable effect occurs; for example, the failure of an engine. When the authority is satisfied that all the necessary criteria are met the aircraft can be issued with a *Type Certificate*, which enables the manufacturer to build a series of aircraft to exactly the same standard as specified in the *Type Record* of the aircraft. Aircraft that are Type Certificated can be produced in series and granted an individual certificate of airworthiness following a shortened flight test programme, or *Production Flight Trial*, which shows that the aircraft does not differ significantly from the fleet and, individually, it meets all the necessary airworthiness criteria.

As a part of the certification process, validated performance data are required for the performance section of the flight manual, known as the *Performance Manual* or the *Operating Data Manual, (ODM)*, which contains the information on the performance of the aircraft needed by the operator for flight planning. Since these data will be used to show that the aircraft can be despatched at a weight that will enable it to meet all the safely criteria with respect to the environmental conditions predicted for its flight, only data verified by flight measurement can be used in the construction of the performance manual. The data for the performance manual need to be presented in a form that enables ready interpretation for any combination of the performance variables, *weight, altitude and temperature (WAT)*.

Data measured in flight are measured under the arbitrary WAT conditions existing at the time. Generally, the weight of the aircraft at the time of test will not correspond to the weight used in the performance estimation process, neither will the weight be constant since the aircraft is burning fuel throughout its flight so that there will be a continuous decrease in its total weight. Also, it is unlikely that the atmosphere in which the test is performed will conform to the ISA model and the actual temperature–height profile will differ from that in the model atmosphere. The flight-measured data will need to be processed to conform to specific combinations of WAT for the construction of the Performance Manual or for direct comparison with the estimated performance in the development trials. Several methods of data processing are available to convert the flight-measured data into standard WAT states; these methods are discussed in Chapter 8.

1.5 Operational performance

Any transport operation carried out by an aircraft must be shown to be safe. The basic requirements for safe flight are that the space required for the aircraft to manoeuvre should never exceed the space available, and that the aircraft carries sufficient fuel for the flight; these fundamental requirements form the basis of *Performance planning* and *Fuel planning*.

Performance planning, which is a part of the flight plan made in advance of the flight, ensures that, at any point in the flight, the aircraft has sufficient performance to be able to manoeuvre within the space available. This criterion must be met in the event of a flight in which an incident occurs causing the performance to be reduced, as well as in a flight in which no such incident occurs. The principal incident that is considered in the flight planning process is the failure of a powerplant. Since this will reduce the performance of the aircraft by a predictable amount, the performance following the failure of the powerplant can be determined in any segment of the flight and the effect on the safety of the aircraft assessed.

The space required for any given manoeuvre is a function of the weight of the aircraft, and the space required increases as the weight increases. Of the performance variables (aircraft weight and the state of the atmosphere defined by its temperature–height profile), the aircraft weight is the only controllable variable that can be used to ensure that the space required for the manoeuvre does not exceed the space available. This determines the maximum weight at which the aircraft can comply with the airworthiness requirements in each segment of the flight and leads to the maximum permissible take-off weight at which the aircraft can comply with all the requirements throughout the flight. The *Maximum permissible take-off weight (MTOW)*, is the end product of performance planning.

Fuel planning ensures that the aircraft carries sufficient fuel for the mission, taking into account reserves for contingencies, diversions and safety. Since the fuel required for the mission will depend on the take-off weight of the aircraft, the fuel planning must follow the flight planning.

The performance data for flight planning are contained in the aircraft performance manual. This is a document produced for each individual aircraft, since aircraft of the same nominal type may differ in detail due to any modifications from the standard design that may have been required during production or later in the life of the aircraft. Also, the performance manual of an aircraft may need to be amended from time to time should the aircraft be modified or repaired during its operational life.

1.6 Conclusions

In this introduction it has been seen that the design of an aircraft is centred on a statement of performance, that its certification as an operational aircraft depends on its ability to meet specified performance criteria relating to its safe operation and that its operational data are based on its demonstrated performance. Performance, therefore, is important to the aircraft throughout its design and development and in its operation as an airborne vehicle. The study of flight-path performance needs to progress through the life cycle of the aircraft from the design concept to its operation as a certificated vehicle if it is to be objective.

This treatment of aircraft performance will start by examining the atmosphere as the operating environment of the aircraft. The state of the atmosphere is fundamental to performance and so an understanding of the structure and characteristics of the atmosphere is essential before the subject of flight-path performance can be discussed.

In designing the aircraft the performance target is achieved by providing the aircraft with the appropriate aerodynamic and propulsive characteristics. The basic layout of the aircraft will be heavily influenced by the payload, for example, the number of passengers or the maximum dimensions and weight of cargo specified in the performance target, and by the range and route over which the aircraft is to operate. Once the basic layout of the aircraft has been determined, the aerodynamic characteristics necessary to achieve the critical elements of the mission – for example, the range and the gradient of climb – can be estimated. The gradient of climb and the range and route structure will affect the maximum installed thrust required to meet the performance target and influence the selection of the powerplant. This is an iterative process in which the design is reviewed and modified until the estimated aerodynamic and propulsive characteristics indicate that the performance target can be achieved. In the design process, algorithms are required for the prediction of the performance of the aircraft in each element of the mission; these need to be in a form that will allow the effects on its performance of changes to the design of the aircraft to be predicted. In Chapter 3, the aerodynamic and propulsive characteristics of the aircraft are discussed, and the algorithms for the design process are developed in Chapters 4 to 7.

Since the design process depends on the estimation of performance from data that have been collated over a period of time, mainly from experimental sources, inevitably it will be inexact. Consequently, the performance estimated from those data would need to be verified by flight measurement to show that the aircraft can meet its design target in practice. This will lead on to the certification process in which the limitations of operation of the aircraft are determined and the data for the performance manual are produced. These aspects of performance measurement, data presentation and its practical application are covered in Chapters 8 to 10 to complete the objective study.

To illustrate the application of the performance theory to aircraft design and operation a series of examples of performance estimation are given in Chapter 11.

Bibliography

Airworthiness Authorities Steering Committee: *Joint Airworthiness Requirements*, JAR-1, Definitions and Abbreviations; JAR-25, Large Aeroplanes (CAA Printing and Publishing Services).

Federal Aviation Administration: *Code of Federal Regulations; Title 14, Aeronautics and Space*. Part 1, Definitions and Abbreviations; Part 25, Airworthiness Standards: Transport Category Airplanes (US Office of the Federal Register, Washington).

Grover, J. H. H. (1989) *Handbook of Aircraft Performance* (BSP Professional Books).

Wagenmakers, J. (1991) *Aircraft Performance Engineering* (Prentice Hall).

2

The atmosphere and air data measurement

The atmosphere that surrounds the Earth can be said to be the working fluid of the aircraft. In this chapter, the characteristics of the atmosphere will be discussed and a model of the atmosphere, which is used as a design reference, will be described. The measurement of the essential air data – altitude, airspeed, Mach number and air temperature – will then be considered.

2.1 Introduction

The atmosphere is a thin, gaseous, layer surrounding the Earth. It is subjected to continuous processes of heating by the Sun's radiated energy by day and cooling by radiation into space at night so that its temperature state is always changing. Furthermore, the convective currents within the atmosphere resulting from its changing temperature state are affected by Coriolis forces as the Earth rotates. This induces swirling motions in the air mass and consequent pressure changes; the atmosphere is, therefore, in a continually changing state of both pressure and temperature. The state of the atmosphere, defined by its temperature and pressure, is fundamental to both the design and the operation of the aircraft. The atmospheric air provides the lift force that supports the aircraft in flight and, through the powerplant, the propulsive forces that are necessary to sustain flight. These forces depend on the properties of the atmosphere, and on the relative motion between the atmosphere and the aircraft. Therefore, the first essential in the estimation, or the measurement, of the performance of the aircraft is to know the state of the atmosphere in which the aircraft is flying and to be able to measure the relative motion between the aircraft and the atmospheric air mass.

2.2 The characteristics of the atmosphere

The atmosphere consists of air, which is a mixture of gases, mainly nitrogen (78%) and oxygen (21%), with traces of argon (0.9%), carbon dioxide (0.03%) and other inert gases in minute quantities. In addition, there are quantities of dust particles,

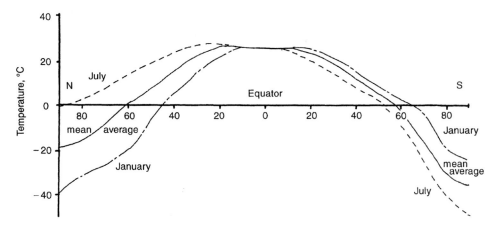

Fig. 2.1 Mean seasonal global temperature distribution.

water vapour and moisture in variable amounts which, although they do not affect the gaseous properties of the air significantly, do play an important part in the development of the structure of the atmosphere.

The atmospheric air can be taken to behave as a neutral gas that obeys the equation of state,

$$p = \rho T R \qquad (2.1)$$

where p = pressure, N/m^2; T = temperature, K; ρ = density, kg/m^3, and R = gas constant, 287.05287 Nm/kg K. Thus, the properties of the atmosphere can be defined in terms of its temperature and pressure.

Over the surface of the Earth, the properties of the atmosphere are not constant but vary with time and geographical position. Geographically, the temperature of the atmosphere near the surface of the Earth varies with latitude; the mean seasonal temperature distribution is shown in Fig. 2.1. At the equator, the seasonal variation of temperature is small and the mean temperature is high, whereas at the poles, the seasonal variation is large and the mean temperature is low. Between the poles and the equator, the mean global temperature roughly follows a sine curve. However, it should be emphasized that local variations in temperature at any latitude caused by land mass distribution, oceans, mountain ranges and deserts, can be very large.

The variation of temperature with time is partly a long period, seasonal, variation, also seen in Fig. 2.1, with the summer period being warmer than the winter. There is also a short period, diurnal, variation – daytime being warmer than night-time. Local weather variations will produce further, random, variation on top of the more regular cycles. The variation of temperature is, therefore, continuous and complex.

The pressure of the atmosphere at the Earth's surface also varies with time. Regions of low pressure, cyclones, and high pressure, anti-cyclones, are formed by the effects of convection currents in the atmosphere as the air is transported from the higher temperature regions to the lower temperature regions. This process would normally take place along lines of longitude but the rotation of the Earth about its polar axis produces Coriolis forces which cause the flow to swirl and to create a series of

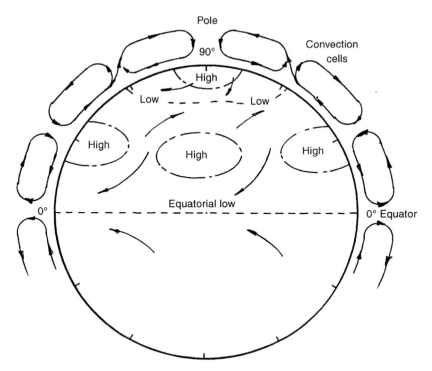

Fig. 2.2 General global atmosphere pressure distribution.

convection current cells. This results in a general pattern of high and low pressure regions over the surface of the Earth (see Fig. 2.2).

However, the pressure pattern is further influenced by ocean and land mass distribution. Since the temperature distribution over the Earth's surface is dependent on its surface characteristics, (oceans and landmasses), convective flows in the local air mass will result. Relatively warm surface temperatures will produce rising air, or low-pressure regions; relatively cool surfaces will produce descending air, or high-pressure areas. Thus, the oceans and landmasses will influence the global pressure pattern with local, seasonally varying, convective flows. The result is a complex pressure distribution over the surface of the Earth which, whilst it has broadly predictable features, is constantly changing.

2.3 Vertical development of the atmosphere

The Earth receives energy by radiation from the Sun and the energy passes through the atmosphere on its way to the surface of the Earth (Fig. 2.3). As it does so, some of the energy is reflected from the atmosphere directly back into space and some is absorbed by the atmosphere; the remainder passes through to the Earth's surface. At the Earth's surface some of the radiation will be absorbed and the remainder reflected back to the atmosphere, the proportion depending on the surface reflectivity, or albedo. The energy that is absorbed heats the surface where it is stored before it is

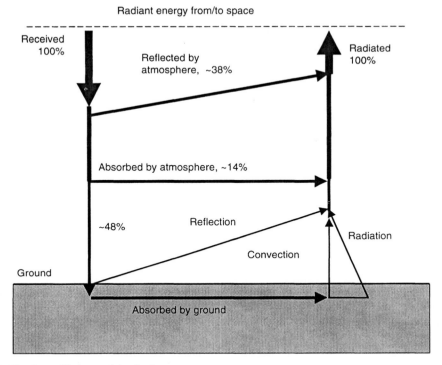

Fig. 2.3 The thermal balance of the Earth.

eventually dissipated by convection to the atmosphere and radiated back into space. The variability of the reflectivity of the ground and the oceans leads to a very complex temperature distribution over the Earth and, consequently, in the convective currents in the atmosphere.

The radiation that is absorbed by the atmosphere is not absorbed uniformly but selectively by different layers, giving rise to a complex temperature–height profile in the atmosphere (Fig. 2.4). At low levels the water vapour and carbon dioxide absorb the terrestrial radiation producing a warm air region near the ground extending upwards to about 11 km. This layer is known as the *troposphere*. In the troposphere, the temperature decreases with increasing height and the tempera-ture–height gradient, which is negative here, is known as the *temperature lapse rate, L.* Above this region there is little water vapour in the atmosphere and its absorbtivity is reduced; this layer is called the *stratosphere* and extends upwards to some 50 km. The ozone content of the atmosphere increases with height up to about 80 km and in the layer between 50 km and 80 km the absorbtivity, particularly of the ultraviolet spectrum, increases to form a further warm air layer, this is the *mesosphere*. Above the mesosphere is a layer of very low pressure, the *thermosphere*, extending to about 800 km and the final layer, the *exosphere*, forms the boundary with space.

The vertical structure of the atmosphere is also affected by latitude, particularly in the polar regions, where the *tropopause*, which is the boundary between the

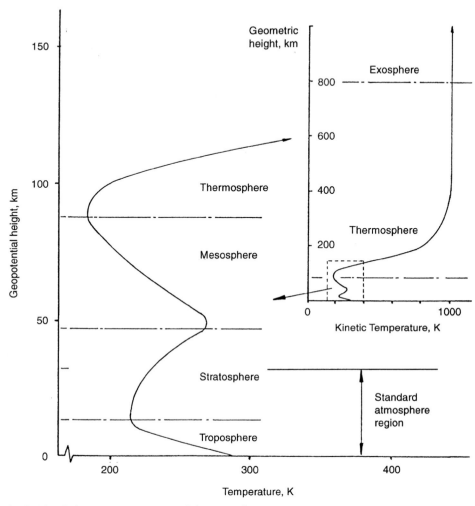

Fig. 2.4 Vertical temperature structure of the atmosphere.

troposphere and the stratosphere, tends to be lower than in the warmer equatorial regions. In addition, the temperature–height profiles at ground level may be modified by the high reflectivity of ice and snow. Figure 2.5 shows a set of temperature–height profiles measured simultaneously at various latitudes in the Northern Hemisphere.

2.4 The standard atmosphere model

The performance of the aircraft depends on the state of the atmosphere in which it is flying. The state of the atmosphere, as defined by its pressure and temperature, is variable, so that the actual performance of the aircraft will depend on its geographical location and on time. In the design of an aircraft, assumptions of the state of the atmosphere will have to be made in order to predict its performance. A model of

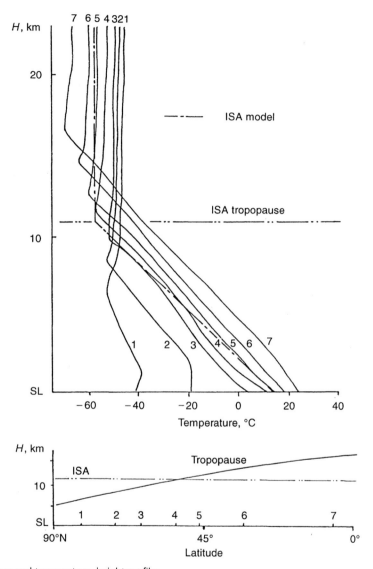

Fig. 2.5 Measured temperature–height profiles.

the structure of the atmosphere is required, first, to act as a standard for the estimation of the performance of an aircraft at the design stage and, secondly, to form the basis for methods of comparison between flight-measured performance in the real atmosphere and the design estimates of performance based on the model atmosphere in order to verify the design methods.

The atmosphere model needs to represent an average atmosphere with respect to geographical and seasonal variations in pressure and temperature and to have a vertical structure, which is similar to that in the real atmosphere. However, it also needs to be simple enough to enable the performance relationships to be developed in an easily usable form. By international agreement, an atmosphere model has been accepted and is used as the basis for all performance work; it is known as the *International Standard Atmosphere* and is generally referred to as *ISA* or, simply, the *Standard Atmosphere*.

The International Standard Atmosphere is constructed on the assumption that the air consists of a perfect gas that obeys the equation of state, (eqn (2.1)), and that the effects on the gas laws of such additions as dust, water vapour and moisture are negligible. The datum of the ISA model is taken to be an atmospheric static pressure of $101\,325\,\mathrm{N/m^2}$ and temperature of $288.15\,\mathrm{K}$. These correspond to the physicists' standard temperature and pressure, $15°C$ and $760\,\mathrm{mm}$ mercury, which approximate to the mean seasonal sea-level values at latitude $45°N$. Because of this, the datum of the atmosphere model is often referred to, incorrectly, as 'sea-level'. It should be remembered that the reference model datum is defined as the height at which the pressure is $101\,325\,\mathrm{N/m^2}$, which may be above or below the actual sea level. The reference datum values of the principal characteristics of the International Standard Atmosphere model are:

$$\text{Reference pressure} \qquad p_0 = 101\,325\,\mathrm{N/m^2}$$

$$\text{Reference temperature} \quad T_0 = 288.15\,\mathrm{K}$$

$$\text{Reference density} \qquad \rho_0 = 1.225\,\mathrm{kg/m^3}$$

The vertical structure of the atmosphere model is defined by the assumption of a series of linear relationships between temperature and height (Fig. 2.6). Up to a height of 32 km, which is the vertical extent of the International Standard Atmosphere model used in connection with aircraft performance, the model consists of

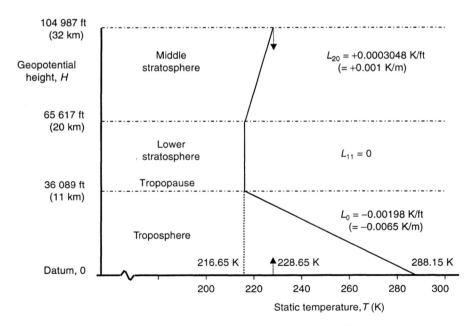

Fig. 2.6 International Standard Atmosphere model; temperature–height profile.

three layers, in each of which the temperature–height profile is given by,

$$T = T_i + L_i(H - H_i) \qquad (2.2)$$

where the subscript i denotes the height of the lower boundary of the layer considered in km. Thus, at the datum level, $i = $ zero, and the temperature lapse rate, L_i, is the rate of change of temperature with height in the layer above H_i.

The first layer in the atmosphere model is the troposphere, which extends to a height of 11 km above the reference datum and in which the temperature decreases with height. At 11 km, the tropopause denotes the level at which the temperature lapse rate changes and the stratosphere region begins. In the lower stratosphere, the temperature lapse rate is zero giving an isothermal layer between 11 km and 20 km. Above 20 km, the temperature lapse rate becomes positive in the middle stratosphere, which extends to a height of 32 km; this is the upper boundary of the ISA model. Above 32 km, the model continues into the upper stratosphere and, eventually, into the higher atmosphere layers that are above the operating heights of conventional aircraft.

Although the International Standard Atmosphere is defined by reference to metric heights, international aircraft operations are currently required to be flown by reference to altimeters calibrated in Imperial units, i.e. feet. Since the performance of the aircraft is associated with practical operations, heights will generally be referred to in feet, rather than in metres, for all performance analysis.

The practical measurement of height in the atmosphere requires the definition of height scales.

2.4.1 Height scales

True height, $-z$, is defined as the vertical, geometric, distance between an object and a datum level. The potential energy of the object above that datum is given, per unit mass, by

$$PE = -\int_0^z g \, dz \qquad (2.3)$$

where g is the acceleration due to gravity. (The negative sign is here used with the true height since, in Earth axes, the positive vertical direction, Oz_e, is downwards, whereas heights are conventionally measured upwards, see Appendix A).

The value of the gravitational acceleration, g, is not constant over the surface of the Earth but varies with latitude, and this is due to two factors. First, the radius of the Earth is not constant but varies between the poles and the equator, the polar radius being some 20 km less than that at the equator. Therefore, the value of the gravitational acceleration, which is inversely proportional to the square of the distance from the centre of the Earth, will be less at the equator than at the poles. Secondly, the centrifugal acceleration due to the rotation of the Earth on its axis also tends to reduce the equatorial value of the gravitational acceleration. Both of these effects are dependent on latitude. Similarly, height above the geoid, which defines the mean surface of the Earth, will produce a decrease in the local value of the gravitational acceleration.

The sum of these effects means that the potential energy of an object above the Earth's surface, if defined by its true (geometric) height, will depend on latitude.

Thus, the integral, eqn (2.3), will have to take into account the variable value of the gravitational field, g. To avoid this complexity, an alternative height scale is defined in which it is assumed that the gravitational field is uniform over the Earth's surface and at all heights and has the standard value, g_0. This leads to the definition of the *geopotential height* scale, H.

When geopotential height is used with the uniform gravitational field, g_0, it gives the same potential energy to a body as the true height used with the variable gravitational field, g. Thus, the height scales are related by potential energy,

$$PE = -\int_0^z g \, dz = g_0 \int_0^H dH = g_0 H \qquad (2.4)$$

The difference between true height and geopotential height is relatively small; geopotential heights of 5000 ft and 40 000 ft would give the same potential energies as true heights of 5001 ft and 40 077 ft respectively. Geopotential height is used for the definition of the properties of the atmosphere.

2.4.2 Pressure–height relationship

Having defined the temperature–height profile in the model atmosphere in terms of geopotential height (eqn (2.2)), a relationship between static pressure and geopotential height can now be determined for the atmosphere model.

The static pressure in a fluid at rest is determined by the hydrostatic law. By considering the vertical pressure gradient in the atmosphere the pressure increment δp over a height increment δH is given by,

$$\delta p = -g_0 \rho \delta H \qquad (2.5)$$

Using eqn (2.1), the equation of state can be expressed as,

$$\frac{\delta p}{p} = -\frac{g_0}{RT} \delta H \qquad (2.6)$$

In the ISA model the temperature is defined as a function of geopotential height (eqn (2.2)), so that eqn (2.6) can be written as,

$$\frac{dp}{p} = -\frac{g_0}{R[T_i + L_i(H - H_i)]} \, dH$$

This can be integrated in each layer of the atmosphere. In the troposphere the pressure–height relationship becomes,

$$\frac{p}{p_0} = \left[1 + \frac{L_0}{T_0} H\right]^{\frac{-g_0}{RL_0}} \qquad (2.7)$$

and in the lower stratosphere where $L = 0$,

$$\ln\left(\frac{p}{p_{11}}\right) = -\frac{g_0}{RT_{11}} (H - H_{11}) \qquad (2.8)$$

where subscript 11 refers to values at the tropopause, 11 km.

Thus, pressure is a unique function of geopotential height in the ISA model. This is a most important relationship since the state of the model atmosphere, in terms of its pressure and temperature, is defined by height only; this will be referred to later in the measurement of air data.

2.4.3 Off-standard and design atmospheres

The International Standard Atmosphere is only a model for use in the estimation and measurement of aircraft performance. The real atmosphere encountered at any particular time and place will generally not conform to the ISA model but will have its own temperature at datum pressure and its own temperature–height profile. The datum of the real atmosphere is usually taken to be an arbitrary height – for example, mean sea-level – at which the pressure will vary with time and the temperature–height profile may range between arctic winter and tropical summer. Any atmosphere that does not conform to the ISA profile is referred to as an *off-standard atmosphere*.

Aircraft are required to operate in a wide range of atmosphere states and their performance will need to be estimated in off-standard conditions. The *design atmospheres* are defined to cover the likely extreme variations in datum level temperatures and temperature–height profiles that typify these operating conditions. Broadly, they consist of temperature–height profiles parallel to the ISA model profile but displaced by an increment in datum temperature; Fig. 2.7(a) shows a set of design atmospheres used in the European airworthiness codes of practice (e.g. JAR 25). The hot climate design atmospheres are formed by adding an increment to the datum temperature and modifying the temperature in the lower stratosphere to give a tropopause level at about 11 km. The cold-climate atmospheres structure has modified temperature lapse rates and lower stratosphere temperatures. In addition, at low levels, additional layers are introduced to give the very low ground level temperatures encountered under these conditions.

The design atmospheres are defined in terms of *pressure heights* (see Section 2.4.4) and, like the ISA, are simply atmosphere models used for the estimation of performance in extreme conditions of operation. Figure 2.7(b) shows the design atmospheres in terms of *geopotential height*. The effect of temperature on the height scale can be clearly seen in the height of the tropopause; the relationship between the pressure height and the geopotential height will be discussed in Section 2.4.4.

2.4.4 Pressure height

In the development of the ISA model it was shown that a unique relationship existed between geopotential height and static pressure; this, in the troposphere, is given by

$$p = p_0 \left[1 + \frac{L_0}{T_0} H \right]^{\frac{-g_0}{RL_0}} \tag{2.7}$$

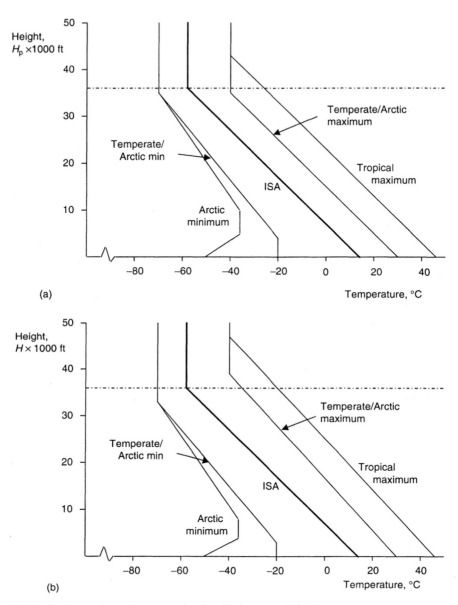

Fig. 2.7 Design Atmospheres. (a) Pressure heights. (b) Geopotential heights.

In an off-standard atmosphere $T \neq T_0$ and, generally, $L \neq L_0$; thus, the relationship between pressure and geopotential height will not be the same as in the standard atmosphere model. To enable the relationship between pressure and height to be applied to all atmosphere states a further height scale needs to be defined, this is the *pressure height*, H_p.

The pressure height, H_p, in an off-standard atmosphere is the geopotential height, H, in the standard atmosphere at which the same pressure occurs. In the standard atmosphere, pressure height and geopotential height are equal by definition.

The difference between pressure height and geopotential height in an off-standard atmosphere can be determined from eqn (2.6). In any atmosphere,

$$\frac{\mathrm{d}p}{p} = -\frac{g_0}{RT}\mathrm{d}H$$

and in the standard atmosphere, since $H = H_p$,

$$\frac{\mathrm{d}p}{p} = -\frac{g_0}{RT_{std}}\mathrm{d}H_p$$

where T_{std} is the standard atmosphere temperature at the pressure height H_p.

It follows that

$$\mathrm{d}H = \left[\frac{T}{T_{std}}\right]_{H_p}\mathrm{d}H_p \qquad (2.9)$$

Thus, a geopotential height increment $\mathrm{d}H$ is related to a pressure height increment $\mathrm{d}H_p$ by a temperature correction. This correction is used to obtain geopotential height intervals from measured (pressure) height intervals for the measurement of gradients of climb and other flight-path related performance characteristics.

2.4.5 Relative properties of the atmosphere

The atmospheric equation of state, eqn (2.1), applies to all points in the atmosphere so that,

$$p = \rho T R$$

and at the ISA datum,

$$p_0 = \rho_0 T_0 R$$

Thus,

$$\frac{p}{p_0} = \frac{\rho}{\rho_0}\frac{T}{T_0}$$

This can be written,

$$\delta = \sigma\theta \qquad (2.10)$$

where $\delta = p/p_0$, the relative pressure; $\sigma = \rho/\rho_0$, the relative density; and $\theta = T/T_0$, the relative temperature.

The relative properties are a convenient means of expressing and manipulating the atmosphere properties and avoiding the need to use the gas constant.

Figure 2.8 shows the relative properties of the standard atmosphere. The relative temperature profile, which is defined by the standard atmosphere model, consists of the three linear segments up to 32 km. The relative pressure profile describes the decrease in pressure with height; it is useful to note some significant points on the

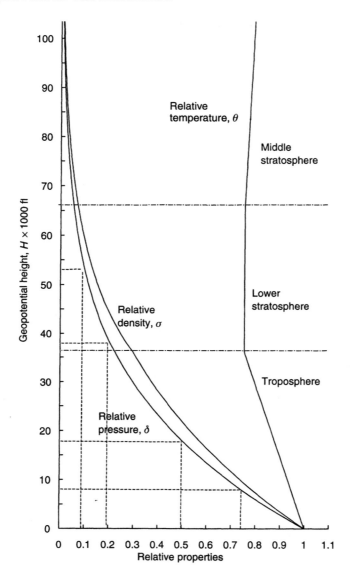

Fig. 2.8 International Standard Atmosphere; relative properties.

pressure height profile;

8000 ft	$\delta = 0.75$	(maximum cabin pressure height for passenger transport)
18 000 ft	$\delta = 0.5$	(short haul operations)
38 000 ft	$\delta = 0.2$	(long range transport operations)
53 000 ft	$\delta = 0.1$	(Concorde and some military operations)
100 000 ft	$\delta = 0.01$	(TR1, SR71 surveillance aircraft, 80–90 000 ft)

The relative density is simply the relative pressure divided by the relative temperature.

2.4.6 Density altitude

It is sometimes more convenient to consider the state of the atmosphere in terms of its density rather than its pressure and temperature separately. In this case, the relationship between the density and height in the standard atmosphere model is used as a datum. In an off-standard atmosphere, the density altitude is defined as the pressure altitude in a standard atmosphere at which the same density would occur. Figure 2.9 shows the relationship between density altitude and pressure height.

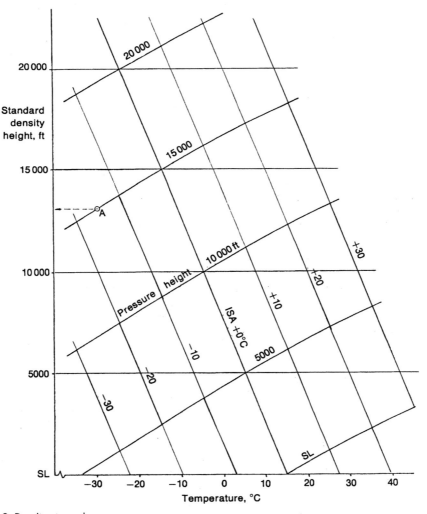

Fig. 2.9 Density atmosphere.

The concept of the density altitude can be illustrated by a simple example. If the observed temperature at a pressure height of 15 000 ft is −30°C (243.15 K, or about ISA-15°C) then, from the pressure–height relationship (eqn (2.8)), the relative pressure at 15 000 ft is

$$\delta = 0.56434$$

and the relative temperature is

$$\theta = 243.15/288.15 = 0.84383$$

giving the relative density to be

$$\sigma = \delta/\theta = 0.66878.$$

Now, from the properties of the standard atmosphere, the pressure height at which a relative density of 0.66878 occurs is 13 120 ft. Thus, the density altitude is 13 120 ft since the standard atmosphere density at this height is equivalent to the actual density at a pressure height of 15 000 ft and a temperature of −30°C; this is shown by point A in Fig. 2.9.

2.4.7 Tables of the International Standard Atmosphere

The properties of the atmosphere are tabulated as functions of height and are shown in Appendix C for reference.

2.5 Measurement of air data

The essential requirements in the measurement of aircraft performance are, first, the knowledge of the state of the atmosphere in which the aircraft is flying and, secondly, the relative motion between the aircraft and the air mass. This information is collected by the air data system

The air data system of an aircraft (see Fig. 2.10) consists of a pitot-static installation to sense the airflow pressures from which height, airspeed and Mach number are derived. An air thermometer, from which the air temperature can be determined, and, in some cases, airflow direction detectors (ADDs), which sense the local flow direction relative to the aircraft body axes, are part of the system. Owing to the relative motion between the aircraft and the atmospheric air mass, and the variation of the state of the atmosphere with height, the determination of the air data is not a straightforward process. Corrections need to be applied to account for systematic errors in the measurement process and for assumptions made in the calibration equations of the instruments.

The primary flight instruments, the altimeter, airspeed indicator and Machmeter, shown as individual instruments in Fig. 2.10, are fed with the airflow pressures from the Pitot-static system. They convert those pressures into the movement of a pointer mechanically, thus making these instruments independent of any electrical power supply. The secondary instruments, the air temperature sensor and ADDs, usually require an electrical power supply for their operation.

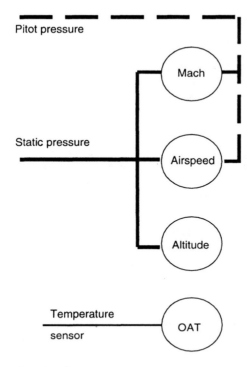

Fig. 2.10 Air data system of an aircraft.

Alternatively, the atmosphere pressures can be fed to an *air data computer* (ADC), which contains pressure transducers to convert the pressures into electrical signals for processing. Since the ADC can be programmed to apply known corrections to the data, it is possible to process the air data and to present it in the most convenient form. The ADC is, however, dependent on electrical power and failure of the power supply could mean the loss of essential information; mechanical instruments are required as a standby system in most cases. The ADC will be discussed further in Section 2.7.

Both the air data computer and the mechanical instruments use the same basic calibration equations to convert the measured data into a suitable form for operational use; the calibration equations will be developed in the subsequent sections. The units used in the display of air data are usually the foot for measurement of height and the knot (or nautical mile per hour) for airspeed, since international regulation requires primary flight instruments to be calibrated in these units.

(NB Although the calibration equations will be developed in SI units it should be appreciated that the instrument displays, and operational data, normally will be in feet (altitude) and knots (airspeed) and reference will be made to these units throughout the text. However, for the sake of clarity, the conversion factors between SI and Imperial units will *not* be included in the calibration equations developed in the subsequent sections.)

2.5.1 Measurement of height

In Section 2.4.2, relationships were found that related pressure to geopotential height in the standard atmosphere (eqns (2.7) and (2.8)). By rearranging these equations, height can be expressed in terms of pressure so that, in the troposphere in which $L \neq 0$,

$$H = \frac{T_0}{L_0} \left[\left(\frac{p}{p_0}\right)^{\frac{-L_0 R}{g_0}} - 1 \right] \tag{2.11}$$

and in the isothermal lower stratosphere, in which $L = 0$,

$$H = 11\,000 - \frac{R T_{11}}{g_0} \ln\left(\frac{p}{p_{11}}\right) \tag{2.12}$$

where subscript 11 refers to conditions at 36 089 ft (11 km).

These expressions form the calibration equations of an instrument that indicates height as a function of barometric static pressure in the atmosphere; this instrument is the *altimeter*. As aircraft operations rarely take place above 65 000 ft (20 km), altimeters are normally only calibrated for use in the troposphere and lower stratosphere.

Since the altimeter calibration is based on the model of the standard atmosphere, it will use the ISA values of temperature and temperature lapse rate to measure altitude in any atmosphere state. The height indicated will be the *pressure height* relative to the ISA datum pressure, 101 325 N/m^2, (usually expressed as 1013 mb or 1013 hectopascals). The concept of pressure height is important in the measurement of air data since it enables a height in any atmosphere state to be related to a pressure in the standard atmosphere. By setting the altimeter to the ISA datum pressure (1013 mb), pressure heights indicated by the altimeter can be converted into static pressures by the calibration equation, thus enabling one of the properties of the atmosphere – static pressure – to be measured.

Although the standard atmosphere model uses a static pressure of 1013 mb as its datum, in practice height is measured with respect to other datum pressures, one of which is Mean Sea Level. Because the sea-level pressure of the atmosphere at any geographical location changes from day to day, the static pressure at sea level will probably not be 101 325 N/m^2. Therefore, the altimeter would not read zero at sea level when set to the 1013 mb datum. This can be overcome by including a facility in the altimeter to enable the pressure datum to be selected so that heights can be measured from any convenient datum; Fig. 2.11 shows the most common altimeter datum settings. When the aircraft is in the take-off or landing phase of flight, heights above the airfield level are required and the local static pressure at the airfield level, p_a, is used as the datum. This setting is known as the QFE. When QFE is set the altimeter will read *heights* above airfield level. When flying across country, height relative to topographical features is required and, since these are charted with respect to mean sea-level, the sea-level pressure, p_{sl}, in that region is set. This is known as the QNH and, when QNH is set, the altimeter reads *altitudes* above mean sea level. The QNH is a regional pressure setting and aircraft on long flights would need a change of the altimeter setting as they crossed the regional boundaries. When

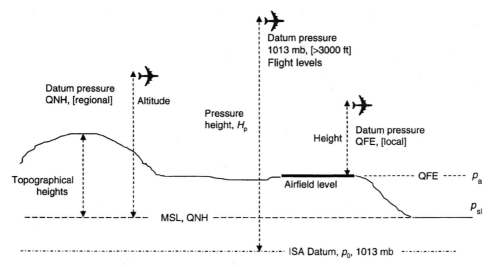

Fig. 2.11 Altimeter reference pressure settings.

flying above 3000 ft (in the UK) all altimeters are set to the standard atmosphere datum pressure, 1013 mb. This eliminates the need to reset the QNH at each regional boundary and ensures that all aircraft fly to a common altimeter setting for vertical separation. When 1013 mb is set, the altimeter readings are referred to as *flight levels (FL)*. A vertical increment of 100 ft is taken to be the increment between flight levels so that height indications of 4000 ft, 10 500 ft and 23 000 ft would be referred to as FL40, FL105 and FL230, respectively.

In the measurement of height, a number of corrections need to be taken into account before the pressure height, and quantities derived from pressure height, can be determined. These can be summarized, working back from the altimeter to the freestream flow.

(a) Altimeter reading; Alt or H_{pI}

This is the reading of an individual instrument. Since the instrument is a mechanical device driven by the static pressure, there will be errors due to mechanical tolerance. The instrument error correction can be evaluated by calibrating the individual instrument against an accurate source of pressure and the correction applied to the altimeter reading to give indicated altitude.

(b) Indicated altitude; H_{pi}

This is the altimeter reading corrected for instrument error. The indicated altitude will be measured with reference to the appropriate altimeter datum pressure setting.

(c) Pressure altitude; H_p

This is the indicated altitude, measured with respect to the appropriate datum pressure setting, corrected for static pressure error. This is the error due to the location of the static pressure source within the disturbed pressure field caused by the presence of the aircraft (see Section 2.6).

(d) Geopotential height interval, dH_p

This is the pressure height interval, dH_p, measured by the altimeter and corrected for temperature difference from the ISA model atmosphere, eqn (2.9).

(e) Static pressure, p, and relative pressure, δ

When the altimeter datum pressure is set to 1013 mb the pressure heights can be converted into atmospheric pressure or relative pressure either by reference to the atmosphere tables or from the ISA pressure–height relationship, eqn (2.8).

2.5.2 Measurement of airflow characteristics

Airspeed is the relative velocity between the aircraft and the air mass in which it is flying. It is one of the most important parameters in aircraft performance since the aerodynamic forces acting on the aircraft, and upon which its performance is based, are functions of airspeed. There are several systems that can measure the speed of the aircraft relative to the Earth (the ground speed), and are used for this purpose in navigation. However, it must be remembered that the air mass is moving relative to the Earth and the airspeed generally is not the same as the ground speed. The airspeed can be measured only by reference to characteristics of the relative motion between the aircraft and the air mass itself. Another characteristic of the airflow that affects aircraft performance is *Mach number*, which is the ratio between the airspeed and the speed of sound in the air mass.

In the measurement of the airspeed, the air mass is stationary and the aircraft is passing through it (although for the purposes of analysis the aircraft is usually considered stationary in a moving airflow). Thus, the flow relationships are adiabatic and the total energy of the flow is constant. Any changes in the energy of the flow must take place as interchanges between the intrinsic energy, the kinetic energy and the potential energy, although the potential energy changes can usually be neglected in practice. This is because there is no significant height change at the point of measurement where the flow is brought to rest in the Pitot tube.

Since the total energy of the flow is constant, the energy relationship, neglecting the potential energy term, can be written in the form,

$$\frac{dp}{\rho} + V\,dV = 0 \tag{2.13}$$

Integrating eqn (2.13) for adiabatic flow in which

$$\frac{p_1}{p_2} = \left(\frac{\rho_1}{\rho_2}\right)^{\gamma}$$

gives,

$$\frac{\gamma}{\gamma - 1}\frac{p}{\rho} + \frac{V^2}{2} = \text{constant} \tag{2.14}$$

relating the flow pressures to the flow velocity, or true airspeed, V.

Alternatively, from the equation of state, eqn (2.1),

$$\frac{p}{\rho} = RT$$

and expressing the gas constant, R, in the form

$$R = C_p \frac{\gamma - 1}{\gamma}$$

eqn (2.14) can be written in the form,

$$C_p T + \frac{V^2}{2} = \text{constant} \tag{2.14a}$$

This relates the flow temperature to the true airspeed.

Equations (2.14) and (2.14a) are alternative statements of the energy equation of the adiabatic flow of an ideal gas and can be used in the measurement of the airflow characteristics.

2.5.3 Measurement of airspeed

The airspeed can be measured by comparing the total and static pressures of the airflow relative to the aircraft. From eqn (2.14), the energy at any two points in the flow are equal, thus,

$$\frac{\gamma}{\gamma - 1} \frac{p_1}{\rho_1} + \frac{V_1^2}{2} = \frac{\gamma}{\gamma - 1} \frac{p_2}{\rho_2} + \frac{V_2^2}{2} \tag{2.15}$$

If point 1 refers to the undisturbed, freestream conditions in which the pressure p_1 is the static pressure p, and V_1 is the true airspeed of the flow V, and point 2 refers to the stagnation conditions in the Pitot tube in which the airflow velocity, V_2, is zero and the pressure p_2 is the total, or Pitot, pressure, p_p, then eqn (2.15) becomes,

$$\frac{\gamma}{\gamma - 1} \frac{p}{\rho} + \frac{V^2}{2} = \frac{\gamma}{\gamma - 1} \frac{p_p}{\rho_2} \tag{2.16}$$

Now the speed of sound, a, is given by

$$a = \sqrt{\gamma RT} = \sqrt{\frac{\gamma p}{\rho}} \tag{2.17}$$

so that eqn (2.16) reduces to

$$p_p = p \left\{ 1 + \frac{\gamma - 1}{2} \left(\frac{V}{a} \right)^2 \right\}^{\frac{\gamma}{\gamma - 1}} \tag{2.18}$$

Comparing the total and static pressures provides the relationship between airspeed and the differential pressure, or impact pressure, p_d,

$$p_d = p_p - p = p \left[\left\{ 1 + \frac{\gamma - 1}{2} \left(\frac{V}{a} \right)^2 \right\}^{\frac{\gamma}{\gamma - 1}} - 1 \right] \tag{2.19}$$

This expression is, in terms of the local atmospheric state, defined by its pressure and temperature, in which the speed of sound and the static pressure are altitude dependent variables. If eqn (2.19) is to be used generally to measure airspeed then it must be independent of the state of the atmosphere. In eqn (2.19), V is the true airspeed, TAS, of the aircraft relative to the air mass and a is the speed of sound in the local flow which, from eqn (2.17), can be expressed in terms of the relative properties as,

$$a^2 = a_0^2 \frac{\delta}{\sigma}$$

so that, from eqn (2.19), $(V/a)^2$ can be written as

$$\left(\frac{V}{a}\right)^2 = \left(\frac{V}{a_0}\right)^2 \frac{\sigma}{\delta} = \left(\frac{V_e}{a_0}\right)^2 \frac{1}{\delta} \tag{2.20}$$

where V_e is the *equivalent airspeed, EAS,* defined by the dynamic pressure, q, such that,

$$q = \tfrac{1}{2}\rho V^2 = \tfrac{1}{2}\rho_0 V_e^2 \tag{2.21}$$

thus

$$V_e \equiv V\sqrt{\sigma}$$

This definition of equivalent airspeed is very important since it enables the dynamic pressure, q, to be found from the EAS alone without the need to calculate the density of the air mass.

Substituting eqn (2.20) into eqn (2.19) gives,

$$p_p - p = p\left[\left\{1 + \frac{\gamma - 1}{2}\left(\frac{V_e}{a_0}\right)^2 \frac{1}{\delta}\right\}^{\frac{\gamma}{\gamma-1}} - 1\right] \tag{2.19a}$$

Now to obtain a unique calibration equation, independent of altitude or temperature, the pressure term, δ, must also be eliminated from eqn (2.19a). The *calibrated airspeed, CAS, or V_c,* is defined on the assumption that the static pressure, p, is equal to the ISA datum pressure, p_0, at all altitudes; that is $\delta = 1$ throughout the atmosphere. Clearly, this assumption will lead to a substantial systematic error that increases with altitude, this is known as the *Scale-Altitude Error.*

Making the assumption that $p = p_0$, eqn (2.20) can now be written as

$$\left(\frac{V}{a}\right)^2 = \left(\frac{V_c}{a_0}\right)^2 \tag{2.20}$$

and when substituted into eqn (2.19) gives,

$$p_p - p = p_0\left[\left\{1 + \frac{\gamma - 1}{2}\left(\frac{V_c}{a_0}\right)^2\right\}^{\frac{\gamma}{\gamma-1}} - 1\right] \tag{2.22}$$

This is known as the *full-law calibration equation* of the airspeed indicator. It is used for the measurement of airspeed by comparison between the Pitot and static pressures of the airflow sensed by the aircraft Pitot-static system.

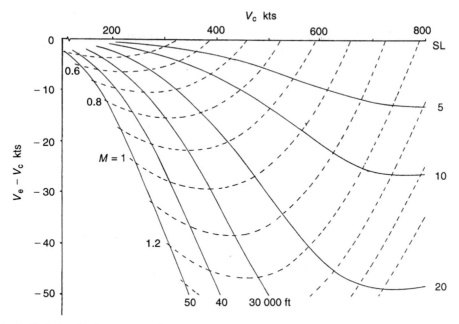

Fig. 2.12 Scale altitude error correction.

The scale-altitude error, resulting from the assumption that $p = p_0$ in the calibration equation, can be accounted for in the measurement of airspeed since the pressure–height relationship in the standard atmosphere model is known. The difference between the EAS, calculated from eqn (2.19a), and CAS, calculated from eqn (2.22), can be determined as a function of CAS and altitude to form the correction in order to obtain EAS from CAS. The correction is shown diagrammatically in Fig. 2.12 in terms of calibrated airspeed and altitude.

A number of definitions of airspeed have been introduced in the derivation of the calibration equation of the airspeed indicator; these can be summarized as follows, working back from the airspeed indicator to the freestream flow.

(a) Airspeed indicator reading; ASIR or V_I

This is the reading of an individual instrument. Since the instrument is a mechanical device driven by the Pitot-static pressures, there will be errors due to mechanical tolerance. These can be evaluated by calibration against an accurate source of pressure and a correction applied to give indicated airspeed.

(b) Indicated airspeed, IAS or V_i

This is the ASIR corrected for instrument error.

(c) Calibrated airspeed, CAS or V_c

This is the IAS corrected for Pitot-static pressure errors due to the location of the Pitot-static sources on the aircraft being within the disturbed pressure field caused by the presence of the aircraft (see Section 2.6).

(d) Equivalent airspeed, EAS or V_e

This is CAS corrected for the scale-altitude error implicit in the calibration equation of the airspeed indicator. The correction is a function of CAS and height, see Fig. 2.12.

(e) True airspeed, TAS or V

This is EAS corrected for air density, eqn (2.21). True airspeed represents the velocity between the aircraft and the air mass in, for example, m/s.

These definitions of airspeed are important and the differences between them need to be fully understood.

Equation (2.22), the full law calibration equation of the airspeed indicator, is a complex function on which to construct an airspeed indicating instrument and a simplified calibration law is often used where only relatively low airspeeds are to be measured. By expanding eqn (2.22), using the binomial theorem, the *simple law calibration* can be deduced, this is

$$p_p - p = \tfrac{1}{2}\rho_0 V_c^2 \left[1 + \frac{1}{4}\left(\frac{V_c}{a_0}\right)^2\right] \tag{2.23}$$

and is used for airspeed indicators for low performance aircraft in which the CAS does not exceed 460 kts. This will be adequate for all subsonic transport operations, but for high performance aircraft the more complex (and therefore more expensive) full law instrument will be required.

2.5.4 Measurement of Mach number

Mach number, M, is the ratio of the true airspeed, V, to the local speed of sound, a,

$$M = \frac{V}{a} \tag{2.24}$$

Although Mach number is often referred to as a speed this is not strictly correct since the speed of sound, given by eqn (2.17), is a function of the state of the atmosphere and can be expressed as,

$$a = a_0 \theta^{\frac{1}{2}} \tag{2.25}$$

where a_0 is the ISA datum value of the speed of sound. It is, therefore, necessary to define the temperature of the air mass before Mach number can be converted into a true airspeed. The relationship between Mach number and airspeed can be expressed as

$$Ma_0 = \frac{V}{\theta^{\frac{1}{2}}} = \frac{V_e}{\delta^{\frac{1}{2}}} \tag{2.26}$$

These alternative expressions occur frequently in statements of performance.

The Mach number of a flow determines certain flow qualities related to changes in flow density. At low Mach numbers, its effects are insignificant and the flow can be regarded as incompressible. As the Mach number increases the flow becomes compressible and density changes become significant. As the Mach number approaches unity, the character of the flow changes and discontinuities in its density appear. These discontinuities are shock waves that dominate the flow at supersonic Mach numbers. Knowledge of the Mach number is important since the change of the flow characteristic with Mach number will affect the drag of the aircraft and hence its performance.

From eqns (2.24) and (2.19), it can be seen that the Mach number can be measured from the flow pressures,

$$\frac{p_p - p}{p} = \left[1 + \frac{\gamma - 1}{2} M^2 \right]^{\frac{\gamma}{\gamma - 1}} - 1 \qquad (2.27)$$

This is the calibration equation of a Machmeter for use in subsonic flight. In supersonic flight, shock waves form at the Pitot tube and the calibration equation for flight at Mach numbers exceeding unity considers the associated pressure effects. The calibration equation, eqn (2.27), combines the driving function of the airspeed indicator with that of the altimeter. This explains why the Machmeter cannot be regarded as a direct indicator of airspeed.

2.5.5 Measurement of air temperature

Any statement of aircraft performance must contain information concerning the static temperature of the air mass; this may be referred to as ambient temperature, outside air temperature or freestream static temperature. The air temperature sensor, however, senses temperature in a moving airstream and, since the flow will be brought to rest by the sensor, it will tend to measure the total, or stagnation, temperature of the flow.

A typical air temperature sensor is usually based on the principle of a ventilated Pitot tube containing a resistance wire sensing element of known temperature–resistance characteristic (see Fig. 2.13). The flow entering the sensor is almost brought to rest at the sensing element, some flow being allowed to leak through a restriction at the rear of the tube to allow the flow sample to be changed continuously. This enables the sensor to respond to changes in the ambient air temperature. The sensor is surrounded by second and third concentric Pitot tubes, also ventilated by flow

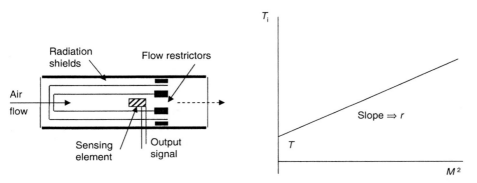

Fig. 2.13 Ventilated Pitot air temperature sensor.

restrictors, which surround the sensing element with air at the (near) total temperature. These act as radiation shields to avoid errors in temperature measurement due to solar radiation or local high temperatures heating the sensing element as well as providing mechanical protection to the sensing element.

Using the energy relationship (eqn (2.14a)) for the flow brought to rest, the total temperature, T_t, is related to the static temperature, T, by

$$C_p T_t = C_p T + \tfrac{1}{2} V^2 \tag{2.28}$$

and from eqns (2.24), (2.25) and (2.26) this becomes

$$T_t = T \left[1 + \frac{\gamma - 1}{2} M^2 \right] \tag{2.29}$$

Due to the design of the air temperature sensor, the flow is not quite brought to rest so that the temperature sensed by the resistive element will tend to be less than the adiabatic total temperature. In addition, there may be convective and conductive heat losses from the sensing element. The ratio between the measured, or indicated, temperature rise, $T_i - T$, and the ideal temperature rise, $T_t - T$, is called the recovery factor, r, of the air thermometer and is similar to an efficiency factor. The recovery factor can be expressed as,

$$r = \frac{T_i - T}{T_t - T} \tag{2.30}$$

which, when included in eqn (2.29), gives the indicated total temperature, T_i, as a function of Mach number,

$$T_i = T \left[1 + r \frac{\gamma - 1}{2} M^2 \right] \tag{2.31}$$

The recovery factor can be found by flying at a number of steady Mach numbers (or airspeeds) at a constant height and plotting the indicated total temperature against M^2, see Fig. 2.13. The static temperature is found by extrapolating to $M =$ zero and the recovery factor is determined from the slope of the line. Once the recovery factor is known, the static temperature can be found from the indicated total temperature and the flight Mach number using eqn (2.31).

2.6 Practical considerations of air data measurement

Although the methods of measurement of the primary air data (altitude, airspeed and Mach number) can be readily deduced from the Pitot and static pressures of the flow, in practice the measurement of these quantities may require consideration of the effect of the aircraft on the local airflow.

When an aircraft is in flight it disturbs the local airflow and creates a pressure field around itself. This is partly because of the wing producing the lift force, but also due to the volume of the aircraft displacing the air mass as it passes through it. The local changes in velocity within the region of disturbed airflow will cause corresponding changes in the local static pressure but, from Bernoulli's equation, the total pressure of the flow will remain constant since the process is adiabatic. For practical reasons, the static pressure has to be measured close to the aircraft. Therefore, it will be measured within the disturbed pressure field and it may differ from the freestream static pressure assumed in the calibration equations of the air data instruments. The indicated values of the altitude, airspeed and Mach number resulting from the measured values of the local, or system, pressures will differ from the values that would occur when using the undisturbed freestream pressures. This error is known as *system pressure error*. The system pressure error can be measured in flight by, for example, trailing a static pressure sensor behind the aircraft in the undisturbed airflow and comparing the system static pressure with the undisturbed, freestream static pressure. The difference between them can be used to evaluate the corrections to the indicated air data, measured by the aircraft system, to give the freestream values of the air data.

The Pitot pressure will not be affected by the disturbance to the flow since, in adiabatic flow, the total pressure remains constant. Provided that the Pitot tube is located outside the boundary layer of the aircraft and is not behind any source of work or heat – for example, a propeller or exhaust – and is aligned closely to the local flow direction, it will register the true total pressure of the flow in subsonic conditions.

The Pitot-static system pressures are not the only air data system parameters that are affected in this way. Since the direction of the local airflow close to the aircraft will be distorted by the presence of the aircraft, the airflow direction detectors, ADDs, will also need to be calibrated. For example, consider the calibration of the ADD measuring angle of attack. The local flow direction relative to aircraft body axes is measured by the ADD and the undisturbed freestream flow direction is derived, for example, from the pitch attitude of the aircraft in straight, steady, level flight. Comparison provides a calibration for the ADD.

2.7 Air data computers

So far, the measurement of the air data has been considered in terms of individual, mechanical instruments, each providing an indication of one air data parameter, Fig. 2.10. In many applications, the air data instruments have been replaced by an air data computer that carries out all the functions of the air data instruments. In addition, it applies the known corrections to give the airspeed, altitude and air temperature in their most convenient form either for display to the pilot or to be transmitted to other systems on the aircraft, see Fig. 2.14.

In the air data computer, total and static pressures from the aircraft Pitot-static system are fed into sensors that are similar to the pressure sensing capsules in the individual airspeed indicator and altimeter instruments. Instead of driving a pointer, the capsules operate transducers that convert the pressures into electrical signals. The signals are transmitted to the processor and used to calculate the indicated values of airspeed, altitude and Mach number using the same analytical functions as the individual instruments. If the Pitot-static system pressure errors are known they can be applied by the computer to give CAS, pressure height and Mach number. In addition, the scale–altitude correction can be applied to the CAS to enable the EAS to be determined. The total air temperature sensed by the air thermometer

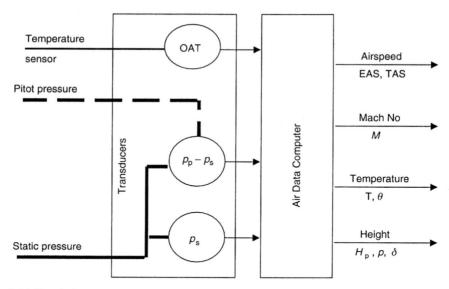

Fig. 2.14 The air data computer system.

can be corrected to static air temperature using the Mach number and the recovery factor of the installation. The air density and true airspeed, TAS, can now be found. It will be noted that the air data computer depends on the calibration of the Pitot-static system for system pressure error, and on the calibration of the air thermometer installation for the recovery factor, in exactly the same way as did the individual air data instruments.

2.8 Conclusions

The simple, linearized, atmosphere model (the International Standard Atmosphere or a design atmosphere) is used as the design standard for the aircraft. It provides a reference for the estimation of its performance in terms of the altitude and temperature states in which the aircraft is intended to operate. It is also used as a datum for the measured performance and the aircraft flight manual will usually refer to operational performance in terms of both the ISA temperature profile (or parallel temperature profiles) and a general atmosphere temperature. The International Standard Atmosphere however, is only a representative model of the atmosphere and should not be taken to be a statement of the conditions of the actual atmosphere.

Knowledge of the state of the real atmosphere in which the aircraft is flying is of prime importance to performance. It will be seen in later chapters that the combination of altitude and temperature affects the performance of the aircraft and may limit its performance to a level that would not allow a sufficient margin for safety. The real atmosphere, however, is a very variable medium and its properties need to be measured in flight so that the actual performance of the aircraft can be interpolated from the operational performance charts contained in the flight manual. Thus, the altimeter (defining the static pressure) and the air temperature sensor – which, between them, determine the atmosphere state – are an essential part of the aircraft instrumentation system relating to performance.

In the following chapters, the optimum performance of an aircraft will be seen to be related to airspeeds defined by the aircraft's aerodynamic characteristics. The airspeed needs to be measured in a form that enables the aircraft to be operated at its best performance.

- Where the performance is a function of dynamic pressure, or expressed in terms of dynamic pressure, equivalent airspeed is the most convenient form of measurement. Equivalent airspeed provides a means of measuring the dynamic pressure of the airflow directly. When equivalent airspeed is used, there is no need to consider the air density so that performance expressed in terms of equivalent airspeed is generally independent of the atmosphere state.
- When high airspeeds are encountered, flow compressibility becomes significant and the aerodynamic characteristics of the aircraft may be affected. Under these conditions the Mach number becomes the dominant speed-related term and the performance of aircraft operating at Mach numbers above about 0.6 is generally expressed in terms of the Mach number rather than airspeed. Since the dynamic pressure derived from the Mach number includes both pressure and Mach number terms, $q = \frac{1}{2}\gamma p M^2$, both altitude and Mach number will normally need to be quoted in any statement of performance expressed in terms of Mach number.
- True airspeed is only needed when the flight path of the aircraft is being considered; for example, in navigation or the calculation of gradients of climb. Since the density of the air is needed to determine true airspeed, measurement of altitude and temperature will be required.

Measurements of atmosphere and air data are fundamental to any statement of performance; this will be seen in later chapters dealing with the operational aspects of aircraft performance.

Bibliography

Barry, R. G. and Chorley, R. J. (1992) *Atmosphere, Weather and Climate* 6th edn (Rouledge).

CAA. *Joint Airworthiness Requirements, JAR-1, Definitions and Abbreviations* (Airworthiness Authorities Steering Committee).

Pogosyan, Kh. P. (1965) *The Air Envelope of the Earth*. NASA TT-F-287.

US Committee on Extension to the Standard Atmosphere (1976) *US Standard Atmosphere, 1976* (US Government Printing Office).

Engineering sciences data unit items
Performance Series, Vols 1–14.

Atmospheric Data for Performance Calculations. Addendum: heights in feet, data in SI units. Vol 2, No 72018, 1972, and Vol 2, No 68046, 1977.

Equations for the Calculation of International Standard Atmosphere and Associated Off-Standard Atmospheres. Vol 2, No 77022, 1986 (Am B).

Height Relationships for Non-Standard Atmospheres. Vol 2, No 78012, 1996 (Am B).

Airspeed Data for Performance Calculations. Vol 2, No 69026, 1992 (Am B).

Introduction to Air Data System Parameters, Errors and Calibration Laws. Vol 11, No 86031, 1991 (Am A).

The Correction of Flight Test Anemometric Data. Vol 11, No 83029, 1992 (Am D).

Pitot and Static Errors in Steady Level Flight. Vol 12, No 86006, 1992 (Am A).

Conversion of Air-data System Pressure Errors into Height and Speed Corrections. Vol 12, No 85036, 1993 (Am B).

The Treatment of Calibrations of Total Air Temperature Probes for use in Flight Test Analysis Work. Vol 12, No 84007, 1991 (Am A).

Aerodynamics Series

Properties of a Standard Atmosphere. Vol 1b, No 77021, 1986 (Am A).

Physical Properties of Design Atmospheres. Vol 1b, No 78008, 1992 (Am C).

3

The force system of the aircraft and the equations of motion

Before the performance of the aircraft can be analysed, it is necessary to consider the system of forces acting on the aircraft and to develop the equations of motion that govern its flight. The characteristics of the forces acting on the aircraft, and their variation with the principal performance variables, also need to be discussed.

3.1 The equations of motion for performance

The equations of motion of the aircraft are statements of Newton's law, $F = ma$, in each of three mutually perpendicular axes. The general force, F, is the sum of the components of a system of forces acting on the aircraft, which results in the inertial force, ma. The system of forces acting on the aircraft can be categorized into four groups; the gravitational forces, F_g, the aerodynamic forces, F_a, and the propulsive forces, F_p, which result in the inertial forces, F_I, so that the statement of Newton's law becomes,

$$F_a + F_p + F_g = F_I \tag{3.1}$$

There will also be a system of moments acting on the aircraft but, as these do not affect the flight path directly, they do not need to be taken into account in the equations of motion for performance. Each group of forces acts in its own axis system and needs to be resolved into the velocity axis system before the equations of motion can be developed. The axis systems are described in full in Appendix A and the full equations of motion for aircraft performance are developed in Appendix B. Only a summary of the characteristics of the forces and the equations of motion will be considered here.

In Appendix B the full development of the equations of motion of the aircraft led to the general equations of motion, eqn (B19):

$$\left.\begin{aligned}
&-D + [T\cos(\alpha + \tau_1)\cos\beta - D_M] - mg\sin\gamma_2 \\
&\qquad = m\dot{V} + \dot{m}V \\
&Y\cos\gamma_1 + L\sin\gamma_1 + T[-\cos(\alpha + \tau_1)\sin\beta\cos\gamma_1 + \sin(\alpha + \tau_1)\sin\gamma_1] \\
&\qquad = mV\dot{\gamma}_3\cos\gamma_2 \\
&Y\sin\gamma_1 - L\cos\gamma_1 + T[-\cos(\alpha + \tau_1)\sin\beta\sin\gamma_1 - \sin(\alpha + \tau_1)\cos\gamma_1] + mg\cos\gamma_2 \\
&\qquad = -mV\dot{\gamma}_2
\end{aligned}\right\} \tag{3.2}$$

In these equations of motion some simplifying assumptions have already been made, these include the assumption that all engines are operating at equal gross thrust. For conventional aircraft, additional assumptions can be made to simplify the equations further, these are:

(i) that the rate of change of aircraft mass is negligible, $\dot{m} = 0$,
(ii) that the aircraft is in symmetric flight so that $\beta = 0$ and $Y = 0$,
(iii) that the gross thrust acts in aircraft body axes, $\tau_1 = 0$,
(iv) that the total net thrust $F_N = [T\cos\alpha - D_M]$, and
(v) that the thrust component $T\sin\alpha$ is small when compared with the lift force.

When these assumptions are made, the equations of motion reduce to a simplified form that can be used for most performance analysis tasks:

$$\left.\begin{aligned}
F_N - D - mg\sin\gamma_2 &= m\dot{V} \\
L\sin\gamma_1 &= mV\dot{\gamma}_3\cos\gamma_2 \\
-L\cos\gamma_1 + mg\cos\gamma_2 &= -mV\dot{\gamma}_2
\end{aligned}\right\} \tag{3.3}$$

The majority of performance analysis is based on the longitudinal equation of motion in which the term, $F_N - D$, is known as the excess thrust and provides the increase in potential energy (climb), or the increase in kinetic energy (acceleration).

The equations of motion stated in eqns (3.2) and (3.3) are written in terms of aircraft with thrust-producing engines. If the aircraft has power-producing engines, which drive propellers to convert the power into thrust, then the equations must be converted into their power form; this will be considered later in the section on propulsive forces.

3.2 The aircraft force system

In the development of the equations of motion, the forces acting on the aircraft are represented as simple force terms and appear as constants. However, the forces stated in eqn (3.1) and in the equations of motion are not simple forces but depend on the performance variables, aircraft weight, airspeed (or flight Mach number) and the state of the atmosphere. In particular, the aerodynamic forces and the propulsive forces are of great importance to the performance of the aircraft. Their characteristics will define, for example, the airspeeds for best climb rate and gradient and for optimum range or endurance in the cruise part of the flight.

Each group of forces can be considered in turn to determine how its characteristics vary with the flight variables.

The inertial forces, F_I

The inertial forces arise from the mass of the aircraft and its acceleration. The accelerations may be linear accelerations or result from the combination of the forward speed of the aircraft with its rates of pitch and turn. The inertial forces act in the velocity axis system, which is discussed fully in Appendix B.

The gravitational forces, F_g

The gravitational force acts downwards in the Earth axis system and is the product of the aircraft mass, m, and the acceleration due to gravity, g. It may be referred to either as weight, W, or as the product mg; each form of reference has its own applications within the theory and practice of aircraft performance.

The aerodynamic forces, F_a

The aerodynamic forces arise from the relative motion between the aircraft and the air mass in which it is flying; they act in the wind axis system. It will be assumed that the reader is familiar with the concepts of aerodynamics and this treatment will only consider the aerodynamic characteristics of the aircraft that are directly applicable to the study of performance.

The dynamic pressure of the airflow, q, may be considered in terms of either airspeed or Mach number,

$$q = \tfrac{1}{2}\rho V^2 = \tfrac{1}{2}\gamma p M^2 \tag{3.4}$$

Whilst either form may be used when considering the non-dimensional aerodynamic forces, the form involving the Mach number is particularly useful when considering operational performance. If the airspeed is considered in terms of the flight Mach number, then the temperature of the atmosphere is implicit in the statement of the Mach number and the atmosphere pressure can be considered independently. Since altitude is related uniquely to the static pressure of the atmosphere, the altitude becomes a basic variable of the aerodynamic forces. Therefore, the forces need to be considered only in terms of their variation with aircraft weight, flight Mach number and altitude rather than in terms of aircraft weight, airspeed, altitude and temperature.

The aerodynamic forces that concern performance are the lift, L, the drag, D, and the sideforce, Y. In the case of an aircraft, the speed of flight is relatively high and the non-dimensional flow variables that characterize the flow are,

(i) The Reynolds number, $R_e = \dfrac{\rho V l}{\mu}$

Typically the flight value of R_e is large, 10^6 to 10^9, and the flow can be treated as continuum flow. If the aerodynamic characteristics of the aircraft have been determined from experimental sources (e.g. wind tunnels), any Reynolds number effects should have been accounted for before being used in any performance estimation process. It is unlikely that the Reynolds number will influence the analysis of the full-scale flight performance of the aircraft significantly, except in extreme cases.

(ii)   The Mach number, $M = \dfrac{V}{a}$

This may vary from almost zero up to a typical maximum of 2.2 for conventional aircraft; higher Mach numbers are possible but raise special problems. Since this treatment of performance is concerned mainly with subsonic flight, the supersonic flow characteristics will not be considered in depth. Only in the transonic region, where the Mach number affects the aerodynamic characteristics of the aircraft, at the higher subsonic Mach numbers, will the effects be considered. In flight up to Mach numbers of 0.5 the flow can be regarded as incompressible and Mach number effects ignored; for $0.5 < M < 0.8$ compressibility becomes significant and may lead to small changes in the lift and drag force characteristics. For most subsonic aircraft the critical Mach number occurs typically around $M = 0.8$; at this Mach number the local flow at points on the aircraft becomes supersonic and shock waves begin to form. This effect starts the change from subsonic to supersonic flow and affects the characteristics of both the lift and drag forces, leading to significant effects on the performance of the aircraft. Mach number is one of the most important variables of performance and its effect on the aerodynamic forces needs to be considered.

3.3 The aerodynamic force characteristics

3.3.1 The lift force, L

The lift force is generated mainly by the wing, but other parts of the aircraft will also produce contributions to the overall lift. The general expression for the lift force relates the lift to the angle of attack of the airflow relative to the aircraft,

$$L = \tfrac{1}{2}\rho V^2 S C_L = \tfrac{1}{2}\rho V^2 S a(\alpha - \alpha_0) \tag{3.5}$$

where α is the lift curve slope, $\mathrm{d}C_L/\mathrm{d}\alpha$, and α is the angle of attack measured from the zero-lift angle of attack, α_0.

The lift characteristic of the plain, cambered aerofoil is shown in Fig. 3.1. There is an angle of attack, α_0, at which the aerofoil produces zero lift; the zero-lift angle of attack is zero if the aerofoil is symmetrical, and negative in the case of a positively cambered aerofoil. As the angle of attack increases, the lift coefficient increases in proportion and the slope of the lift characteristic is known as the lift curve slope. The lift curve slope has a theoretical value of 2π per radian if the aerofoil is a flat plate of infinite span, but this is increased by the thickness of the aerofoil section and reduced as the aspect ratio decreases. A typical range of values for the lift curve slope is between 4 and 6 per radian depending on the aerofoil section and wing geometry. A straight wing of aspect ratio around 10 and an aerofoil with a thickness of about 12% will have a lift curve slope of about 5.7/rad.

As the angle of attack increases, so the lift coefficient increases until the pressure distribution over the aerofoil section starts to cause separation of the flow. This

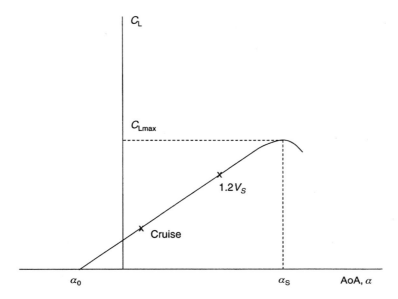

Fig. 3.1 Lift characteristic of a plain, cambered aerofoil.

causes the lift curve slope to decrease as the angle of attack increases and a point is reached when the slope becomes zero; this is the point of maximum lift coefficient, $C_{L\,max}$, which denotes the stall. The angle of attack at the stall, α_s, is known as the stalling angle of attack and is the greatest angle of attack at which the aircraft can be maintained in steady, '1g' flight. Any further increase in angle of attack will produce a decrease in lift coefficient and the lift force is then less than the weight of the aircraft. In this state, the aircraft will sink and, usually, pitch nose-down in the stall. The stall denotes the boundary of controlled flight and defines the low speed limit of the performance envelope of the aircraft. The stall is normally preceded by aerodynamic buffeting caused by the separation of the flow. This acts as a natural stall warning and the stall buffet boundary is sometimes used as the low speed limit to performance; the airworthiness requirements contain a number of definitions of the stall and stall boundaries. Since the stall is an uncontrollable state of flight, all speeds scheduled for operational manoeuvres will have a margin of safety over the stall speed.

The lift characteristic can be modified by leading edge and trailing edge flaps (and other devices), so that the aerodynamic properties of the wing are better suited to the different performance regimes. Figure 3.2 shows the general effects of leading and trailing edge flaps.

- The basic plain aerofoil is optimized for cruising flight; it has low drag and cruising flight takes place at a low angle of attack and hence a low lift coefficient. However, the stalling lift coefficient of the plain aerofoil would be too low for the take-off and landing manoeuvres and would result in speeds for these manoeuvres that would be too high. Assuming a safety margin of speed over the stall, the minimum speed in a manoeuvre will be typically $1.2V_s$ and the speed scheduled for take-off or landing will be based on a lift coefficient of $0.7C_{L\,max}$ (Fig. 3.1).

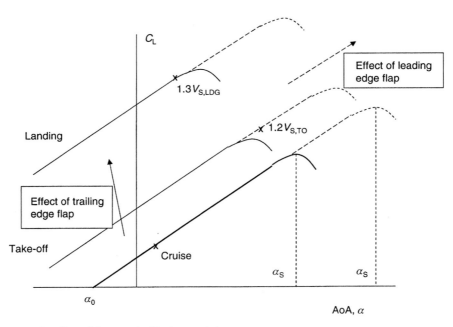

Fig. 3.2 The effect of flaps on the lift characteristic.

- Leading edge flap deflection has the effect of extending the lift curve to a higher stalling angle of attack, and hence lift coefficient. This would enable the take-off and landing speeds to be reduced, but it would result in a high nose-up attitude because of the large stalling angle of attack. The leading edge flap will also increase the drag, particularly at a low angle of attack.
- The deflection of the trailing edge flaps has the effect of increasing the camber of the aerofoil section and thus shifting the lift characteristic upwards as the zero-lift angle of attack becomes more negative. There is also a tendency to decrease the stalling angle of attack slightly. The trailing edge flap allows higher lift coefficients to be achieved at lower angle of attack and, thus, at lower pitch attitudes. The deployment of the trailing edge flap is often made in several stages. First, a rearward translation of the flap without significant deflection extends the wing area. Effectively, this decreases the wing loading and permits an increase in lift coefficient. Secondly, deflection of the extended flap increases the aerofoil camber. Effectively, this shifts the lift curve upwards and increases the lift coefficient for a given angle of attack. There may be a number of stages of deflection optimized for take-off, climb, descent, approach and landing.

Flap systems are often combined with slats and slots, and a flap extension may open a slot between the flap and wing, or expose a slat, to assist the flow over the aerofoil. A combination of leading edge and trailing edge flap can be found that permits the take-off and landing manoeuvres, and other manoeuvres, to be carried out at reasonable speeds and safe pitch attitudes. Figure 3.2 shows typical flap and angle of attack combinations for the principal states of flight.

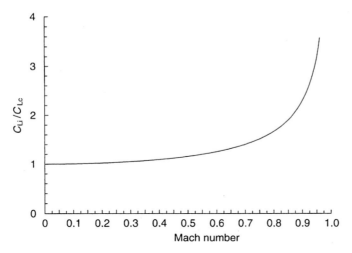

Fig. 3.3 The compressible lift coefficient.

The effect of Mach number on lift

The main flight variable that affects the characteristic of the lift force is the Mach number. As the Mach number of the airflow increases, so the characteristics of the flow change from those of an incompressible fluid to those of a compressible fluid. This modifies the pressure coefficients, and hence the force coefficients, generated by the aircraft. The compressible flow coefficients are related to the incompressible flow coefficients by the Prandtl–Glauert factor, β, so that the compressible lift co-efficient is given by,

$$C_{Lc} = C_{Li}/\beta \qquad\qquad (3.6)$$

where $\beta = \sqrt{(1 - M^2)}$ for $M < 1$.

The ratio between the compressible and incompressible lift coefficients is shown in Fig. 3.3.

Whilst this effect appears to be very significant when seen in terms of the lift coefficient, its real effect is felt on the angle of attack of the aircraft. Since the aircraft flies at (almost) constant weight, the lift coefficient decreases with Mach number squared and, at high subsonic Mach numbers, the angle of attack of the aircraft will be small. Figure 3.4 shows the typical effect of Mach number on the angle of attack required for steady, level, flight at constant aircraft weight in compressible flow when compared with incompressible flow. It can be seen that the effect of Mach number on the angle of attack is relatively small. Therefore, it is not likely to produce very significant effects on angle of attack dependent variables in the normal, subsonic, range of the operating Mach number.

3.3.2 The sideforce, Y

The aerodynamic sideforce generated by the aircraft arises from sideslipping flight. It can be regarded as a 'lateral lift' due to the sideslip angle, β, which acts as a 'lateral

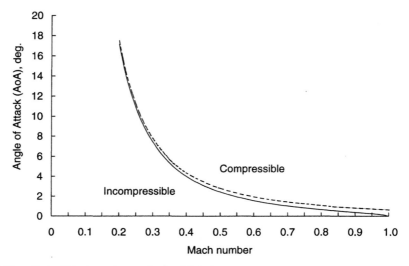

Fig. 3.4 The effect of Mach number on Angle of Attack.

angle of attack'; the comments on the lift force can be generally applied to the sideforce. In symmetric flight there is no sideslip and the aerodynamic sideforce will be zero. Except in special cases in which the aircraft is in asymmetric flight, for example – flight with asymmetric thrust following an engine failure – the sideforce has little significance on performance.

3.3.3 The drag force, *D*

The drag force is the most important aerodynamic force in aircraft performance. In subsonic flight, it is made up of several components, each of which has its own characteristics. The components are the lift independent drag, D_z, the lift dependent drag, D_i, and, at high subsonic Mach numbers, a volume dependent wave drag, D_{wv}. The sum of the drag components makes up the total drag of the aircraft.

It is usually assumed in the analysis of subsonic performance that the drag polar of the aircraft is parabolic and represented by the lift dependent and lift independent terms only, the drag coefficient being given by,

$$C_D = C_{Dz} + KC_L^2 \tag{3.7}$$

where C_{Dz} and K are constants.

Whilst this approximation is used to develop the basic functions of aircraft performance it should be remembered that the real drag characteristic will not be purely parabolic but will contain terms dependent on Mach number. Moreover, particularly at the higher subsonic Mach numbers, the drag characteristic of the aircraft may deviate considerably from the parabolic approximation. In the following subsections, each element of the drag force will be considered separately and the

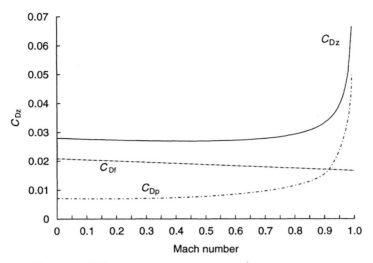

Fig. 3.5 The zero-lift drag coefficient.

effect of the flight variables, Mach number, weight and altitude, will be assessed on each element.

The lift independent drag, D_z

The lift independent drag coefficient can be broken down into two parts, the surface friction drag and the profile drag. The surface friction drag coefficient, C_{Df}, usually accounts for about 75% of the lift independent drag and tends to decrease slightly as the Mach number increases, as the result of a Reynolds number effect. The profile drag coefficient, C_{Dp}, which accounts for the other 25% of the lift independent drag, is a pressure dependent drag. This is affected by the Prandtl–Glauert factor in the same manner as the lift coefficient, increasing rapidly as the Mach number approaches unity, see Fig. 3.5.

Here, it can be seen that the value of C_{Dz} remains almost constant up to a Mach number of about 0.7; this is typical for a conventional subsonic aircraft.

When the compressible, zero-lift, drag coefficient is multiplied by the dynamic pressure, q, to turn it into a force, the effect of the Mach number can be seen when compared with the assumption of the constant C_{Dz} from the parabolic drag polar, see Fig. 3.6. There is good agreement between the predicted drag forces up to a Mach number of about 0.8, above which the compressible flow drag force increases significantly.

The forces are expressed here as *Drag Area, D/S*, which is a convenient way of expressing the drag without involving the scale of the aircraft:

$$D/S = qC_D = \tfrac{1}{2}\gamma p M^2 C_D$$

The zero-lift drag force is directly proportional to the atmospheric pressure, p, since the drag force is proportional to the dynamic pressure, q, (eqn (3.4)). Thus, for flight at a given Mach number, the zero-lift drag force will decrease as altitude increases since the atmospheric pressure decreases as a function of altitude (see Chapter 2).

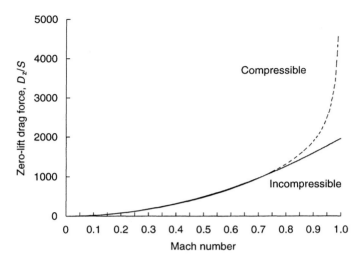

Fig. 3.6 Effect of Mach number on the zero-lift drag force.

Aircraft weight has no effect on the zero-lift drag force.

The lift dependent drag D_i

The lift dependent, or vortex, drag coefficient, C_{Di}, is a function of the angle of attack, α, and is usually taken to be

$$C_{Di} = KC_L^2 \qquad (3.8)$$

where K is generally assumed to be $1/\pi Ae$ in incompressible flow.

This approximation is based on the aspect ratio of the wing, A, and the span efficiency factor, e, which is a function of the spanwise wing load distribution. However, there may be contributions to the lift force from parts of the aircraft other than the wing, notably the tailplane, and basing the lift dependent drag factor, K, on the wing alone is likely to be optimistic. Flow separation at low airspeeds may also contribute to the effective value of the lift dependent drag factor, although it may not be strictly dependent on the lift force itself. In addition, the vortex drag is a function of angle of attack, α, and the Mach number effect on α, shown in Fig. 3.4, will produce a further contribution to the value of K. The value of the lift dependent drag factor, K, will usually have to be determined experimentally but it can be generally accepted as being reasonably constant over the working range of the lift coefficient.

The lift dependent drag force, D_i, is given, as a drag area, by,

$$D_i/S = qKC_L^2 = \frac{K(W/S)^2}{\frac{1}{2}\gamma p M^2} \qquad (3.9)$$

and is shown in Fig. 3.7, for a given weight and altitude combination.

Since the lift dependent drag force is inversely proportional to the dynamic pressure, q, it will decrease with Mach number squared and increase with increasing altitude. Increasing aircraft weight will also increase the lift dependent drag force.

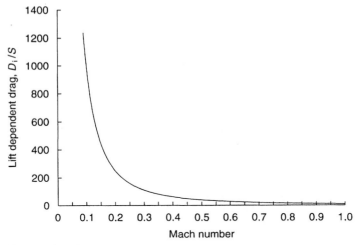

Fig. 3.7 The lift dependent drag.

The volume dependent wave drag, D_{wv}

As the aircraft passes through the air mass its volume displaces the flow and produces local disturbances in flow velocity. At the critical flight Mach number, M_{crit}, the local flow at points on the aircraft becomes supersonic and shock waves begin to form, growing in strength as the flight Mach number increases. The energy required to sustain these shock waves manifests itself as a drag force that increases rapidly as the flight Mach number exceeds its critical value. There is no simple expression for the volume dependent wave drag. However, experimental results indicate that, above the critical Mach number, the volume dependent wave drag coefficient is related to the volume, and other dimensions, of the aircraft by a relationship – based on the slender body theory – of the form,

$$C_{Dwv} \propto K_0 \text{Vol}^2 \tag{3.10}$$

where K_0 is a shaping factor, which is a function of Mach number. A first-order approximation to K_0 is that K_0 increases as Mach number squared above M_{crit} in the transonic region. In supersonic flight beyond the transonic region, K_0 tends to decrease. On this assumption, the volume dependent wave drag can be expected to increase as the fourth power of Mach number in the transonic region. This indicates the significance of the wave drag term in the drag characteristic of the aircraft above the critical Mach number, as shown in Fig. 3.8.

As in the case of the zero-lift drag, the volume dependent wave drag will decrease as altitude increases for a given Mach number and is independent of aircraft weight.

The overall drag force, D

The overall drag force is the sum of the components of the drag force, the zero-lift drag, the lift dependent drag and the volume dependent wave drag. Each component has been shown to be a function of Mach number, altitude (or pressure) and, in the case of the lift dependent drag, aircraft weight. The drag characteristic is shown in Fig. 3.9 for a given weight and altitude combination.

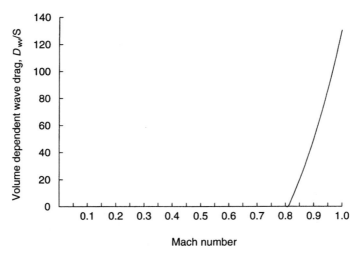

Fig. 3.8 The volume dependent wave drag.

Figure 3.9 shows that, below the critical Mach number, there is a reasonable comparison between the compressible flow drag characteristic and the incompressible approximation. This justifies the use of the simple, incompressible, parabolic drag polar in the development of the basic expressions of performance. However, it should be remembered that the parabolic drag polar is an approximation and that any performance characteristics estimated on the assumption of a parabolic drag polar will not be exact. In practice, it will be necessary to measure the performance of the aircraft in flight to define the actual performance achieved. At Mach numbers above the critical value, the drag force increases rapidly and the approximation becomes invalid; any estimation of the aircraft performance above M_{crit} will need consideration of the full drag characteristic of the aircraft.

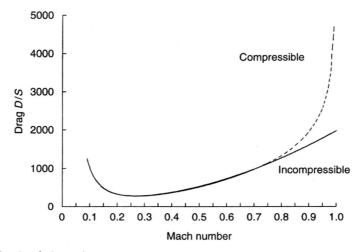

Fig. 3.9 The aircraft drag polar.

The minimum drag speed

In eqn (3.7), the drag characteristic was taken to be

$$C_D = C_{Dz} + KC_L^2 \tag{3.7}$$

and has been shown reasonably to represent the subsonic aircraft at Mach numbers below M_{crit}. If the drag characteristic is factored by $\frac{1}{2}\rho_0 V_e^2 S$ to convert it into force units, eqn (3.7) becomes

$$D = YV_e^2 + \frac{ZW^2}{V_e^2} = D_z + D_i \tag{3.11}$$

where $Y = \frac{1}{2}\rho_0 S C_{Dz}$ and $Z = K/\frac{1}{2}\rho_0 S$, both of which are constants. Figure 3.10 shows the total drag force, and its two components, for a given aircraft weight, W.

Differentiating eqn (3.11) with respect to EAS leads to the expression for the minimum drag speed,

$$V_{emd} = \sqrt[4]{\frac{ZW^2}{Y}} \tag{3.12}$$

This occurs when the two components of the drag force are equal. The minimum drag speed is important to performance and it will be seen in later chapters that it determines the best operating speeds of aircraft with thrust producing engines. The relative magnitudes of the zero-lift drag, D_z and the lift dependent drag, D_i, will affect the minimum drag force and the minimum drag speed. If the zero-lift drag is reduced then the total drag will be reduced but the minimum drag speed will be increased. If the lift-dependent drag is reduced then the total drag will be reduced and the minimum drag speed will be reduced. The ability to adjust the minimum drag speed in this way is an important tool in the design of the aircraft performance characteristics for different regimes of flight.

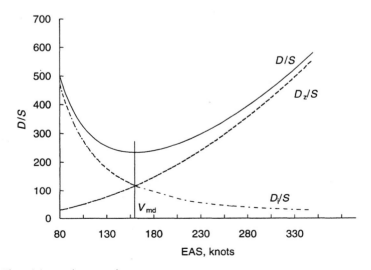

Fig. 3.10 The minimum drag speed.

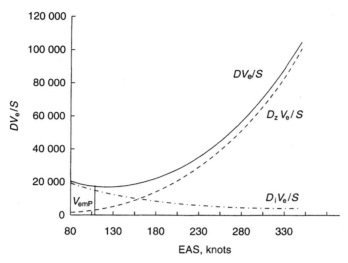

Fig. 3.11 The minimum power speed.

The minimum power speed

Power is the product of force and velocity and the performance equation can be considered in terms of the thrust-power, P, required, rather than the thrust force required. The drag-power is given by multiplying the drag force, D, eqn (3.11), by the true airspeed, V, so that, in terms of EAS the drag power equation becomes,

$$P\sqrt{\sigma} = DV_e = YV_e^3 + \frac{ZW^2}{V_e} \qquad (3.13)$$

and is shown in Fig. 3.11. (Here it should be noted that the true airspeed, V, has been converted to equivalent airspeed, $V_e/\sqrt{\sigma}$, before the multiplication.)

Differentiating eqn (3.13) with respect to EAS leads to the expression for the minimum power speed,

$$V_{emP} = \sqrt[4]{\frac{ZW^2}{3Y}} \qquad (3.14)$$

The minimum power speed is important to the performance of aircraft with power producing engines. It will be seen in later chapters that it determines the best operating speeds of aircraft with power producing engines in the same way that the minimum drag speed determines the optimum performance of aircraft with thrust producing engines. The relative magnitudes of the zero-lift drag, D_z, and the lift dependent drag, D_i, will affect the minimum drag power and the minimum power speed in the same general way as they affected the minimum drag force and minimum drag speed.

Although the minimum drag speed and minimum power speed are related by a simple numerical factor, $\sqrt[4]{3} = 1.316$, they should not be considered to be simply related in their application to aircraft performance. The minimum drag speed relates to the performance of aircraft with thrust producing engines, whilst the minimum power speed relates to aircraft with power producing engines.

Some further relationships of the drag characteristic will be summarized in Section 3.5.

3.4 The propulsive forces

There are two basic forms of powerplant used for aircraft propulsion:

- The thrust-producing powerplant, which produces its propulsive force directly by increasing the momentum of the airflow through the engine, and
- The power-producing powerplant, which produces shaft power that is then turned into a propulsive force by a propeller.

Each form of powerplant has different characteristics and needs to be considered separately.

3.4.1 The thrust-producing powerplant

The usual form of thrust-producing engine is the turbojet or turbofan, although rockets could be included in this category.

The turbojet or turbofan engine uses atmospheric air as its working fluid and, with the addition of fuel, burns the air to increase its energy. The high-energy air is then expelled through a nozzle with increased momentum to produce the thrust force. The principle is shown in Fig. 3.12. Atmospheric air flows into the intake where it is slowed down to a velocity that can be accepted by the compressor. After compression, fuel is mixed with the air and the mixture is burned in the combustion chamber. The hot gas produced is passed through a turbine that extracts energy to drive the compressor and the exhaust is expelled through a nozzle that converts its remaining energy into thrust.

In simplified terms, the turbojet engine can be considered to produce thrust by increasing the momentum of its internal flow stream. The *net propulsive force*, F_N, is the difference between the stream force entering the engine and the stream force exiting the engine. The thrust produced at the exit plane of the nozzle is known as the *gross thrust*, F_G, and is equal to the rate of change of momentum of the exhaust gas flow, $F_G = \dot{m}V_j$. The flow into the intake also contributes to the engine thrust. In this case, the momentum of the flow is lost as the air enters the engine. The

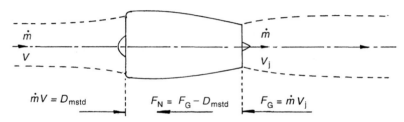

Fig. 3.12 The thrust-producing powerplant.

force due to the intake flow, known as the *momentum drag*, D_m, is equal to the rate of change of momentum in the intake airflow, $D_m = \dot{m}V$. The net propulsive thrust is given by

$$F_N = F_G - D_m = \dot{m}V_j - \dot{m}V \tag{3.15}$$

Turbofan engines may have more than one flow path, a core or hot flow and a bypass or cold flow. Strictly, each needs to be considered separately but, in this treatment, a mean, gross thrust will be assumed for the engine. It will also be assumed in the simple analysis that the intake and exhaust mass flows are equal. This is reasonable since the increase in mass flow at the exhaust due to the addition of the fuel mass flow may well be offset by compressor air bleeds for aircraft pneumatic services, e.g. pressurization and anti-icing.

The gross thrust, F_G, acts in 'thrust axes', which may not be parallel to the aircraft body axes. Thus, there may be a need to resolve the gross thrust into aircraft body axes before it can be used in the equations of motion. An example is seen in the case of the vectored thrust engine in which the thrust axes are variable with respect to the aircraft body axes (Appendix B).

The momentum drag, D_m, acts in velocity axes since it represents a change of momentum of the airflow in the direction of flight. The momentum drag is the product of the engine air mass flow and the aircraft true airspeed. Although referred to as a 'drag' force, the momentum drag is part of the engine thrust as it results from the engine internal flow stream. Any forces resulting from the external flow to the engine will be included in the airframe drag (Appendix B). The allocation of flow forces to the airframe drag or to the propulsive thrust is known as thrust-drag accounting. It is important to distinguish between these contributions since the optimum operating airspeeds of the aircraft are determined by its drag characteristic. Allocation of a force contribution into the wrong side of the thrust-drag 'balance sheet' will result in inaccurate estimations of the performance of the aircraft.

The net thrust of the powerplant will be affected by the flight Mach number and altitude. It is not possible to postulate any precise function that will relate thrust to Mach number or altitude for all thrust-producing powerplants. However, simple relationships can be developed that will enable the general characteristic of the thrust variation with Mach number and altitude to be deduced. From eqn (3.15) the net thrust can be expressed as,

$$F_N = \dot{m}(V_j - V) \tag{3.16}$$

The turbine engine is a volumetric device and the air mass flow, $\dot{m}$, is the product of the volume of air passed by the engine per second (which is controlled by the engine rotational speed), and the density of the air entering the engine. Since the airflow needs to be slowed down to a Mach number of about 0.5 before it can be accepted by the compressor there will be an isentropic change to the density of the flow as it enters the engine intake. The density of the air entering the engine, ρ_i, will be given by

$$\rho_i = \rho[1 + 0.2(M^2 - M_i^2)]^{\frac{1}{\gamma-1}} \tag{3.17}$$

where subscript i refers to the conditions at the engine compressor face.

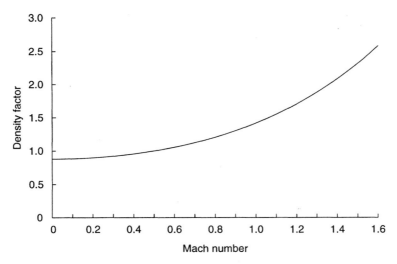

Fig. 3.13 The effect of Mach number on intake density.

The rise in density at the compressor face is shown in Fig. 3.13. The increased density will increase the air mass flow through the engine at any given engine rotational speed and hence the thrust will tend to increase with increasing flight Mach number.

From eqn (3.16), it can be seen that the net thrust is also proportional to the difference between the velocity of the engine exhaust flow, V_j, and the aircraft true airspeed, V. The velocity of the engine exhaust flow is a function of the temperature of the exhaust gas and will be determined by the engine throttle setting. For any given engine thrust setting the exhaust gas velocity can be considered constant. The airspeed of the aircraft is independent of the engine thrust setting so that, as the flight Mach number increases, the difference between the exhaust velocity and the true airspeed, and hence the net thrust, decreases, Fig. 3.14.

The overall effect of Mach number on the net thrust is the product of the two functions, eqns (3.16) and (3.17), and is shown in Fig. 3.15. Here it can be seen that the thrust characteristic is substantially influenced by the temperature of the exhaust gas. If the exhaust gas is relatively cool, as for example in the case of a high bypass ratio turbofan, then the exhaust gas Mach number will be low and the effect of the density function will be small. The thrust will decrease almost linearly with flight Mach number. A pure turbojet, which has no bypass flow, will have a relatively hot exhaust gas flow. Therefore, the density function will tend to dominate the thrust function and help to maintain the thrust level as the flight Mach number increases. Figure 3.15 shows the form of the thrust characteristics of low, medium and high bypass ratio engines.

It is emphasized that the thrust characteristics shown in Fig. 3.15 have been developed to show the likely variation of thrust with Mach number and do not represent a means of calculating or estimating the thrust of an engine.

The thrust produced by an engine decreases with altitude. Empirical data show that the decrease in thrust can be reasonably approximated by a function of

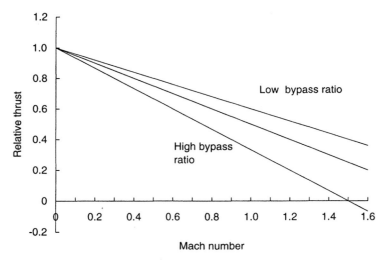

Fig. 3.14 Thrust decrease due to flight Mach number.

the form,

$$\frac{F_N}{F_{N0}} = \sigma^x \tag{3.18}$$

where the exponent x has a value of about 0.7 in the troposphere and unity in the stratosphere. These values may vary with characteristics of the engine cycle, particularly the bypass ratio.

The specific fuel consumption, C, is similarly affected, in this case as a function of temperature. The values of the exponent y is about 0.5 and may be influenced by bypass ratio.

$$\frac{C}{C_0} = \theta^y \tag{3.19}$$

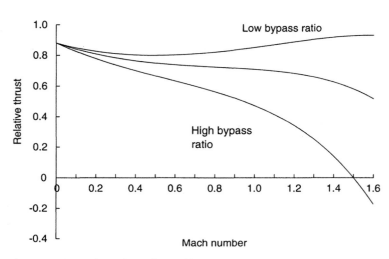

Fig. 3.15 Thrust variation with Mach number and bypass ratio.

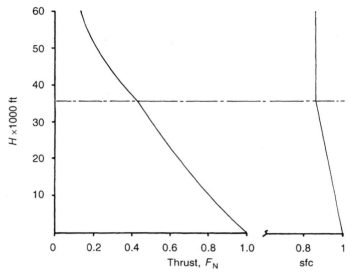

Fig. 3.16 The effect of altitude on thrust and specific fuel consumption.

These functions are shown in Fig. 3.16.

From eqn (3.3), the longitudinal performance equation of motion for aircraft with thrust-producing engines is given by,

$$F_N - D - W \sin \gamma_2 = m\dot{V} \qquad (3.20)$$

The excess thrust, $F_N - D$, which is available for climb or acceleration is the difference between the drag characteristic and the thrust characteristic of the aircraft, see Fig. 3.17.

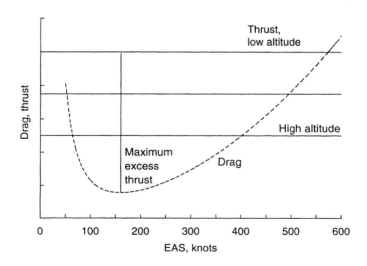

Fig. 3.17 Excess thrust.

Figure 3.17 is drawn in terms of equivalent airspeed so that the drag characteristic is independent of altitude. However, the thrust decreases with increasing altitude so that the excess thrust tends to decrease as altitude increases and the aircraft will eventually reach a performance ceiling at which the excess thrust is zero. This occurs at an airspeed equal to the minimum drag speed of the aircraft.

In Fig. 3.17, the thrust is shown as being independent of airspeed for the purpose of illustration. This is not generally the case and the thrust characteristic will be of the form shown in Fig. 3.15. This will lead to a maximum excess thrust close to, but not necessarily equal to, the minimum drag speed.

3.4.2 The power-producing powerplant

The power-producing powerplant delivers its power through a rotating shaft to a propeller that converts the power into propulsive thrust. The powerplant may be either a piston engine or a gas turbine that converts the energy of its gas flow into shaft-power rather than into thrust. In either case the shaft-power output is not greatly affected by airspeed and, to a first-order approximation, the power can be regarded as independent of airspeed.

The shaft-power, P, is converted into thrust, T, by the propeller. In the process, losses occur and the thrust-power produced will be less than the shaft-power delivered. The propeller efficiency, η, is the ratio of the thrust-power output to the shaft-power input so that,

$$\eta P = TV \qquad (3.21)$$

where V is the true airspeed.

This implies that the propulsive thrust increases as airspeed decreases at constant engine power and that the thrust will become infinite at zero airspeed. In practice, the propeller efficiency will vary with airspeed and a constant engine shaft-power produces a finite thrust at zero airspeed, known as the *Static Thrust*, which decreases as the airspeed increases. The propeller efficiency, η, is a characteristic of an individual propeller and depends on the Advance ratio, J, of the propeller, and the Power coefficient, C_P, of the engine.

The propeller Advance ratio is given by

$$J = V/nD$$

and the engine Power coefficient by,

$$C_P = \frac{P}{\frac{1}{2}\rho n^3 D^5}$$

where n is the rotational speed in rad/s and D is the propeller diameter.

The propeller efficiency generally can be regarded as being reasonably constant in the cruising range of airspeed in the case of a variable pitch propeller. A typical relationship between the thrust and drag of an aircraft with power-producing engines is of the form shown in Fig. 3.18.

The longitudinal performance equation of motion for aircraft with power-producing engines differs from that for aircraft with thrust-producing engines since

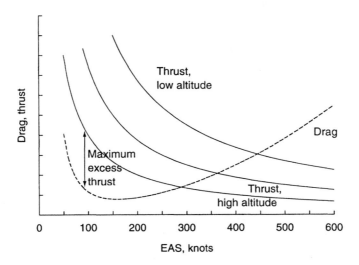

Fig. 3.18 The thrust characteristic of a power-producing engine.

the propulsive thrust is a function of airspeed, eqn (3.21). The equation is now written in the form,

$$\frac{\eta P}{V} = D + W \sin \gamma_2 + m\dot{V}$$

or, rearranging and writing in terms of equivalent airspeed gives,

$$\eta P \sqrt{\sigma} - D V_e = W V \sqrt{\sigma} \left[\sin \gamma_2 + \frac{\dot{V}}{g} \right] \qquad (3.20a)$$

which leads to the conclusion that the excess power is the difference between the equivalent-thrust power, $\eta P \sqrt{\sigma}$, and the equivalent-drag power, $D V_e$, see Fig. 3.19.

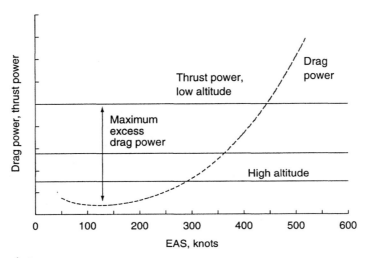

Fig. 3.19 Excess thrust power.

Figure 3.19 is drawn in terms of equivalent airspeed so that the equivalent-drag power, DV_e, is independent of altitude. However, the equivalent-thrust power, $\eta P\sqrt{\sigma}$, decreases with increasing altitude so that the excess thrust power tends to decrease as altitude increases and the aircraft will eventually reach a performance ceiling at which the excess thrust power is zero. This occurs at an airspeed equal to, or very close to, the minimum power speed of the aircraft.

The simplified powerplant characteristics described in Section 3.3 are intended to give a general understanding of the manner in which propulsive thrust may vary with the performance variables airspeed (or Mach number), altitude and temperature. They are not intended to be a means of calculating powerplant performance or of scaling performance for the effects of those variables.

3.5 Aerodynamic relationships

If it is assumed that the aircraft has a parabolic drag polar, eqn (3.7), then a number of relationships can be deduced that can be used in the derivation of some simplified expressions for the performance of the aircraft. These relationships, which were addressed in Section 3.2, are well known and will be quoted without proof.

Figure 3.7 shows that the parabolic drag polar has a minimum value that is important in determining the airspeeds for optimum performance. At minimum drag, it can be shown that

$$C_{Dz} = KC_L^2 \tag{3.22}$$

which implies that the lift coefficient for minimum drag is given by,

$$C_{Lmd} = [C_{Dz}/K]^{\frac{1}{2}} \tag{3.23}$$

and the minimum drag speed, and minimum drag Mach number, are given by,

$$V_{md} = [2W/\rho S]^{\frac{1}{2}}\{K/C_{Dz}\}^{\frac{1}{4}} \quad \text{and} \quad M_{md} = [2W/\gamma p S]^{\frac{1}{2}}\{K/C_{Dz}\}^{\frac{1}{4}} \tag{3.24}$$

respectively.

The power required for level flight is given by the product of the drag force and the true airspeed, DV, since in steady level flight $T = D$. The airspeed for minimum power, V_{mp}, can be shown to be given by,

$$V_{mp} = \frac{1}{\sqrt[4]{3}}V_{md} \tag{3.25}$$

Although there is a simple numerical constant linking the minimum drag speed and the minimum power speed these terms should only be applied to aircraft with the appropriate powerplant. The minimum power speed relates to the performance of aircraft with power-producing engines whereas the minimum drag speed relates to the performance of aircraft with thrust-producing engines. Only in the case of the glider, which has no engines, do both minimum power speed and minimum drag speed have significance.

The lift–drag ratio, L/D, is a measure of the aerodynamic efficiency, E, of the aircraft and has a maximum value at the minimum drag speed so that,

$$\frac{L}{D_{min}} = E_{max} = \frac{1}{2[KC_{Dz}]^{\frac{1}{2}}} \tag{3.26}$$

The relative airspeed, u, is the ratio between the airspeed and the minimum drag speed,

$$u = V/V_{md} \tag{3.27}$$

Using the relative airspeed, the ratio of the drag to minimum drag can be expressed as,

$$\frac{D}{D_{min}} = \frac{1}{2}\left[u^2 + \frac{1}{u^2}\right] \tag{3.28}$$

In addition, the powerplant propulsive thrust can be expressed in terms of the minimum drag, D_{min}, of the aircraft; this form of expression will be used in the generalized performance equations. In the case of thrust-producing engines the dimensionless thrust, τ, is given by,

$$\tau = F_N/D_{min} \tag{3.29}$$

and, in the case of power-producing engines, the dimensionless power, λ, is given by,

$$\lambda = \eta P/D_{min} V_{md} \tag{3.30}$$

Using eqns (3.26) to (3.30) the performance equation, eqn (3.3), can now be written in a generalized, dimensionless, form

$$\left[\frac{\lambda}{u} + \tau\right] - \frac{1}{2}\left[u^2 + \frac{1}{u^2}\right] = E_{max}\left\{\sin\gamma_2 + \frac{\dot{V}}{g}\right\} \tag{3.31}$$

This equation can be used to determine the performance characteristics of aircraft with either form of powerplant, or a mixture of thrust- and power-producing engines, as in the case of the turbo-prop.

These relationships will be used in the following chapters for the development of expressions for the estimation of the performance of the aircraft.

3.6 Conclusions

The purpose of the equations of motion is to enable expressions for performance to be developed that will reasonably represent the actual performance characteristics of the aircraft. However, no estimation process can ever be totally accurate and there will always be some discrepancies between the estimated performance and the performance achieved by the aircraft in flight. The question is – is the benefit of the effort required to improve the quality of the estimation process worth the cost?

By assuming the parabolic form of the drag polar and a constant lift curve slope, it has been shown that simple, generalized, expressions can be determined that will provide an adequate estimate of performance for most design study purposes. If the drag polar is not parabolic, and cannot be represented by a reasonably simple equation, the performance expressions will become less general and may require 'point-to-point' calculation to produce a performance characteristic. At Mach numbers above the critical value, for example, it is necessary to do this since the deviation of the drag characteristic from the parabolic form renders the general expressions invalid. The process for the estimation of the forces acting on the aircraft could be extended to include the full effects of Reynolds number and Mach number on the lift and drag characteristics. In addition, it could take into account any other variables that may cause deviation from those simplified characteristics. However, the effects may only be significant in extreme cases. In the normal working range, any difference between the simplified force characteristics and the actual force characteristics should be small enough to be acceptable, and the estimation of the performance good enough for general design studies. Only when detailed estimation of a critical condition is required should it be necessary to include the secondary effects of all variables on the aerodynamic forces.

The force system of the aircraft and the equations of motion developed in this chapter, and in Appendices A and B, have been simplified to represent the aircraft in normal, symmetric, all-engines operating, flight. If it were necessary to consider the performance of an aircraft in any other flight condition then the full equations would need to be used. For example, flight with one engine inoperative involves an asymmetric state using sideslip or angle of bank, either of which will require additional terms to be included in the force equations.

The benefits of the simplified expressions in enabling the expressions for performance to be developed in their most general form, and the optimum performance to be determined, will be seen in the following chapters.

Bibliography

Küchemann, D. (1978) *The Aerodynamic Design of Aircraft* (Pergamon Press).
MIDAP Study Group (1979) *Guide to In-flight Thrust Measurement of Turbojets and Fan Engines.* AGARD-AG-237.
Mattingly, J. D. *et al.* (1987) *Aircraft Engine Design* (AIAA Education Series).
Whittle, F. (1981) *Gas Turbine Aerothermodynamics* (Pergamon Press).

Engineering sciences data unit items
Performance Series, Vols 1–14.
Introduction to Equations of Motion for Performance. Vol 1, No 78038, 1979 (Am A)
Simplified Forms of Performance Equations. Vol 1, No 80032, 1980.
Representation of Drag in Aircraft Performance Calculations. Vol 3, No 81026, 1997 (Am C).
Examples of Performance Analysis using Data Obtained Concurrently in Air-path, Body and Earth Axes. Vol 13, No 79018, 1979.
Introduction to the Measurement of Thrust in Flight (Air Breathing Ducted-flow Engines). Vol 13, No 69006, 1981 (Am A).

Aerodynamics Series
Introductory Sheet on Subcritical Lift-dependent Drag of Wings. Vol 2e, No 66031, 1995 (Am C).
Drag Due to Lift for Plane Swept Wings, Alone or in Combination with a Body, up to High Angle of Attack at Subsonic Speeds. Vol 2e, No 95025, 1997 (Am A).
Lift-curve Slope of Wing–body Combinations. Vol 4b, No 91007, 1995 (Am D).

4

Cruising performance

Since an aircraft usually spends the greater part of its mission in cruising flight, the cruising performance has a strong influence on the overall mission performance. At the design stage of the aircraft, or when considering a modification to an existing type of aircraft, the designer needs the ability to assess the effect of a design change on the overall performance. An easily applied and flexible method of estimation of cruising performance analysis is a most important tool in aircraft design.

4.1 Introduction

The cruising performance of an aircraft is one of the fundamental building blocks of the overall mission. In the cruising segment of the mission, both height and airspeed are essentially constant and the aircraft is required to cover distance in the most expedient manner. Usually, the majority of the fuel carried in the aircraft will be used during the cruise. The distance that can be flown, or the time that the aircraft can remain aloft, on a given quantity of fuel are important factors in the assessment of the cruise performance.

In civil transport operations, the cruising performance of the aircraft has a strong influence on the economics of the operation of the aircraft. The cost of fuel contributes to the cost of the operation, but so does the cost of time. The aircraft needs to be flown at a speed and in a manner that will optimize the overall operating cost; cruising performance is important to the overall balance between fuel consumption and the time of flight.

In military operations, the cruise performance determines the radius of action or the endurance of the aircraft; both of which are important parameters in its operational effectiveness. The lower the fuel consumption per unit distance in cruising flight the greater will be the radius of action for a given fuel load. The aircraft will be able to reach targets at a greater range or operate from bases further back from the action in relative safety. Similarly, surveillance or patrol missions depend on the endurance of the aircraft, and operation at the correct speed for minimum fuel consumption is an essential part of any patrol mission.

In the analysis of cruise performance, the aircraft is considered to be in steady, level, straight, symmetric flight with no acceleration or manoeuvre. Under those conditions the aircraft can be taken to be in a state of equilibrium in which the forces and moments are in balance; this is referred to as being *in Trim*. In practice, cruising flight may involve very low levels of climb or acceleration. In addition, it may be required to make gradual manoeuvres associated with the mission, for example turning to change track and to correct errors in its course, or climbing to change cruising altitude. Usually these manoeuvres can be neglected in design estimations or, if it is considered necessary, corrections applied to account for the errors produced. In practice, an aircraft carries a fuel contingency allowance over and above the estimated cruising fuel requirement to allow for such unscheduled manoeuvres. When the aircraft is cruising in trim the equations of motion of a conventional aircraft, eqn (3.3), can be reduced to the simple statements,

$$\left.\begin{array}{c} F_N = D \\ L = W \end{array}\right\} \tag{4.1}$$

The development of the basic expressions for cruising flight in this chapter is based on these simplified statements.

(It should be remembered that the simplified statements given in eqn (4.1) contain a number of assumptions that must be fulfilled if the expressions developed from them are to be used to estimate the performance of the aircraft.)

4.2 Specific air range and specific endurance

Cruising efficiency can be measured in terms of either the range or the endurance of the aircraft. The Specific Air Range (SAR) is defined as the horizontal distance flown per unit of fuel consumed and the Specific Endurance (SE) is defined as the time of flight per unit of fuel consumed.

The distance travelled, x, in still air is given by the time integral of the true airspeed, V, so that

$$V = dx/dt \tag{4.2}$$

and in cruising flight the true airspeed is usually quoted in knots, or nautical miles per hour (nm/hour).

In addition, during cruise, fuel is burned and the fuel mass flow, Q_f, will determine the rate of change of mass of the aircraft,

$$Q_f = -dm/dt \tag{4.3}$$

The fuel mass flow is usually quoted in kg/hour and is negative since the mass of the aircraft decreases with time as fuel is burned.

The specific air range is an expression of the instantaneous distance flown per unit of fuel consumed and can thus be expressed as,

$$SAR = -dx/dm = V/Q_f \tag{4.4}$$

and has units of length/mass. It would normally be quoted in nautical miles per kilogram (nm/kg).

The specific endurance is an expression of the instantaneous flight time per unit of fuel consumed and can be expressed as,

$$SE = -dt/dm = 1/Q_f \qquad (4.5)$$

and would normally be quoted in hr/kg.

Since the drag of the aircraft is a function of the aircraft weight, which is continuously decreasing as fuel is burned, the specific air range and the specific endurance will be *point performance* parameters, relating to the range and endurance at that point on the cruise path. To find the cruise range or endurance, the SAR or SE must be integrated over the cruise flight path as functions of the aircraft weight. Neither the SAR nor the SE is conveniently formulated for integration in the form given in eqns (4.4) and (4.5). They need to be written in terms of the performance variables before they can then be integrated to give the range or endurance of the aircraft. Because of the fundamental differences of the propulsive characteristics of thrust-producing and power-producing engines, the performance of the aircraft with each type of engine must be considered separately.

4.3 Range and endurance for aircraft with thrust-producing engines

If the aircraft is powered by thrust-producing engines (turbojets or turbofans), the fuel flow is seen to be a function of engine thrust which, in cruise, is equal to aircraft drag (eqn (4.1)).

The specific fuel consumption, C, is defined as the fuel flow per unit thrust

$$C = Q_f/F_N \qquad (4.6)$$

and has units of kg/N hr.

(NB *It should be noted that, in the subsequent analysis of the cruising performance of the aircraft, dimensional consistency of the expressions might not be strictly observed. This is particularly the case where the specific fuel consumption is used, since the units in which it is quoted may vary. The expressions that are developed here will be kept in their simplest possible form and may not include all the terms necessary to maintain their strict dimensional consistency. Therefore, it may be necessary to include the gravitational constant, g, or other constants, to make the units consistent, a check on the units of the expressions will show when this is needed.*)

Although eqn (4.6) suggests that the fuel flow is proportional to thrust, the specific fuel consumption (sfc), may be a function of other performance-related parameters and a number of alternative fuel flow laws may be considered. Examples of some of the commonly assumed laws are,

(i) $$\qquad\qquad\qquad\qquad C = C_1 \qquad (4.7)$$

Assuming a constant value for sfc is the simplest fuel flow law. It implies that the fuel flow is directly proportional to thrust. This law is usually accepted in the determination of the general expressions for range and endurance. In practice, it does not reflect the effects of changes in engine operating conditions or in flight conditions and so it should only be applied when variations in thrust or Mach number are small and cruising conditions are constant.

(ii)
$$C = C_2 \theta^{\frac{1}{2}} M^n \tag{4.8}$$

This is a reasonable approximation to the fuel flow law of a turbojet or turbofan engine. It takes into account variations in the temperature of the atmosphere, θ, and of the effects of flight Mach number. The exponent n may vary and empirical data indicate values ranging between about 0.2 for a turbojet and about 0.6 for a high bypass ratio turbofan. This law is not particularly difficult to apply in the integration of SAR or SE.

(iii)   or
$$\left. \begin{array}{l} C = C_3 + C_4 M \\ C = C_5 + C_6 F_N \end{array} \right\} \tag{4.9}$$

These are further attempts to produce approximations to empirical fuel flow data over a range of thrust and Mach numbers but they tend to be more cumbersome when introduced to the range and endurance equations.

In the following analysis of the cruise performance, the simple fuel law, eqn (4.7), will be used. The effects of using eqn (4.8) as an alternative law will be discussed later.

Using eqns (4.1) and (4.6) in the expressions for SAR and SE, eqns (4.4) and (4.5) lead to expressions in a form suitable for integration,

$$\text{SAR} = \frac{V}{C} \frac{L}{D} \frac{1}{W} \tag{4.10}$$

and

$$\text{SE} = \frac{1}{C} \frac{L}{D} \frac{1}{W} \tag{4.11}$$

Since these point performance characteristics include the aircraft lift–drag ratio, they will have maximum values at airspeeds related to the minimum drag speed of the aircraft. Writing eqns (4.10) and (4.11) in coefficient form, and substituting for airspeed, gives,

$$\text{SAR} = \frac{1}{C} \left[\frac{2}{\rho W S} \right]^{\frac{1}{2}} \frac{C_L^{\frac{1}{2}}}{C_D} \tag{4.12}$$

and

$$\text{SE} = \frac{1}{C} \frac{C_L}{C_D} \frac{1}{W} \tag{4.13}$$

If it is assumed that the aircraft has a parabolic drag polar and that the simple fuel flow law for thrust-producing engines, eqn (4.7), applies, then, for the instantaneous or point performance of the aircraft, the maximum SAR would be obtained by flying at an angle of attack corresponding to $\{C_L^{\frac{1}{2}}/C_D\}_{\max}$. This gives an optimum airspeed for maximum range of $\sqrt[4]{3}V_{\text{md}}$, or $1.316V_{\text{md}}$. Similarly, the maximum SE would be obtained by flying at the minimum drag speed, V_{md}. These results apply to any point along the flight path but, in some methods of cruising, do not necessarily apply continuously along the flight path.

Equations (4.10) and (4.11) are in a form that can be integrated to give the *path performance* of the aircraft in cruising flight. As the aircraft cruises, and fuel is consumed, the weight of the aircraft decreases. It can be seen that the cruise performance is a function of, firstly, the quantity of fuel available for cruise and, secondly, of the effect of the change of weight on the minimum drag speed. The range and endurance are found, as path performance functions, by integrating the SAR and SE over the change in weight between the beginning and end of cruise. If the initial weight of the aircraft is W_i, and the final weight is W_f, then the fuel ratio, ω, can be defined as,

$$\omega = W_i/W_f \qquad (4.14)$$

so that the weight of fuel consumed can be related to the initial, or final, cruise weight,

$$W_i - W_f = W_i(1 - 1/\omega) \qquad (4.15)$$

In the cruise, lift equals weight so that,

$$W = \tfrac{1}{2}\gamma p M^2 S C_L \qquad (4.16)$$

As weight decreases during the cruise the variables on the right-hand side of eqn (4.16) must vary to compensate. These are air pressure (which can be controlled by the cruise altitude), flight Mach number and lift coefficient (controlled by angle of attack). Three methods of cruise can therefore be considered, in each of which one of the variables is varied to compensate for the decrease in weight and the other two are maintained constant. Each method produces a different result and has its particular application in aircraft operations.

4.3.1 Cruise method 1. Constant angle of attack, constant Mach number

In this method, the air pressure must be allowed to decrease to allow for the decrease in aircraft weight as fuel is consumed, thus

$$\frac{W}{p} = \tfrac{1}{2}\gamma M^2 S C_L \qquad (4.17)$$

This implies that the aircraft must be allowed to climb during the cruise to maintain the parameter W/p constant and the method is known as the *Cruise–Climb* technique. Since the angle of attack is constant throughout the cruise the lift coefficient, and hence lift–drag ratio, L/D, will be constant. The constant Mach number implies flight at constant true airspeed.

Let the range under cruise method 1 be R_1, then from eqn (4.10),

$$R_1 = \frac{V}{C}\frac{L}{D}\int_{W_i}^{W_f} \frac{dW}{W} \qquad (4.18)$$

which becomes,

$$R_1 = \frac{V}{C}\frac{L}{D}l_n\omega \qquad (4.19)$$

This expression is the best known expression for the range of an aircraft and is known as the *Breguet Range* expression. Although it offers the optimum performance in terms of distance flown on a given fuel load there are practical reasons that make its application to flight operations difficult, and further consideration of this cruise method is necessary.

Substituting for the true airspeed, V, and writing eqn (4.19) in coefficient form gives,

$$R_1 = \frac{1}{C}\left[\frac{2W}{\rho S}\right]^{\frac{1}{2}} \frac{C_L^{\frac{1}{2}}}{C_D} l_n \omega \tag{4.20}$$

This has a maximum value that occurs at an airspeed corresponding to $1.316 V_{md}$. However, it may not be possible, or expedient, to cruise at the optimum airspeed and the effect of cruising at an alternative airspeed needs to be considered.

Since the cruise–climb is flown at constant angle of attack the ratio C_L/C_{Lmd} will be constant and, therefore, the relative airspeed, $u = V/V_{md}$, will be constant throughout the cruise. Also, since the airspeed is constant in this method of cruise, $V = V_i = V_f$ and therefore the initial and final relative airspeeds in the cruise are given by $u_i = V/V_{mdi}$ and $u_f = V/V_{mdf}$. Thus, $V = u_i V_{mdi}$ or $u_f V_{mdf}$ so that eqn (4.20) can be written in terms of either the minimum drag speed at the beginning of cruise, V_{mdi}, or at the end of cruise, V_{mdf}. Using eqns (3.20), (3.22), (3.23) and (3.24) in eqn (4.19) gives,

$$R_1 = \left[\frac{V_{mdi}}{C} E_{max}\right]\left\{\frac{2u^3}{u^4 + 1}\right\} l_n \omega \tag{4.21}$$

This consists of three parts.

The square bracket is known as the *range factor*. It contains a term that is a function of the airframe–engine combination and can be regarded as a constant scaling factor, although it contains two variables – the initial cruise weight and air density. This term can be used to determine the effect of modifications to the aircraft (in respect of either the airframe or the powerplant) on the cruise performance through the drag characteristic, the specific fuel consumption and the aircraft weight.

The curly bracket is a function of the relative airspeed. It acts as a shaping factor, which is characteristic of the method of cruise.

The third term is a function of the fuel ratio. This is also a characteristic of the method of cruise and the magnitude of this term will determine the relative range of the cruise methods.

The product of the curly bracket and the function of the fuel ratio is known as the *range function* of the cruise method.

Figure 4.1 shows the range function of the cruise–climb method for three values of fuel ratio representing the cruise fuel required for short, medium and long-range aircraft. Maximum range is obtained by flying at a relative airspeed $u = 1.316$ and the range penalty for operation at any airspeed other than this can be determined. The range of the aircraft, in navigational units, can be found by multiplying the range function by the range factor.

The values of the fuel ratio shown in Fig. 4.1, and subsequent figures, are typical of very long range aircraft, $\omega = 1.5$, medium to long range aircraft, $\omega = 1.3$, and short-range aircraft, $\omega = 1.1$.

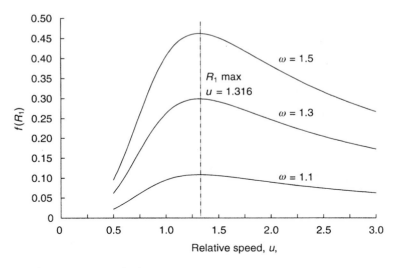

Fig. 4.1 Range function cruise method 1.

The endurance of the aircraft can be found by integrating eqn (4.11) over the weight change during the cruise, in the same manner as the integration of the SAR. This leads to the expression for the endurance of the aircraft under cruise method 1, E_1,

$$E_1 = \frac{1}{C} \frac{C_L}{C_D} l_n \omega \tag{4.22}$$

This shows that maximum endurance is obtained by flying at the minimum drag speed.

Writing eqn (4.22) in terms of the relative airspeed, u, gives,

$$E_1 = \left[\frac{E_{max}}{C}\right] \left\{\frac{2u^2}{u^4 + 1}\right\} l_n \omega \tag{4.23}$$

where the square bracket is the *endurance factor* for the airframe–engine combination and the relative airspeed and fuel ratio terms provide the *endurance function*. The product of these gives the endurance of the aircraft in hours. The endurance function is shown in Fig. 4.2.

Although the cruise–climb method gives the best possible range for a given quantity of fuel, its use is limited by practical restrictions to aircraft operations. Due to the constraints of the control of air traffic, and the need to provide vertical separation between flights in different directions, aircraft cannot be allowed to change height freely. This means that, in practice, it is unusual to be able to take advantage of the benefits of the cruise–climb unless the aircraft is operating in airspace with no conflicting traffic. As a compromise, aircraft are sometimes allowed to 'step-climb' during the cruise; this involves discrete increments in height, compatible with the requirements for vertical separation, at intervals along the route to keep W/p close to the required value. In a typical transport flight, cruising at about 30 000 ft a step climb of 2000 ft would be required after about 2 hours flying to bring the value of W/p back to its initial value.

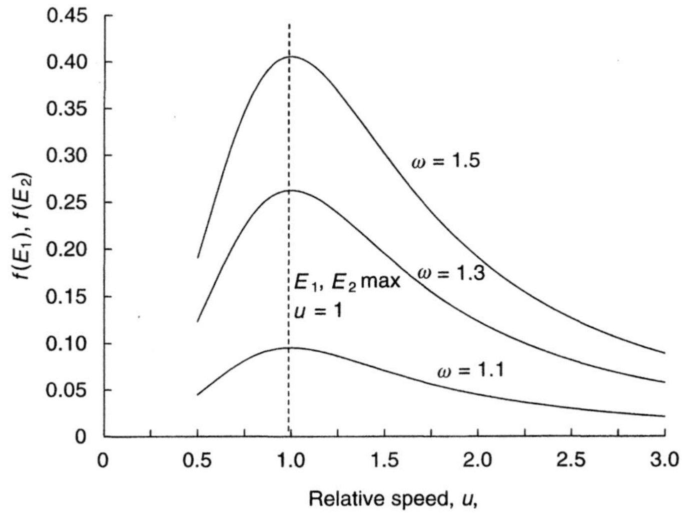

Fig. 4.2 Endurance function cruise methods 1 and 2.

The operation of the aircraft in the cruise–climb differs in the troposphere and in the stratosphere because of the effect of the structure of the atmosphere on the engine thrust characteristic. This can be explained by considering the thrust–drag balance, in parametric form (see Chapter 8). In the cruise–climb the aircraft is cruising at constant Mach number and constant angle of attack, thus the drag coefficient is constant, and

$$\frac{F_N}{\delta} = \frac{D}{\delta} = \tfrac{1}{2}\gamma p_0 M^2 S C_D \tag{4.24}$$

which are constants under these cruise conditions.

Now the parametric form of the engine thrust (Chapter 8) shows that the thrust is related to the engine rotational speed, N, and the flight Mach number by a functional relationship of the form,

$$\frac{F_N}{\delta} = f\left[\frac{N}{\theta^{\frac{1}{2}}}, M\right] \tag{4.25}$$

In the troposphere, the relative temperature, θ, decreases with height so that, during the cruise–climb, the parametric engine rotational speed, $N/\theta^{\frac{1}{2}}$, increases and hence the parametric thrust increases and will cause the aircraft to climb or to accelerate. This tendency will need to be checked by a continuous reduction in engine rotational speed to maintain the parametric rotational speed constant during the cruise–climb so that, with the constant cruise Mach number, the parametric thrust–drag balance is maintained.

In the isothermal stratosphere, where the parametric engine rotational speed will remain constant if the actual engine rotational speed is constant, the rates of change of the parametric thrust and parametric drag as height increases are equal. The cruise–climb is achieved by trimming the aircraft to the required angle of attack and setting the thrust, governed by the engine rotational speed, to give the required Mach number. The aircraft will then cruise–climb as weight decreases without the need for any further correction to thrust setting or to trim other than to account for any shift of the centre of gravity (CG) as fuel is burned. Cruise–climb is the ideal cruise method for operation in the stratosphere.

(The parametric form of the performance variables referred to above is discussed fully in Chapter 8.)

4.3.2 Cruise method 2. Constant angle of attack, constant altitude

In this case, the Mach number, or true airspeed, must be reduced during cruise, so that

$$\frac{W}{M^2} = \tfrac{1}{2}\gamma p S C_L \tag{4.26}$$

This implies that the lift coefficient, and lift–drag ratio, will be constant during the cruise and that, substituting for airspeed in eqn (4.10), the range under cruise method 2, R_2, will be given by the integral,

$$R_2 = \frac{1}{C}\left(\frac{2}{\rho S C_L}\right)^{\frac{1}{2}}\frac{L}{D}\int_{W_i}^{W_f}\frac{\mathrm{d}W}{W^{\frac{1}{2}}} \tag{4.27}$$

which, on integration, becomes,

$$R_2 = \frac{1}{C}\left[\frac{2W_i}{\rho S}\right]^{\frac{1}{2}}\frac{C_L^{\frac{1}{2}}}{C_D}2(1 - \omega^{-\frac{1}{2}}) \tag{4.28}$$

As in cruise method 1, this gives maximum range when the cruising airspeed is $1.316V_{md}$ since the angle of attack is maintained constant.

As in method 1, this is a constant angle of attack cruise method so that the relative airspeed is constant throughout the cruise and $u = u_i = u_f$. Writing eqn (4.28) in terms of the relative airspeed gives the range factor and range function for cruise method 2,

$$R_2 = \left[\frac{V_{mdi}}{C}E_{max}\right]\left\{\frac{2u^3}{u^4 + 1}\right\}2\left(1 - \omega^{-\frac{1}{2}}\right) \tag{4.29}$$

The range factor is identical to that found in cruise method 1. The range function has a similar general form to that of cruise method 1, but its value is smaller for a given value of fuel ratio, ω, so that the overall range, R_2, is less than R_1. The range function is shown in Fig. 4.3.

The endurance of the aircraft, found by integrating eqn (4.11) over the weight change during the cruise, is seen to produce the same expression as that found under cruise method 1, if the constant fuel flow law is assumed. This is because it is also a constant angle of attack method.

Cruise method 2 has the disadvantage that the cruise Mach number, and hence true airspeed, is continuously reduced to compensate for the decrease in aircraft weight. This will increase the time of flight and the cost of the time penalty incurred is likely to far outweigh any fuel advantage the method may have. This method of

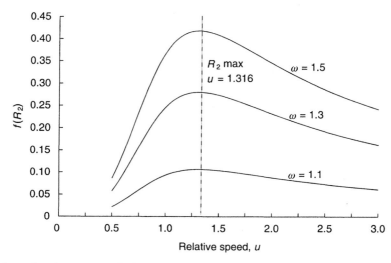

Fig. 4.3 Range function cruise method 2.

cruise can be considered for patrol or surveillance operations, in which endurance is more important than distance travelled, but it will require a constant reduction in engine thrust to maintain the cruise conditions. It has the advantage of being a constant attitude cruise method that may be favourable to some surveillance sensors.

4.3.3 Cruise method 3. Constant altitude, constant Mach number

In this case, the angle of attack must be allowed to decrease as the weight decreases to maintain W/C_L constant.

$$\frac{W}{C_L} = \tfrac{1}{2}\gamma p M^2 S \qquad (4.30)$$

Assuming that the aircraft is cruising at a speed greater than its minimum drag speed, then the decrease in lift coefficient during the cruise will cause the drag coefficient, and hence the drag force, to decrease. This will require a progressive decrease in the thrust required to maintain the Mach number or true airspeed constant. Since the lift–drag ratio will not be constant, the range will need to be found by integrating the drag over the weight change during cruise.

$$R_3 = \frac{V}{C} \int_{W_i}^{W_f} \frac{dW}{D} \qquad (4.31)$$

Now for an aircraft with a parabolic drag polar,

$$D = qSC_{Dz} + \frac{KW^2}{qS} \qquad (4.32)$$

where $q = \tfrac{1}{2}\rho V^2$.

Substituting in eqn (4.31) and integrating gives,

$$R_3 = \frac{V}{C} \frac{1}{(KC_{Dz})^{\frac{1}{2}}} \left\{ \tan^{-1} \left[\frac{W_i}{qS} \left(\frac{K}{C_{Dz}} \right)^{\frac{1}{2}} \right] - \tan^{-1} \left[\frac{W_f}{qS} \left(\frac{K}{C_{Dz}} \right)^{\frac{1}{2}} \right] \right\} \quad (4.33)$$

Now,

$$\frac{W_i}{qS} \left(\frac{K}{C_{Dz}} \right)^{\frac{1}{2}} = \frac{C_{Li}}{C_{Lmdi}} = \frac{V_{mdi}^2}{V_i^2} = \frac{1}{u_i^2}$$

and, similarly,

$$\frac{W_f}{qS} \left(\frac{K}{C_{Dz}} \right)^{\frac{1}{2}} = \frac{1}{u_f^2}.$$

Since the airspeed is constant in this method of cruise, $V = V_i = V_f$ and, therefore, the initial and final relative airspeeds in the cruise are given by $u_i = V/V_{mdi}$ and $u_f = V/V_{mdf}$; thus $V = u_i V_{mdi}$ or $u_f V_{mdf}$. Also, $V_{md} \propto W^{\frac{1}{2}}$ so that $u_f = \omega^{\frac{1}{2}} u_i$, and writing eqn (4.33) in terms of the relative airspeed gives,

$$R_3 = \left[\frac{V_{mdi}}{C} E_{max} \right] 2u_i \left\{ \tan^{-1} \left[\frac{1}{u_i^2} \right] - \tan^{-1} \left[\frac{1}{\omega u_i^2} \right] \right\} \quad (4.34)$$

This expression for the range of the aircraft is considerably more complex than those found in the other two cruise methods, and the cruising speed for maximum range is found to be a function of the fuel ratio, ω. This can be seen by considering the relative airspeed, u, at the beginning and end of the cruise. The cruising speed is maintained constant so that $V_i = V_f$, but the minimum drag speed will decrease with aircraft weight causing the relative airspeed, u, to increase during the cruise. This means that the optimum airspeed for maximum range will be a function of the weight change during the cruise and, therefore, of the fuel ratio, ω. Figure 4.4 shows the range function

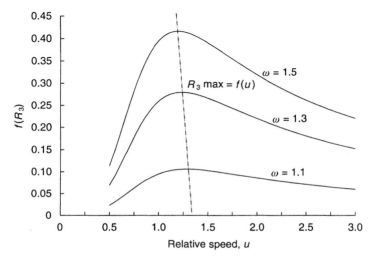

Fig. 4.4 Range function cruise method 3.

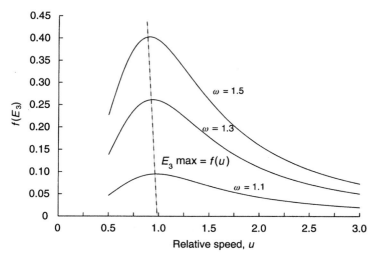

Fig. 4.5 Endurance function cruise method 3.

for cruise method 3 and the variation of the initial relative airspeed for maximum range as the fuel ratio increases.

Following the integration of eqn (4.11), the endurance under cruise at constant altitude and Mach number is given by,

$$E_3 = \left[\frac{E_{max}}{C}\right] 2 \left\{ \tan^{-1}\left[\frac{1}{u_i^2}\right] - \tan^{-1}\left[\frac{1}{\omega u_i^2}\right] \right\} \tag{4.35}$$

and the relative airspeed for best endurance will, similarly, be a function of fuel ratio. Figure 4.5 shows the endurance function that indicates that the optimum endurance can only be achieved by commencing cruise at an airspeed less than the minimum drag speed. This will require cruise on the backside of the drag curve, which tends to be speed unstable. This cruise method, therefore, is not ideal for missions to be flown for endurance.

4.3.4 Comparison of cruise methods

It has been possible to write the range and endurance attained by each method of cruise in the form of a product of a range factor and a range function, and of an endurance factor and an endurance function, giving,

$$R = \left[\frac{V_{mdi}}{C} E_{max}\right] \{ f_R(u_i, \omega) \} \tag{4.36}$$

and

$$E = \left[\frac{E_{max}}{C}\right] \{ f_E(u_i, \omega) \} \tag{4.37}$$

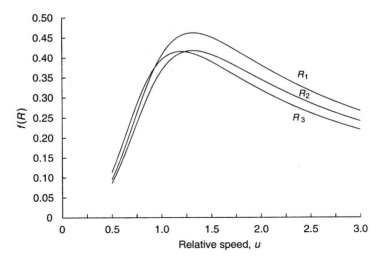

Fig. 4.6 Comparison of cruise methods for range.

This enables the methods of cruise to be compared in terms of the relative magnitudes of the range and endurance functions.

The range functions are compared in Fig. 4.6 for a fuel ratio of 1.5 and show that the cruise–climb is the optimum method of cruise, indicating that, at its best, it gives about 10% better range than the other methods. However, operational considerations generally demand the constant altitude, constant Mach number cruise, which tends to be the least efficient in terms of fuel consumption.

The comparison between the cruise methods for endurance, Fig. 4.7, shows less disparity but favours the constant angle of attack methods. This is particularly the case when flying for endurance, as the cruise would generally be performed at an

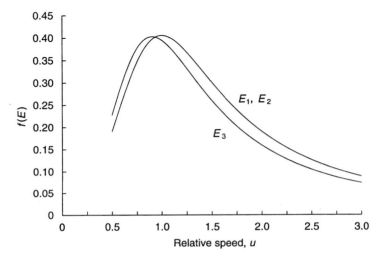

Fig. 4.7 Comparison of cruise methods for endurance.

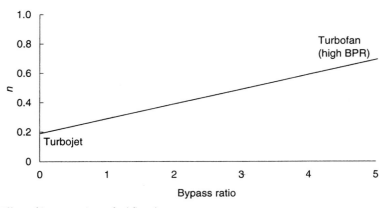

Fig. 4.8 Effect of bypass ratio on fuel flow law.

airspeed slightly above the minimum drag speed to avoid the backside of the drag curve and to give flight path stability.

4.3.5 The effect of alternative fuel flow laws

The range and endurance functions developed above have used the simple fuel flow law, eqn (4.7), which assumes that the fuel flow is proportional only to net thrust. It was accepted that this fuel flow law is idealized, probably not fully describing the characteristics of the engine, and that a more realistic law should be applied. If an alternative fuel flow law is considered, for example eqn (4.8), then it may produce different optimum operating airspeeds.

Equation (4.8) states a fuel flow law of the form,

$$C = C_2 \theta^{\frac{1}{2}} M^n \tag{4.8}$$

where n may vary from about 0.2 for a turbojet to about 0.6 for a high bypass ratio turbofan, see Fig. 4.8.

When the SAR and SE, eqns (4.10) and (4.11), are written in terms of Mach number and the alternative fuel flow law substituted they give,

$$\mathrm{SAR} = \frac{a_0 \theta^{\frac{1}{2}}}{C_2 \theta^{\frac{1}{2}}} M^{1-n} \frac{L}{D} \frac{1}{W} \tag{4.38}$$

and

$$\mathrm{SE} = \frac{1}{C_2 \theta^{\frac{1}{2}} M^n} \frac{L}{D} \frac{1}{W} \tag{4.39}$$

When these expressions are differentiated, they give the airspeeds for maximum SAR and SE to be,

$$V_{\mathrm{max\ SAR}} = \left[\frac{3-n}{1+n}\right]^{\frac{1}{4}} V_{\mathrm{md}} \tag{4.40}$$

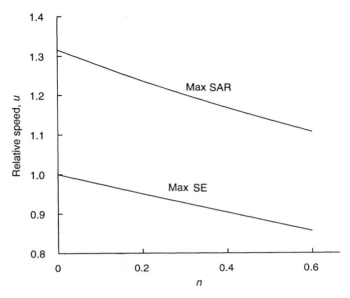

Fig. 4.9 Effect of bypass ratio on optimum speeds for range and endurance.

and

$$V_{\text{max SE}} = \left[\frac{2-n}{2+n}\right]^{\frac{1}{4}} V_{\text{md}} \tag{4.41}$$

Figure 4.9 shows how the optimum speeds vary with the bypass ratio of the engine. This implies that as the exponent, n, increases (i.e. as the bypass ratio of the engine increases), the relative airspeed for best range or endurance will decrease. This may be advantageous when cruising for maximum range but when cruising for endurance it will demand a cruising speed below the minimum drag speed with the consequence of flight path instability.

The SAR and SE, eqns (4.38) and (4.39), can be integrated to produce the full range and endurance functions under the alternative fuel flow laws. These will be similar in form to those produced by the assumption of the simple law but will show different relative speeds for optimum cruise performance.

4.3.6 The effect of weight, altitude and temperature on cruise performance

The range factor determined for each of the cruise performance expressions can be written as,

$$\frac{V_{\text{mdi}}}{C} E_{\text{max}} = \frac{1}{C}\left[\frac{W_i\theta}{\delta_i}\right]^{\frac{1}{2}}\left\{\frac{2RT_0}{p_0 S C_{\text{Lmd}}}\right\}^{\frac{1}{2}} E_{\text{max}} \tag{4.42}$$

in which the square bracket contains the variable elements of the expression.

From eqn (4.42) the range factor is seen to be proportional to the square root of the initial cruising weight which appears to suggest that the range of the aircraft will increase with its weight. It should be remembered that the fuel ratio is the ratio of the initial to final cruise weights, eqn (4.14), so that

$$W_i = \omega W_f$$

Therefore, for a given final weight of the aircraft, the increase in initial weight implies an increase in the fuel available for the cruise; it is this that extends the range not the weight of the aircraft itself. If the additional weight consists only of payload or of aircraft zero-fuel weight – that is, an increase in the final weight – then the fuel ratio will be decreased and hence the range will be reduced.

An increase in air temperature increases the range since the TAS is increased and the aircraft flies further in a given time, during which it burns the same quantity of fuel. However, in eqn (4.42) the specific fuel consumption is assumed to be a simple constant, which may not be the case. If the specific fuel consumption is a function of air temperature, as in eqn (4.8), then the effect of the temperature may be lessened and the range may even decrease as temperature increases.

Increasing altitude will produce an increase in the range as the ambient relative pressure decreases. In the troposphere, the effect will be reduced by the accompanying decrease of temperature with altitude and further affected by any dependency of the specific fuel consumption on air temperature. Generally, cruising at higher altitude will lead to better range performance. However, it has been seen that the optimum subsonic performance of an aircraft, in terms of the range or endurance it can attain from a given quantity of fuel, is a function of its minimum drag speed. Therefore, operation at airspeeds other than its optimum airspeed will incur a range or endurance penalty. The minimum drag speed in terms of equivalent airspeed is unaffected by altitude. However, as the altitude of operation increases, so the true airspeed, and Mach number, of the minimum drag speed increase and the cruise Mach number will approach its critical value. The aircraft drag will then increase due to the onset of the wave drag, and the parabolic drag polar no longer describes the drag characteristic of the aircraft.

The optimum altitude for cruise will be determined by combining the optimum cruise airspeed and the critical Mach number, M_{crit}, so that best range is flown at the highest true airspeed. This gives the greatest economy of operation by minimizing the fuel consumed and the time of flight. Assuming that the best range is obtained by flying at $1.316V_{md}$ then, from eqn (3.20),

$$M_{crit} = 1.316 \left[\frac{2}{\gamma S}\right]^{\frac{1}{2}} \left[\frac{K}{C_{Dz}}\right]^{\frac{1}{4}} \left[\frac{W}{p}\right]^{\frac{1}{2}} \tag{4.43}$$

and the parameter W/p can be evaluated. From this, the optimum cruise altitude can be found for a given aircraft weight, this is shown in Fig 4.10.

If the aircraft is flown at a higher than optimum altitude then the critical Mach number is exceeded and the increase in drag will reduce the range. To avoid this, the aircraft must be flown at an airspeed less than the optimum, again with a range penalty. If the cruise is at lower than optimum altitude then it would be usual practice to cruise at the critical Mach number to take advantage of the higher airspeeds. In this

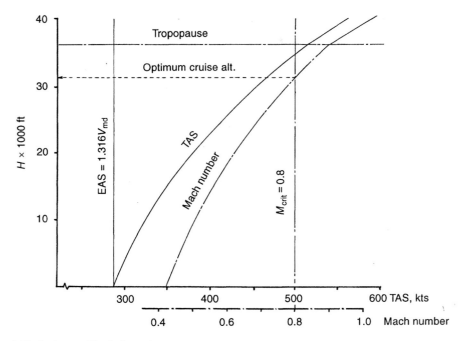

Fig. 4.10 Optimum altitude for cruise.

case, it would be necessary to accept that the relative airspeed will be greater than the optimum.

It is also assumed in eqn (4.42) that the specific fuel consumption does not vary with altitude or Mach number. In practice, the fuel flow law may be a function of the atmospheric variables, and of Mach number, and contribute to the altitude effects within the range factor. A further effect can be seen in eqn (4.40) in which the optimum relative airspeed is seen to be a function of the bypass ratio of the engine. In this case, the fuel flow law produces an optimum relative airspeed that is less than $1.316V_{md}$. Therefore, from eqn (4.43), the optimum cruise altitude at which the best range speed and the critical Mach number are balanced will be increased and better cruise economy can be achieved.

The endurance factor of the aircraft shows that the endurance is unaffected by altitude, other than through any dependence of the specific fuel consumption on the atmospheric variables.

4.4 Range and endurance for aircraft with power-producing engines

If the aircraft has power-producing engines, which produce shaft power with negligible residual thrust, the power is converted into propulsive thrust by a propeller. In this case, the performance equation is written in terms of the thrust power available and drag power required for cruising flight,

$$\eta P = DV \tag{4.44}$$

where η is the propeller efficiency.

The specific fuel consumption is defined in power terms as,

$$C = \frac{Q_f}{P} \tag{4.45}$$

This is usually quoted in kg/kW hr.

(NB *In the subsequent analysis of the cruise performance of the aircraft with power-producing engines, particularly where the specific fuel consumption is used, it may be necessary to include the gravitational constant, g, and other constants, to make the units of the expressions consistent. A check on the units of the expressions will show when this is needed.*)

The specific air range and specific endurance of the aircraft with power-producing engines are given by,

$$\mathrm{SAR} = \frac{V}{CP} = \frac{\eta}{C}\frac{L}{D}\frac{1}{W} \tag{4.46}$$

and

$$\mathrm{SE} = \frac{1}{CP} = \frac{\eta}{CV}\frac{L}{D}\frac{1}{W} \tag{4.47}$$

respectively.

From these expressions it can be deduced that the SAR is a maximum when the aircraft is flying at its minimum drag speed, V_{md}, and that the SE is a maximum when flying at the minimum power speed, V_{mp}. Now, since $V_{md} = 1.316V_{mp}$, it can be further deduced that the optimum airspeeds for the performance of aircraft with power-producing engines are related to the minimum power speed in the same way

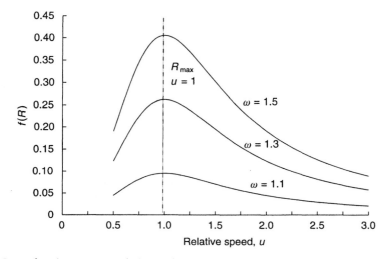

Fig. 4.11 Range function, power-producing engines.

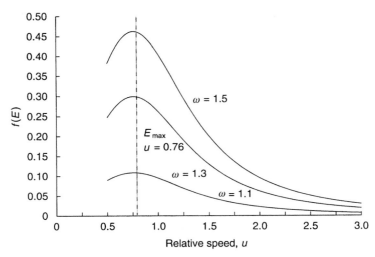

Fig. 4.12 Endurance function, power-producing engines.

as the optimum airspeeds of aircraft with thrust-producing engines are related to the minimum drag speed.

Equations (4.46) and (4.47) can be integrated to give expressions for the cruise performance of the aircraft with power-producing engines in the same way as eqns (4.10) and (4.11) were used to give the cruise performance for aircraft with thrust-producing engines. For example, in the case of the cruise–climb the range and endurance would be given by,

$$R = \left[\frac{\eta E_{\max}}{C}\right]\left\{\frac{2u^2}{u^4+1}\right\}l_n\omega \tag{4.48}$$

and

$$E = \left[\frac{\eta E_{\max}}{CV_{\mathrm{md}}}\right]\left\{\frac{2u}{u^4+1}\right\}l_n\omega \tag{4.49}$$

These are shown in Figs 4.11 and 4.12.

Since the best endurance speed is very low, it may be too close to the stalling speed for safe operation. It is likely that the aircraft would have to be flown at an airspeed in excess of the theoretical optimum when it is being operated for patrol missions in which maximum endurance is the primary requirement.

4.5 Aircraft with mixed powerplants

Some aircraft have powerplants with both thrust-producing and power-producing characteristics. Although there have been aircraft designed with a combination of turbojet engines and piston engines, the most common example is that of the turbo-prop. Here, the shaft power of the turbine engine is converted into thrust through the propeller and the residual energy in the exhaust gas is converted into thrust by the exhaust nozzle. The cruising performance characteristic of an aircraft with mixed powerplants lies between those of the aircraft with pure thrust- or pure power-producing powerplants. It needs to be estimated by taking the proportion of direct thrust to thrust power produced by the engine.

Using the cruise–climb range expression as an example, the principle can be demonstrated. From eqns (4.21) the range of the aircraft with thrust-producing engines is given by,

$$R_T = \left[\frac{V_{mdi}}{C_T} E_{max}\right] \left\{\frac{2u^3}{u^4 + 1}\right\} l_n \omega \tag{4.21}$$

and from eqn (4.28) the range of the aircraft with power-producing engines is given by,

$$R_P = \left[\frac{\eta E_{max}}{C_P}\right] \left\{\frac{2u^2}{u^4 + 1}\right\} l_n \omega \tag{4.48}$$

where subscript T refers to the thrust-producing engine and subscript P refers to the power-producing engine.

As an approximation to the cruise performance of an aircraft with a mixed powerplant, these can be combined into a common equation,

$$R = E_{max} \left[\Pi \frac{\eta}{C_P} + (1 - \Pi) \frac{u V_{md}}{C_T}\right] \left\{\frac{2u^2}{u^4 + 1}\right\} l_n \omega \tag{4.50}$$

where Π is the proportion of the thrust derived from the shaft power in the overall thrust of the powerplant and C_P and C_T are the specific fuel consumptions based on the shaft power and net thrust respectively. It can be seen that the expression in the square brackets proportions the thrust and power terms and, since it also contains a term in u, modifies the range function in the curly brackets. (The same expression can be applied to the endurance equation.)

In the case of the turbo-prop powerplant, the specific fuel consumption is usually based on the equivalent shaft horsepower, ESHP, of the engine. ESHP is the combination of the thrust output with the shaft power output to give the total output in power form as if the engine was a pure power-producing engine. In effect, the expression in the square bracket in eqn (4.50) describes the combination of thrust and power into ESHP so that the performance can be estimated as if the aircraft had a pure power-producing engine. However, it is unlikely that the proportions of thrust and power will be independent of speed or engine output, and so the expression will need to be calculated for each combination of engine power setting and aircraft speed. Because of this, cruise performance calculations for turbo-prop aircraft will usually need to be performed in a 'point-to-point' manner rather than by a continuous function.

4.6 Conclusions

The expressions for the cruising performance developed in this chapter have been based on the assumption that the aircraft has a parabolic drag polar and that the engine specific fuel consumption law is a simple constant. In practice, this is unlikely to be the case and the actual cruising performance of the aircraft will differ from the performance predicted by the simplified expressions.

The purpose of the simplified expressions is to provide a reasonable means of estimating the cruising performance at the design stage of the development of the aircraft. They may also be used to estimate the effect on the performance of a modification that affects the drag characteristic of the airframe or specific fuel consumption of the engine. For this purpose, the expressions for the range and endurance were developed in the form of the product of a [range or endurance factor], a {relative speed function} and fuel ratio function (e.g. eqn (4.21)). The breakdown into this form enables the cruising performance to be analysed in several ways.

- The range or endurance factor allows the effect on the cruise performance of a change in the drag characteristic or the specific fuel consumption to be determined so that the cost-benefit of a design change can be seen and assessed.
- The relative speed and fuel ratio functions enable the best method of operation to be determined and the quantity of fuel required for a given mission to be assessed.

In the development of the expressions for the range and endurance the integration has been taken from the initial to the final cruise weights of the aircraft and the expressions developed in terms of the initial weight, W_i, at the beginning of the cruising phase. In practice, the determination of the fuel required for a mission may require the integration to be performed from the weight at the end of the cruising phase of the flight, the final weight, W_f, instead of the initial weight. This is because the only known weight datum is the weight at the end of the mission (this is discussed further in Chapter 10). This results in the range and endurance factors and the fuel ratio functions being expressed in terms of the final conditions of the cruise phase rather than in terms of the initial conditions. Otherwise, the expressions for cruising flight and their applications are unaffected.

It is possible to develop more exact expressions for cruising flight which include alternative drag characteristics and fuel flow laws. However, they are generally more difficult to handle and to break down into forms that show the relative effects of design changes. Since the operational performance data used for flight planning are required to be derived from validated, flight measured, performance data, the cost-benefit of attempting to produce an exact cruise performance model needs to be carefully considered.

Bibliography

Bert, C. W. (1981) Prediction of range and endurance of jet aircraft at constant altitude. *AIAA Journal of Aircraft*, **18**(10), and **19**(8) (Tech Comment).
Miele, A. (1962) Flight Mechanics, Vol 1, *Theory of Flight Paths* (Pergamon Press).
Peckham, D. (1973) Range performance in cruising flight. *RAE Tech Rept 73164.*

Engineering sciences data unit items
Performance Series, Vols 1–4.
Introduction to Estimation of Range and Endurance. Vol 8, No 73018, 1980 (Am A).

Approximate Methods for Estimation of Cruise Range and Endurance: Aeroplanes with Turbojet and Turbofan Engines. Vol 8, No 73019, 1982 (Am C).

Estimation of Cruise Range: Propeller-driven Aircraft. Vol 8, No 75018, 1975.

Lost Range, Fuel and Time due to Climb and Descent: Aircraft with Turbo-jet and Turbo-fan Engines. Vol 8, No 74018, 1977 (Am A).

Lost Range, Fuel and Time due to Climb and Descent: Propeller Driven Aeroplanes. Vol 8, No 77015, 1977.

5

Climb and descent performance

The ability to control the change of the kinetic and potential energies of the aircraft, and the optimization of the energy management process, is fundamental to its safe and efficient operation.

5.1 Introduction

In the overall mission of the aircraft, there will be a climb phase in which the aircraft increases its height to the required cruising level and a descent phase from the end of the cruise to the landing. In these phases of the flight, the difference between the propulsive thrust and the airframe drag is used to change the potential energy, and the kinetic energy, of the aircraft. If the thrust exceeds drag, the aircraft will climb and if the drag exceeds thrust it will descend; the rate at which this occurs will depend on the relative magnitudes of the thrust and drag forces. Although climb and descent imply changes in height, they may also involve changes in true airspeed since the air density decreases with altitude. If the rates of climb or descent are high, the acceleration of the aircraft implicit in the climbing manoeuvre will have to be taken into consideration.

Climb performance is important from both economic and flight safety points of view. In a climb, the potential energy of the aircraft is increased and fuel energy must be expended to achieve this. The fuel required to climb to a given height can be minimized by the use of the correct climb technique and optimum economy of operation can be attained. Economy, however, is not the only criterion of operation. The safety of the aircraft depends on its ability to climb above obstructions at all points on the flight path. Sufficient excess thrust must be available to ensure that the aircraft can meet certain minimum gradients of climb in any of the safety critical segments of the flight.

The descent is less critical than the climb economically since the aircraft will be operating at low thrust and hence low fuel flow. However, several safety-related considerations will affect the choice of the flight path in the descent. Among them are the attitude of the aircraft, the rate of change of cabin pressure and the need for the engines to supply power for airframe services. The descent strategy will need to consider all of these.

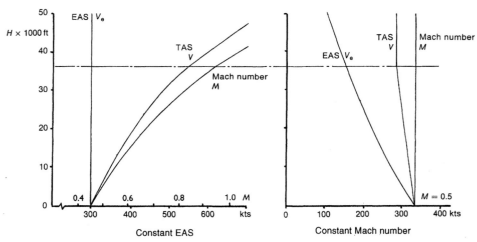

Fig. 5.1 Speed relationships in climbing flight.

A climb, or a descent, will usually be performed with reference to an indication either of airspeed or of Mach number. If the climb were based on an airspeed reference then, strictly, it would be the calibrated airspeed, assuming that any instrument error and Pitot-static pressure errors had been accounted for (Chapter 2). At the typical climbing speeds of a subsonic transport aircraft the scale-altitude correction is small and the calibrated airspeed (CAS) is close to the equivalent airspeed (EAS). Therefore, for all practical purposes, the climb can be assumed to be performed at constant EAS. This implies that, as the aircraft climbs, the ambient air density will be decreasing and the true airspeed (TAS) will be increasing, thus the aircraft will be accelerating throughout the climb. If the climb is based on Mach number then, in the troposphere, as altitude increases, the ambient temperature will be decreasing and with it the speed of sound. This implies that the true airspeed will be decreasing as the aircraft climbs. In the isothermal stratosphere, a climb at constant Mach number results in a constant TAS and there is no acceleration. Figure 5.1 shows the relationship between true airspeed, equivalent airspeed and Mach number in climbing flight. It is evident that if a climb is performed at a constant EAS then the Mach number will increase with height and the critical Mach number will eventually be reached. Alternatively, if the climb is performed at constant Mach number then the EAS will decrease towards the stalling speed as height increases.

In practice, an aircraft climbing to a height at which the Mach number would approach its critical value would usually start the climb at a constant EAS and the Mach number would be allowed to increase. In this state, the angle of attack is constant and the climb can be made at a constant, and possibly optimum, lift–drag ratio. As the Mach number increases, it becomes necessary to avoid the drag rise as the critical Mach number is reached. The climb would be converted into a constant Mach number climb allowing the EAS to decrease as the climb continues. The aircraft will no longer be climbing at the optimum aerodynamic efficiency but the drag rise

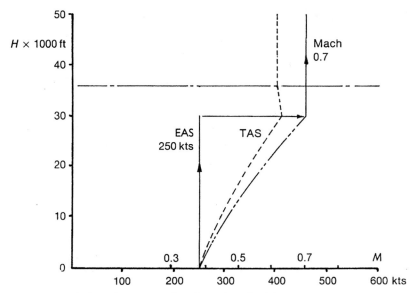

Fig. 5.2 Typical climb or descent speed/mach number schedule.

will be avoided and a good compromise can be achieved between climb performance, EAS and Mach number, see Fig. 5.2. The descending flight path is usually structured in a similar manner.

In the case of a subsonic aircraft, with a normal thrust-to-weight ratio at take-off, the rate of climb is usually low enough to allow the acceleration term in the performance equation to be neglected since the rate of change of air density, and hence TAS, with time is small. Under these conditions, the climb can be treated as being quasi-steady for the purposes of performance analysis. Similarly, the rates of descent involved with subsonic aircraft operations are usually low enough to allow the same assumption to be made in descending flight. This chapter will deal principally with the climb and descent performance of aircraft with the moderate thrust-to-weight ratios of a transport aircraft, typically around a maximum of 0.3 at take-off, as a steady-state analysis.

If the thrust-to-weight ratio is high, as in the case of a military combat aircraft, then the acceleration during climb cannot be omitted and the simultaneous change of potential and kinetic energies must be taken into account. This is known as the *total energy* climb and requires a quite different approach to its analysis, it is considered briefly in Section 5.7.

Steady-state climb and descent can be measured in terms of rate or gradient. The rate of climb is the vertical velocity, dH/dt, and is usually quoted in feet/min. The gradient of climb, γ_2, is defined by,

$$\gamma_2 = \sin^{-1}\left[\frac{dH/dt}{V}\right] \tag{5.1}$$

where the height, H, is geopotential height.

The gradient is often expressed as a *percentage gradient*, or grad%, which is defined as,

$$\text{grad}\% = 100\tan\gamma_2 \tag{5.2}$$

5.2 Climb and descent performance analysis

The powerplant of the aircraft may be thrust producing or power producing. Thrust-producing engines, turbojets and turbofans, produce thrust that is relatively constant with small change of airspeed in subsonic flight. Power-producing engines, piston engines or turbo-shaft engines, produce shaft power, which is relatively constant with change of airspeed, and which needs to be converted into propulsive thrust by a propeller. The differing characteristics of these powerplants lead to different criteria for optimum climb performance and each needs to be considered separately.

5.2.1 Aircraft with thrust-producing engines

The equations of motion of an aircraft with thrust-producing engines in a climb (or a descent) can be taken from eqn (3.3). In a straight, wings level, climb in which the flight path gradient is constant the equations of motion can be written,

$$\left. \begin{array}{r} F_N - D = W \sin \gamma_2 + m\dot{V} \\ L = W \cos \gamma_2 \end{array} \right\} \tag{5.3}$$

(NB. It should be remembered that these equations of motion contain simplifying assumptions and can only be used when those conditions apply.)

If the aircraft has a normal take-off thrust-to-weight ratio of about 0.3 then the rates of climb will be low enough to assume that the acceleration associated with the rate of climb is negligible. The climb can then be assumed to be made either at constant airspeed or at constant Mach number. Also, the gradient of climb and descent will be low enough to allow the assumption that $\cos \gamma_2 = 1$ in eqn (5.3), and the equation can be simplified further to the form,

$$\left. \begin{array}{r} F_N - D = W \sin \gamma_2 \\ L = W \end{array} \right\} \tag{5.4}$$

These equations will be used to derive the climb and descent performance expressions for the quasi-steady flight path.

From eqn (5.4) the excess thrust $[F_N - D]$ provides the gradient of climb,

$$[F_N - D]\frac{1}{W} = \sin \gamma_2 \tag{5.5}$$

so that, if the thrust is constant, the best gradient of climb will be obtained by flying at the minimum drag speed. Figure 5.3 shows the ideal thrust and drag relationship (relative to the minimum drag) in which the thrust does not vary with airspeed and the maximum excess thrust occurs at the minimum drag speed. It should be noted, however, that, in practice, the airspeed is likely to influence the thrust to some extent. Therefore, the airspeed for optimum climb gradient will be found to be close to, but not necessarily at, the minimum drag speed.

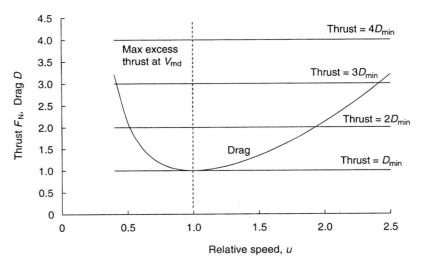

Fig. 5.3 Excess thrust, thrust-producing engines.

Using eqn (5.1) in eqn (5.5) leads to an expression for the best rate of climb,

$$[F_N - D]\frac{V}{W} = \frac{dH}{dt} \tag{5.6}$$

This indicates that the airspeed for the best rate of climb occurs when the excess thrust power, $F_N V$, over drag power, DV, is a maximum. Since the ideal thrust power increases linearly with true airspeed, the best rate of climb is predicted to be at an airspeed greater than the minimum drag speed; this is seen in Fig. 5.4. In this case, there is no simple solution for the airspeed for best rate of climb, this will occur

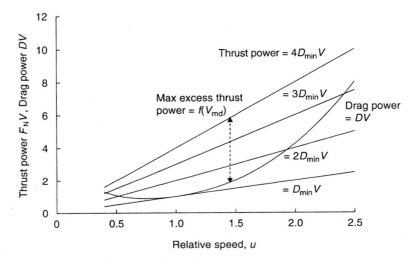

Fig. 5.4 Excess thrust power, thrust-producing engines.

when the difference between the thrust power and drag power is a maximum and is a function of the excess thrust-power.

Generalized climb performance for rate and gradient of climb or descent

The climb and descent performance can best be analysed in a general manner by considering a dimensionless form of the performance equation, eqn (3.27). The generalized climb performance equation can be written for thrust-producing engines as,

$$\tau - \tfrac{1}{2}[u^2 + u^{-2}] = E_{\max} \sin \gamma_2 = E_{\max} \left(\frac{v}{u}\right) \tag{5.7}$$

where the dimensionless rate of climb, v, is defined as,

$$v = \frac{\mathrm{d}H/\mathrm{d}t}{V_{\mathrm{md}}} \tag{5.8}$$

Equation (5.7) can be applied to any aircraft for which the drag characteristic, the thrust and the weight are known. By differentiating eqn (5.7), the relative airspeeds for best climb or descent performance can be found.

Climb gradient

From eqn (5.7) the gradient of climb is given by,

$$E_{\max} \sin \gamma_2 = \tau - \tfrac{1}{2}[u^2 + u^{-2}] \tag{5.9}$$

For maximum gradient $\mathrm{d}\gamma_2/\mathrm{d}u = 0$, which occurs when $u = 1$ if τ is constant, and confirms that the steepest climb occurs at the minimum drag speed (eqn (5.5)). Figure 5.5 shows the dimensionless climb gradient as a function of relative airspeed for several values of dimensionless thrust, τ; combinations of τ and u that give positive values of $\sin \gamma_2$ produce climbing flight. Descending flight occurs when the

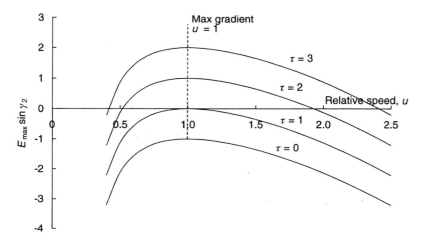

Fig. 5.5 Gradient of climb, thrust-producing engines.

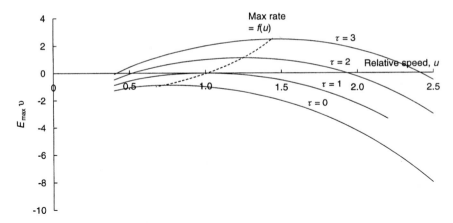

Fig. 5.6 Rate of climb, thrust-producing engines.

combination of τ and u give negative values of $\sin \gamma_2$. Typical values of τ for a transport aircraft at take-off thrust usually lie between 3 and 4.

In the special case of gliding flight, in which $\tau =$ zero, flight at the minimum drag speed will give the shallowest glide angle, which will give the greatest range of glide; this speed is used when cruising between thermals. The minimum glide angle will be,

$$\gamma_2 = \sin^{-1}[1/E_{\max}] \qquad (5.10)$$

Climb rate

From eqn (5.7) the rate of climb is given by,

$$E_{\max}v = \tau u - \tfrac{1}{2}[u^3 + u^{-1}] \qquad (5.11)$$

and for maximum rate of climb $\mathrm{d}v/\mathrm{d}u =$ zero. In this case, there is no simple solution and the relative airspeed for best rate of climb is found to be a function of the dimensionless thrust,

$$u = \frac{1}{\sqrt{3}}[\tau \pm (\tau^2 + 3)^{\frac{1}{2}}]^{\frac{1}{2}} \qquad (5.12)$$

Figure 5.6 shows the dimensionless rate of climb as a function of relative airspeed for several values of dimensionless thrust, and the relative airspeed for best rate of climb is seen to increase with dimensionless thrust.

In gliding flight, the minimum sink rate is attained by flying at a relative airspeed of $u/\sqrt[4]{3}$ which is the minimum power speed of the aircraft. Flying at this speed will maximize the time (or endurance) of gliding flight and is the speed used for climbing in thermals.

5.2.2 Aircraft with power-producing engines

Using the same assumptions that were used in the case of the aircraft with thrust-producing engines, the equation of performance for the aircraft with power-producing

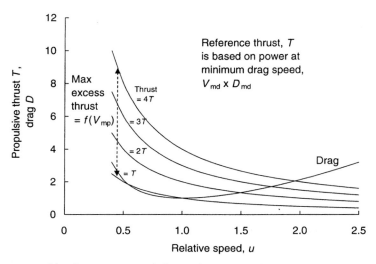

Fig. 5.7 Excess propulsive thrust, power-producing engines.

engines (eqn (3.17)) can be written, in parallel with those for aircraft with thrust-producing engines, as

$$\frac{\eta P}{V} - D = W \sin \gamma_2 \qquad (5.4a)$$

where $\eta P/V$ is the propulsive force developed by the engine–propeller combination.

The gradient of climb is given by,

$$\left[\frac{\eta P}{V} - D\right] \frac{1}{W} = \sin \gamma_2 \qquad (5.5a)$$

which has a maximum value when the excess propulsive force is a maximum. This occurs at an airspeed less than the minimum drag speed in climbing flight. Figure 5.7 shows the excess propulsive force (relative to the power at the minimum drag speed, $V_{md} \times D_{md}$). It indicates that the maximum occurs at an airspeed that is less than the minimum drag speed and which tends to decrease as the power available increases.

The rate of climb is given by,

$$[\eta P - DV] \frac{1}{W} = \frac{dH}{dt} \qquad (5.6a)$$

and is a maximum at the minimum power speed. Figure 5.8 shows the excess thrust power that occurs at the minimum power speed of the aircraft, assuming that the thrust power is independent of airspeed.

Generalized performance

The generalized performance equation for climb and descent is given, from eqn (3.27), for aircraft with power-producing engines as,

$$\frac{\lambda}{u} - \tfrac{1}{2}[u^2 + u^{-2}] = E_{max} \sin \gamma_2 = E_{max} \frac{v}{u} \qquad (5.13)$$

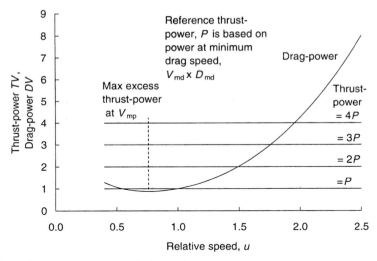

Fig. 5.8 Excess thrust power, power-producing engines.

Climb gradient

From eqn (5.13) the gradient is given by,

$$E_{max} \sin \gamma_2 = \frac{\lambda}{u} - \tfrac{1}{2}\left[u^2 + u^{-2}\right] \tag{5.14}$$

By differentiating eqn (5.14) the relative airspeed for best climb gradient is found to occur when $d\gamma_2/du = 0$ which gives,

$$u^4 + \lambda u - 1 = 0 \tag{5.15}$$

This has no simple solution but shows that the relative airspeed for maximum gradient of climb is a function of engine power. The dimensionless climb gradient is shown in Fig. 5.9 as a function of relative airspeed for several values of dimensionless power, λ. In gliding flight, λ = zero, and the shallowest glide angle is given by flying at the minimum drag speed. As the power increases, the airspeed for maximum climb gradient decreases and, for values of dimensionless power greater than unity, the best gradient is attained by flying at airspeeds less than the minimum power speed. In practice, this may be impractical since the aircraft may be operating too close to the stalling speed for safety; this will be referred to later.

Climb rate

From eqn (5.13) the rate of climb is given by,

$$E_{max}v = \lambda - \tfrac{1}{2}\left[u^3 + u^{-1}\right] \tag{5.16}$$

For maximum rate dv/du = zero which occurs when $u = 1/\sqrt[4]{3}$, which is the minimum power speed, eqn (3.21). Figure 5.10 shows the dimensional rate of climb as a function of relative airspeed for several values of dimensionless power. The best rates of climb are attained by flying at the minimum power speed at all levels of power. In gliding flight, the minimum sink rate occurs at the minimum power speed.

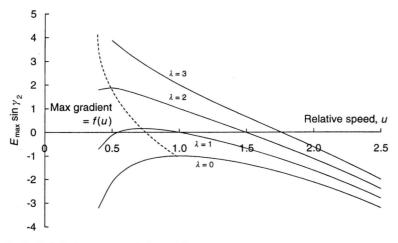

Fig. 5.9 Gradient of climb, power-producing engines.

5.2.3 Aircraft with mixed powerplants

Aircraft may have powerplants that do not fall easily into either the thrust-producing or power-producing categories; (a classic example of such an aircraft was the Convair B36 that had six piston engines and four turbojets). Turbo-prop engines usually produce some residual thrust from their exhaust gas and, in some cases, it forms a substantial contribution to the overall propulsive force. The unducted fan engine combines the principle of the propeller with the high bypass ratio turbofan, and its propulsive thrust characteristic is neither that of a pure power-producing engine nor that of a pure thrust-producing engine.

In these cases, the powerplant cannot be regarded as purely thrust-producing or power-producing and it will have a characteristic somewhere between the two.

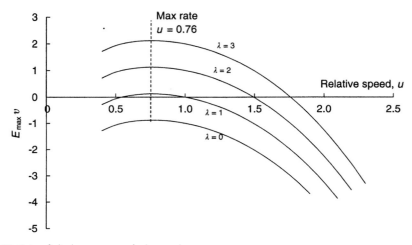

Fig. 5.10 Rate of climb, power-producing engines.

Where the thrust and power elements can be separated, the climb performance of the aircraft with such a powerplant can be analysed by using the generalized expression for performance, derived from eqn (3.27), and including both the power and thrust terms.

$$\left[\frac{\lambda}{u} + \tau\right] - \tfrac{1}{2}[u^2 + u^{-2}] = E_{max} \sin \gamma_2 = E_{max} \frac{v}{u} \qquad (5.17)$$

This leads to the conclusion that the best gradient of climb will be attained by flying at an airspeed below the minimum drag speed and the best rate of climb will be found at an airspeed above the minimum power speed. The actual airspeeds will be functions of the excess propulsive thrust and the relative proportions of thrust and power developed by the powerplant.

5.3 Measurement of best climb performance

In the foregoing analysis it has been assumed that the output of the engine, either thrust-producing or power-producing, is not affected by airspeed and that the optimum airspeeds for best rate or gradient of climb have been determined on this assumption.

In practice, the thrust of a turbojet or turbofan engine will be affected by airspeed and the excess thrust will not necessarily occur at the minimum drag speed. Similarly, the thrust power of the power-producing engine–propeller combination will depend on airspeed since there may be a direct effect of airspeed on the intake airflow of the engine, producing a 'supercharging' or ram effect. In addition, there may be a dependence of the propeller efficiency on the advance ratio, J ($= V/nD$), which is proportional to airspeed. This implies that the excess thrust power will not necessarily occur at the minimum power speed of the aircraft. Therefore, although the theoretical analysis of the climb predicted airspeeds for optimum climb performance related to the minimum drag and minimum power speeds of the aircraft, the actual optimum climb speeds may not occur at precisely the predicted airspeeds and will need to be determined by flight trials.

One of the usual ways of determining the best climbing performance is by partial climbs. This technique consists of flying a number of short climb segments through a datum height. Each of the climbs is flown at a different airspeed around the predicted speed for best climb performance. The measured rates of climb are corrected for change of aircraft weight due to the consumption of fuel during the series of climbs (Chapter 8), so that the performance relates to a common standard aircraft weight. The measured climb rate and climb gradient can then be plotted against airspeed to give the actual airspeeds for best performance under the weight, altitude and temperature conditions of the test. This process will need to be repeated for all airframe configurations that will be used in flight, for example, all expected flap settings and with landing gear extended and retracted, and at altitudes appropriate to the configuration. From the measured data, the climb performance of the aircraft can be determined in terms of both rate and gradient of climb. From the flight measured data the airspeeds at which the optimum climb performance is achieved can be found. These data will be used in the development of the aircraft performance manual.

An alternative method is to measure the maximum excess thrust or power by level accelerations. In this technique, the aircraft is flown as slowly as possible in level flight; maximum thrust or power is selected and the airspeed recorded in a level acceleration to maximum airspeed. From the acceleration, the excess thrust or power can be deduced and thus so can the speeds for best climb performance. The level acceleration method is best suited to aircraft with thrust-producing engines. The partial climb method is best suited to aircraft with power-producing engines since the best climb speeds tend to be towards the lower end of their speed range.

5.4 Climb performance in aircraft operations

In practice, the climb may be performed to give either a steep gradient of climb or a high climb rate, the choice will depend on the most critical consideration of the phase of flight.

5.4.1 Climb gradient

In the take-off and initial climb phase, the most critical consideration is that of flight safety and the need to ensure that the aircraft can avoid all known obstructions along its flight path. In the licensing of the airfield, a departure path is defined along which no obstructions are permitted and the aircraft is guaranteed a clear flight path. The definition of the departure path is complex and depends on the size of the airfield and the type of aircraft operations that are intended. For large, international, airports, the obstacle limitation surface – which defines the safe departure path – is a surface, of gradient 2%, extending from the end of the take-off distance available on the runway to a distance of 15 000 m. (A full definition can be found in ICAO International Standards and Recommended Practices, Annex 14, Aerodromes.) Therefore, to guarantee a safe departure from the airfield the aircraft must be capable of climbing at a gradient of at least 2% under all conditions, including emergency conditions with one engine inoperative. This will be discussed further in Chapter 9. Clearly, in this phase of flight the aircraft needs to be operated at an airspeed that will produce the best gradient of climb so that the departure flight path will be steep enough to exceed the minimum safe gradient specified. Therefore, the airspeed chosen for the after-take-off climb should be that for maximum gradient. However, the airspeed for best gradient is usually a low speed and may be close to other critical operating airspeeds, such as the stalling speed or minimum airspeeds for lateral-directional control. Restrictions on the airspeed scheduled for the climb are based on a safe margin over the stalling speed and the ability to maintain lateral-directional control in the event of a sudden loss of propulsive thrust on an engine. This will often result in the scheduled airspeed for the climb being higher than that for optimum climb gradient, this is particularly true in the case of aircraft with power-producing engines.

One of the most critical parts of the flight path is the after-take-off climb. This is made with the aircraft in the take-off configuration, initially with landing gear extended, and with flaps set to optimize the take-off speed and runway distance requirement. In this state, the climb gradient with one engine inoperative will often be the critical limiting factor in determining the maximum allowable take-off weight of the aircraft.

5.4.2 Climb rate

Once the aircraft has climbed to a safe height, usually taken to be 1500 ft above the airfield, the need to avoid ground-based obstacles is no longer critical and the climb can continue in the most expedient manner. In the case of transport operations, this will usually be the most economic climb. This will be based on either the minimum time to climb to operating height, the minimum fuel consumed in the climb or some compromise between these which will give the best overall economy.

The maximum rate climb will enable the aircraft to reach its operating height in the minimum time so that the cruise phase can commence. The airspeed for best climb rate is higher than that for best gradient. Therefore, following the after-take-off climb, the aircraft can be accelerated to its climb speed for best climb rate in its en-route configuration and continue to climb to cruise altitude following a convenient schedule of airspeed and Mach number.

Aircraft with power-producing engines will usually climb at their airspeed for best rate of climb, which will be close to their minimum drag speed. The climb will then continue to the cruising height where the aircraft will accelerate to its cruising speed.

Aircraft with thrust-producing engines have an airspeed for best rate of climb that is a function of their excess thrust; the greater the excess thrust the higher will be the airspeed for best climb rate, eqn (5.12). As the aircraft climbs, the thrust will decrease and with it the optimum airspeed for climb rate. The airspeed used in the climb will generally be a compromise based on the excess thrust, which will be a function of the weight, altitude and temperature (WAT) conditions at the start of the climb. It will take into account the anticipated WAT changes during the climb to give the best average climb performance throughout the climb. As the climb continues, the flight Mach number will increase as the relative pressure of the atmosphere decreases. It may become necessary to convert the climb to constant Mach number to avoid the drag rise that would reduce the climb performance (Fig. 5.2).

5.4.3 Minimum fuel climbs

The fuel consumed in the climb can be expressed in terms of the specific climb, SC, as,

$$SC = \frac{dH/dt}{Q_f} \tag{5.18}$$

and is expressed in terms of ft/kg.

If the fuel flow is measured during the flight trials to determine the optimum climb speed, then the specific climb function can be formed to give the airspeed for a minimum fuel climb. In the simple analysis, in which the assumption of constant specific fuel consumption is made, the thrust or power is set to the maximum continuous setting. The minimum fuel climb will then occur at the airspeed for best rate of climb. However, in practice, the specific fuel consumption, together with the output of the powerplant, may vary with air temperature and Mach number and an airspeed–Mach number schedule may be found that will optimize the climb for minimum fuel consumption. Any analysis of the economic benefits of a minimum fuel climb will depend on the direct operating costs of the aircraft and therefore, cannot be conclusive on performance grounds alone.

5.4.4 Noise limitations

Transport aircraft are required to conform to stringent noise regulations on take-off and climb-out from airports; operators of aircraft that exceed the noise limits may be subjected to heavy penalties. To conform with the regulations it may be necessary to reduce the thrust or power in the after-take-off climb before the aircraft has reached the noise measuring station that will be positioned at a point under the departure path. The thrust reduction will reduce the rate, and gradient, of the climb and will extend the time taken to reach the cruising altitude. Engines with a low noise signature, or which do not require a large thrust reduction to comply with the noise regulations, will provide the aircraft with a better performance in the climb. This is because they can operate at higher thrust levels without exceeding the noise limits.

5.5 Descent performance in aircraft operations

If the propulsive thrust is less than the airframe drag then the aircraft will decelerate or descend, as can be seen from the generalized climb performance characteristics, Figs 5.5 and 5.6. The descending flight path can be varied from a shallow descent to a very steep descent either by reducing the engine thrust or by increasing the airframe drag. The drag can be increased either by aerodynamic means or by varying the airspeed. Thus, the aircraft has a very wide range of descent path profiles available to it. In the special case of gliding flight, in which there is no propulsive thrust, the descent will be determined by the lift–drag ratio, E. In this case, the minimum rate of descent occurs at the minimum power speed and the minimum gradient occurs at the minimum drag speed.

Although in Fig. 5.5 it is apparent that a descent can be produced by flying at an airspeed less than the minimum drag speed, the aircraft will not have flight path stability in this condition. Flight path stability occurs when the flight path gradient can be controlled by the use of the elevator control only, Fig. 5.11. If the aircraft is flying at an airspeed greater than the minimum drag speed then the flight path gradient of descent can be increased (steepened) by increasing the airspeed. This is achieved by a nose down pitch change with no need to adjust the engine thrust setting. Conversely, a decrease in descent gradient can be made by a nose up pitch change that will decrease the airspeed; this control can be achieved by using the elevator control alone. However, on the backside of the drag curve (that is, at airspeeds less than the minimum drag speed), the rate of change of drag with airspeed is negative and the flight path gradient cannot be controlled by the elevator alone. To maintain precise

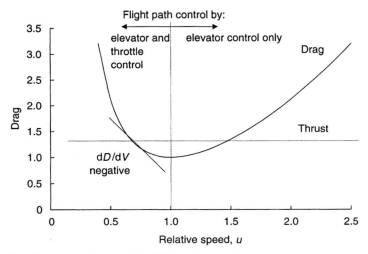

Fig. 5.11 Effect of minimum drag speed on flight path control.

control of the flight path gradient, changes in thrust setting will be necessary in addition to the elevator control inputs to control the descent gradient, otherwise large excursions from the flight path will occur. Descents at airspeeds less than the minimum drag speed (or, in the case of aircraft with power-producing engines, below the minimum power speed), are generally to be avoided. Unless an auto-throttle is employed to maintain airspeed, accurate manual control of the flight path will be difficult.

In practice, limitations to the descent performance may be necessary. In transport operations it would be undesirable to make a very steep, high airspeed descent since this would entail a steep nose-down attitude that could be uncomfortable, if not dangerous, to persons in the cabin. In addition, the rate of increase of cabin pressure during descent must be kept to a reasonably low value to prevent discomfort due to the re-pressurization of the passengers' ear passages. The rate of change of cabin pressure should not exceed the equivalent of 300 ft/min at sea level. This implies that, if the cabin is pressurized to the equivalent of 8000 ft pressure height, the descent to sea level should take not less than 24 minutes regardless of the pressure height from which the aircraft commenced its descent. An exception to this general rule is the emergency descent following the loss of cabin pressure. In this case, the aircraft must descend to a safe altitude as quickly as possible and the highest rate of descent must be used.

The optimization of the descent is not as straightforward as the optimization of the climb. Since the engines will be operating at a low thrust or power, the fuel consumption will be low and the optimization based on fuel consumed is not usually considered to be the most critical condition. Normally, the powerplant will produce some propulsive force that will contribute to the performance equation and the aircraft cannot be considered to be in a true glide. A turbojet or turbofan engine will usually produce a residual thrust, even when operating at the flight idle setting. There will be a small, but not negligible, thrust contribution, which will reduce both the descent rate and the gradient that would occur in the glide. It is not

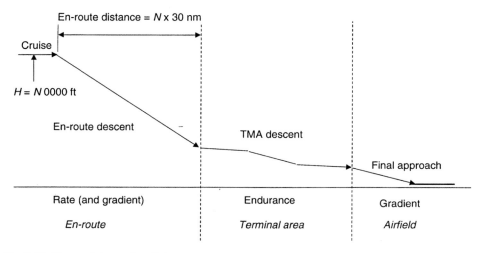

Fig. 5.12 Phases of descending flight.

uncommon to have to increase the drag of the aircraft to enable it to descend at a sufficiently high rate or gradient. Flaps, spoilers, airbrakes and landing gear are all used as means of increasing the drag to obtain a suitable descent performance. Propeller driven aircraft can usually develop sufficient drag from the propeller to avoid the need for airbrakes or spoilers; the flaps and landing gear together with the propeller will normally produce a steep enough descent for all practical purposes. The engine, however, must be operated in such a way that all essential aircraft systems remain fully operative. Pressurization must be maintained and generators, hydraulic pumps and engine air bleeds will need to provide the necessary output. Therefore, the minimum engine settings in the descent may be determined, for example, by the need to provide power for the anti-icing systems.

There are several phases of a typical descent flight path from cruising altitude down to landing. Each phase of the descent has different criteria that govern the manner in which the aircraft is flown. Figure 5.12 shows the phases of a typical descent.

From cruising height, the aircraft will descend towards the terminal area of the airfield in the en-route descent phase. In this phase, the descent will usually be made at a high airspeed and a rate of descent commensurate with the requirement for re-pressurization of the cabin, and engine power settings necessary to keep the aircraft systems operative. Typically, this phase may be flown by reference to a simple rule of thumb, such as commencing the descent at a distance of 30 nm from the destination for each 10 000 ft of altitude. This type of strategy will often produce a good working compromise between rate of descent and airspeed for large transport aircraft. The strategy could be improved by using a flight management system to provide a more exact optimization of the descent. It would adjust the airspeed and rate of descent to complete the descent to the required height over the reporting point at the required time. This is referred to as 4-D navigation since it combines area navigation with height and time. During the en-route descent, it may be necessary to fly a schedule of Mach number and airspeed, as in the climb, to avoid the critical Mach number and stalling speed boundaries of the flight envelope.

From the boundary of the terminal area the aircraft will normally be subject to air traffic control restrictions that demand that it flies at a given airspeed to maintain traffic separation as it is manoeuvred onto the final approach. The descent is likely to be at a low rate and priority given to the positioning manoeuvres. Since the airspeed is now constrained, the aircraft performance must be optimized by changes in its configuration. Flaps and other aerodynamic devices can be used to ensure that the necessary safety margins of airspeed are complied with and that the aircraft is being flown in the most economic manner. In this phase, the aircraft should be flown at a speed close to its maximum endurance speed for the best economy as it is manoeuvred onto the final approach.

On the final approach, the gradient of the flight path is the main criterion. The gradient must be steep enough to exceed the slope of the minimum obstacle limitation surface, but not so steep that the flare to touchdown requires an excessive pitch attitude change. Typically, the gradient of the descent flown by large transport aircraft will be about 3°, which is equivalent to a 5% gradient. Smaller transport aircraft are often capable of using steeper final approach gradients for approaches into airports with restricted approach paths or to assist in the separation of airport traffic arrivals. During the final approach to the landing, the aircraft will need to be flown at the lowest airspeed at which the safety margins can be met and at a pitch attitude that allows for a smooth flare and touchdown. In this phase of flight, the handling qualities must be such that the aircraft can be flown with accurate flight path control. This implies that the airspeed should not be less than the minimum drag speed to maintain flight path stability. The minimum drag speed is determined by the drag characteristic of the aircraft and by its weight (eqn (3.20)).

$$V_{\mathrm{md}} = [2W/\rho S]^{\frac{1}{2}}\{K/C_{\mathrm{Dz}}\}^{\frac{1}{4}} \tag{3.20}$$

By increasing the zero-lift drag coefficient, C_{Dz}, the minimum drag speed will be reduced although the overall drag force will be increased; doubling the zero-lift drag will usually decrease the minimum drag speed by almost 20%. The zero-lift drag increase can be produced by lowering the landing gear and flaps and by using airbrakes, spoilers or other devices specifically designed to produce high zero-lift drag forces. The final approach will usually be flown in a high drag configuration with landing gear extended, flaps fully extended and probably with airbrakes deployed. In this way, the minimum drag speed is reduced to its lowest value and the aircraft will have the necessary flight path stability at its minimum approach speed. A further benefit is that the increased drag will require the engines to be operating at a fairly high thrust setting. In this state, should the aircraft have to abandon the approach and go around, the engines will respond quickly to the demand for maximum climb thrust. At the same time, the high drag devices can be retracted and an excess of thrust over drag for the climb can be achieved in the minimum time. If the aircraft was operating at low engine thrust without the high drag devices, the engine response time to achieve climb thrust would be much longer and the aircraft drag could not be reduced. This would lead to a delay in achieving the necessary climb gradient.

The emergency descent is used, for example, when the aircraft needs to descend rapidly to recover cabin pressure should the pressurization system fail. In this case, the descent to a safe altitude, below 10 000 ft, at which the ambient pressure of the atmosphere is high enough to breathe without the assistance of supplementary oxygen, must be made in the shortest possible time. Two strategies can be considered. First, a high airspeed descent using minimum thrust and with the aircraft in a clean configuration. This strategy will be limited by the maximum Mach number that can be achieved before the onset of Mach buffet or handling problems occur. If the cruising Mach number was already close to the limiting Mach number, the excess drag that could be achieved may not be sufficient to produce a high enough rate of descent. Furthermore, the high airspeed may lead to restrictions on the manoeuvring of the aircraft. Secondly, a low airspeed, high drag descent can be used. In this strategy, the aircraft must first be slowed down to an airspeed at which the flaps, landing gear and other high drag devices can be extended. The descent will then be made at the highest structural limiting airspeed and minimum thrust. In this case, time is lost in the deceleration process and the aircraft may have a very steep nose-down attitude in the descent, which could cause difficulties in passenger or cargo restraint. There is no absolute rule for the emergency descent strategy, a procedure will have to be developed for each aircraft type that will minimize the descent time and keep the aircraft within its design limitations.

5.6 The effect of wind on climb and descent performance

Wind is the relative velocity between the general air mass and the ground. Usually the wind can be assumed to have only a horizontal component of velocity. However, in close proximity to the ground, it will tend to follow the ground profile and so may have considerable vertical velocity components where sloping or undulating ground occurs; in some cases this effect may extend to considerable heights. For the purposes of this analysis, the aircraft will be taken to be operating over level ground so that only the horizontal component of the wind velocity will be considered. Close to the ground the relative velocity of the wind produces a boundary layer in which the wind speed decreases as height decreases, this will produce further effects on the flight path of the aircraft.

The flight path of the aircraft is calculated relative to the air mass and, so far, the development of the theory covering the climb and descent performance achieved by the aircraft has been assumed to occur in still air conditions. In a moving air mass the *actual performance* of the aircraft relative to the air mass is not affected since the reference axes for performance are velocity axes. Since their origin is at the centre of gravity (CG) of the aircraft and moves with the aircraft, it assumes a zero velocity datum within the air mass. The aircraft will achieve the same rates and gradients of climb, and descent, relative to the air mass regardless of the wind. However, the performance relative to the ground will be affected by the wind. This is the *perceived performance* seen by the observer, from either the aircraft or the ground, and which affects the ability of the aircraft to clear ground-based obstructions.

Figure 5.13 shows the effect of a headwind or tailwind on the climb and descent gradients. The rate of climb, or descent, relative to the air mass is unaffected by the wind but the horizontal component of the true airspeed is increased by the tailwind

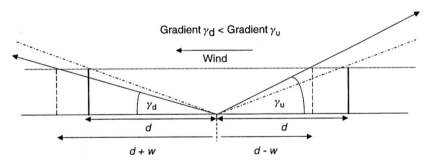

Fig. 5.13 Effect of wind on climb gradient.

or decreased by the headwind. The perceived gradient of the flight path, γ_p, relative to the ground thus becomes,

$$\gamma_p = 100 \tan \left[\frac{\sin \gamma_2}{\cos \gamma_2 - \dfrac{V_w}{V}} \right] \tag{5.19}$$

where V_w is the headwind velocity component, and γ_2 is the actual gradient of climb or descent relative to the air mass.

The perceived gradients are reduced relative to the actual gradients in a tailwind and increased in a headwind. Thus, if an aircraft climbs downwind, the ability to clear obstructions will be reduced although the aircraft is still producing its predicted climb gradient with respect to the air mass. The effect of wind on the perceived climb gradient has caused incidents to occur, particularly in cases where the aircraft has encountered a tailwind in a critical climb situation. Similarly, the downwind approach is a well-established cause of landing incidents due to the reduction in perceived gradient of descent. The situation is made more difficult as the aircraft enters the boundary layer region in which there is a wind velocity gradient caused by the rate of change of wind-speed with height. Because of these effects, it is normal practice when taking the wind component into account to use a factor of 50% for the headwinds and 150% for tailwinds. This factor will be seen in the wind-speed correction in the take-off and climb performance data of the aircraft performance manual (Chapter 10).

An extreme case of the wind effect on the flight path is that of *Wind-shear*, in which the rate of change of wind velocity is very large. Wind-shear is caused by severe meteorological conditions associated with rainstorms that create very strong local downdraughts, known as a microburst, which spread out rapidly as they contact the ground. This results in localized changes in wind speed and direction that may be large and occur very suddenly. Since wind-shear occurs close to the ground, aircraft on final approach or making their after-take-off climb are particularly at risk. On encountering wind-shear, a sudden change in airspeed will occur, together with a change in the flight path gradient through the mechanism described above. To recover to its former state of flight the aircraft will need a rapid response in both engine thrust and angle of attack which, in severe cases, may be beyond the performance capability of the aircraft. Wind-shear warning systems, which sense changes in airspeed and vertical motion, can enable an early response to be made, which will minimize the effect of a wind-shear encounter.

5.7 High-performance climb

Although this treatment of performance is not intended to cover supersonic flight in depth, the climb performance of aircraft with a high thrust-to-weight ratio should be mentioned.

In the case of high-performance combat aircraft, with afterburners operating, the thrust-to-weight ratio at take-off may be about unity and the climb rates achieved by such aircraft are very high, of the order of 50 000 ft/min. This leads to a different approach to the climb performance since it is not possible to treat the climb as a pure change in potential energy. As the aircraft climbs, the ambient air density decreases and, if the climb is flown at a constant EAS, the TAS and Mach number will increase during the climb (Fig. 5.1). In a climb from sea level to 40 000 ft, in ISA conditions, the relative density decreases from unity to about 0.25 and, therefore, the TAS will be doubled. A climb to 40 000 ft by an aircraft with a take-off thrust-to-weight ratio of unity would take about one minute. Therefore, if the EAS were 300 kts there would be an associated acceleration of about 300 kts/min during the climb. The climb is thus a combination of the increase in potential energy and kinetic energy.

The combination of the potential and kinetic energies of the aircraft, per unit mass, is known as the *specific total energy*, E_s. This is sometimes referred to as the *energy height* since it represents the height the aircraft would attain if all its kinetic energy were to be converted into potential energy. The specific total energy is given by,

$$E_s = \left\{ H + \frac{V^2}{2g} \right\} \tag{5.20}$$

From the performance equation, eqn (5.3), the *excess thrust* is given by,

$$[F_N - D] = W \sin \gamma_2 + m\dot{V} \tag{5.21}$$

This can be written in terms of the *specific excess power (SEP)*, which is the excess power per unit weight, as,

$$[F_N - D]\frac{V}{W} = \left\{ \frac{dH}{dt} + \frac{V}{g}\frac{dV}{dt} \right\} = \frac{d}{dt}\left\{ H + \frac{V^2}{2g} \right\} \tag{5.22}$$

From eqn (5.22), it can be seen that the SEP is equivalent to the rate of change of specific total energy and can be used to increase the total energy of the aircraft in a combination of climb and acceleration.

A fighter aircraft has to climb to its operational height and accelerate to its operational Mach number in the shortest time possible if it is to gain the advantage in an air-to-air combat mission. Minimizing the time required to increase the total energy of the aircraft to its combat height and Mach number is fundamental to the operational technique. By maximizing the SEP, the time taken to increase the specific total energy will be minimized. Therefore, the speed schedule flown in the climb is chosen to maximize the difference between the thrust-power, $F_N V$, and the drag-power, DV. In the transonic region, there is an increase in drag that reduces the excess thrust. Acceleration across the transonic region can be assisted by reducing the rate of climb, or even diving, to give the priority to the increase in kinetic energy until the aircraft is established in supersonic flight.

Although the high performance climb is associated with supersonic aircraft, some subsonic aircraft may have rates of climb that are great enough to make the kinetic energy increase significant. In these cases, it may be necessary to consider the acceleration term in the analysis of the climbing performance. In a climb made at a steady EAS, the rate of climb can be corrected by including the effect of the acceleration in the performance equation, thus eqn (5.21) can be written in the form,

$$[F_N - D] = \frac{W}{V} \frac{dH}{dt} \left\{ 1 - \frac{V_e^2}{2g\sigma^2} \frac{d\sigma}{dH} \right\} \qquad (5.23)$$

in which the curly brackets account for the proportion of the excess thrust used in acceleration.

5.8 Conclusions

Climb performance is one of the most critical areas in both the design and the operation of an aircraft. To conform to the airworthiness requirements it has been seen that a multi-engined aircraft needs to be able to climb with one engine inoperative in the take-off configuration and exceed a minimum climb gradient requirement. This condition will usually determine the minimum installed thrust or power requirement for take-off. Economically, since the climb is performed at high thrust or power, the fuel consumed in the climb is a significant proportion of the total fuel for the mission. Optimization of the climb for minimum fuel use may be an important factor in the overall performance of the aircraft.

Whilst the performance in the descent phase of the flight is less critical than in the climb phase, considering both the power required and the safety aspects of the flight path, it is nevertheless important. Descent profiles based on rate and on gradient, and in various airframe configurations of flap setting, airbrake setting and with landing gear retracted and extended, will be required on every descent from the cruise to the landing. In this aspect of performance, which often attracts scant attention, there are many opportunities to optimize the performance and to save fuel or time.

There are, however, constraints on both the climb and descent performance that may be set by considerations of safety or by practical operational factors. Stalling speeds, minimum control speeds, noise limitations, aircraft attitudes, rates of change of cabin pressure and the operation of aircraft systems all need to be considered and may override the optimum aerodynamic performance criteria. In such cases, it may be possible to consider re-designing some aspect of the aircraft so that the limitations on the performance are reduced, but the penalties, which will inevitably occur in other areas, need to be considered. From the generalized analysis, the effect of airframe design modifications that affect the drag characteristic, or the effect of engine thrust or power changes, can be seen, and evaluated, in terms of their effect on the climb and descent performance.

Bibliography

Miele, A. (1962) *Flight Mechanics, Vol 1, Theory of Flight Paths* (Pergamon Press).

Rutowski, E. S. (1954) Energy approach to the general aircraft performance problem. *Journal of the Aeronautical Sciences*, **21**.

Engineering sciences data unit items

Performance Series, Vols 1–14.

Estimation of Rate of Climb. Vol 1, No 92019, 1992.

Acceleration Factors for Climb and Descent Rates at Constant EAS, CAS, M. Vol 1, No 81046, 1992 (Am A).

Effects of Cabin Pressure on Climb and Descent Rates. Vol 1, No 94040, 1994.

Energy–Height Method for Flight Path Optimization. Vol 9, No 90012, 1990.

Variability of Standard Aircraft Performance Parameters. Vol 10, No 91020, 1991.

The Measurement and Analysis of Climb and Excess Power Performance. Vol 13, No 70023, 1982 (Am A).

6

Take-off and landing performance

In this chapter, there will be reference to definitions of many terms associated with take-off and landing performance. These definitions are often complex and may depend on a particular airworthiness code of practice. The definitions given in this chapter may not be rigorous and are given to illustrate principles. If a rigorous definition is required, the appropriate airworthiness code of practice should be consulted.

6.1 Introduction

All conventional aircraft flights start at the point of departure with a take-off and end at the destination with a landing; these are known as the *terminal phases* of the flight.

In the take-off phase, the aircraft is transferred from its stationary, ground-borne, state into a safe airborne state. Similarly, in the landing phase, the aircraft is transferred back from the airborne state to the ground-borne state and brought to a halt. Since these manoeuvres take place in close proximity to the ground, and at a low airspeed, there is a relatively high risk to the safety of the aircraft. The manoeuvres must be carried out in a manner that will reduce the risk of an incident occurring to an acceptably low level of probability; this is the basis of *scheduled performance*, which will be addressed fully in Chapter 9. In this chapter the theory of the flight path of the aircraft will be addressed, but will need to be considered with the regulated performance in mind.

The terminal phases of the flight path consist of two parts. In the first part, the ground-run distance, the aircraft is in contact with the ground and its weight is supported, at least partly, by the landing gear. In the second part, the airborne distance, the aircraft is in transition between the ground-borne state and safe airborne flight. There are several aspects of the terminal manoeuvres that need to be taken into account in the analysis of the flight path. These include the handling qualities of the aircraft, which will not be considered in detail here since they do not directly affect the flight path of the aircraft. However, they may need to be mentioned in the determination of certain limiting airspeeds and distances. Further consideration needs to be given to the possibility of an emergency occurring during the terminal phase, one such foreseeable emergency being the failure of an engine. It is required to be shown that, for example, in the case of a large public transport aircraft, the performance must be such that, at whatever time an engine fails, including during take-off, a forced landing should not be necessary. This is known as *engine failure accountability*.

The definition of the take-off and landing flight paths of an aircraft depends on the purpose of the aircraft and on the level of flight safety that is required for that purpose. If, for example, the aircraft is intended for public transport then the highest level of flight safety is required and the take-off and landing performance must reflect this in the ability to account for a foreseeable emergency. In military combat operations, the need to carry out a mission may outweigh the need to guarantee that the aircraft can take-off with full emergency accountability. In these cases, the aircraft may be committed to a forced landing in the event of an emergency occurring.

Take-off and landing performance can be classified into a number of groups depending principally on the length of airfield required. Whilst the terminology for this classification has fallen into common usage, its definition is sometimes imprecise since take-off and landing distances are related, to some extent, to the size of the aircraft. To clarify the terminology, the definitions given here are intended as a classification based on both the length of airfield required and on the design characteristics of the aircraft.

(i) CTOL – Conventional Take-Off and Landing

This defines the take-off and landing flight paths of conventional aircraft with moderate thrust-to-weight ratios and no extreme aerodynamic characteristics in terms of sweepback, aspect ratio, wing section thickness or high-lift devices. The performance is such that the requirements for engine failure accountability can be fully met; CTOL describes the airfield performance of most transport aircraft, either civil or military, operating from a normal airfield with a paved runway. The take-off and landing distances required for conventional aircraft operations are relatively long and may restrict the number of airfields into and out of which the aircraft can operate safely. If those distances can be reduced then the number of airfields available to the aircraft will be increased; this may make the aircraft more effective both economically and operationally. This leads to:

(ii) RTOL – Reduced Take-Off and Landing

By increasing the thrust-to-weight ratio of the aircraft, and introducing more powerful aerodynamic high-lift devices, high-drag devices and thrust reversal, the take-off and landing distances required can be reduced and shorter airfields made available to the aircraft. For RTOL, the thrust-to-weight ratio at take-off is usually increased by installing larger engines to provide a greater propulsive thrust. In the past, systems for additional thrust that have been installed for use during the take-off, to increase the available take-off thrust-to-weight ratio, include water–methanol injection to increase the engine mass flow, auxiliary engines or rocket assistance packs. Generally, such systems are not economical as they incur a parasitic weight penalty or generate unacceptable noise levels. High lift devices can be used to increase the maximum lift coefficient and lower the lift-off speed. These include full-span flaps on both leading and trailing edges of the wing, drooped ailerons and passive boundary layer control systems to prevent flow separation at low airspeeds. Relatively high thrust, combined with high lift devices, which produce a good lift increment without a high drag penalty, will produce very significant reductions in the take-off distances. If the high lift devices can also produce high drag, or separate high-drag devices are installed, then steep approaches to landing, and powerful aerodynamic braking, can result in very much shortened landing distances. RTOL aircraft with such performance are often referred to, justifiably, as STOL aircraft, this is, however, a misnomer since RTOL performance does not depend on power-induced lift and full engine failure accountability is achievable.

(iii) STOL – Short Take-Off and Landing

The take-off and landing distances can be further reduced only by using the engines to contribute to the aerodynamic lift required to support the aircraft in flight. This can be achieved by power-induced wing lift augmentation, active boundary layer control or by deflected or partially vectored thrust. Power-induced lift can be produced by installing the propeller in front of the wing and deflecting the slipstream with large flaps. In this way the wing area immersed in the propeller slipstream can generate a very considerable lift force, even at zero forward speed. Similarly, the efflux from a turbo-jet or turbo-fan engine can be deflected by wing flaps or the engine nozzle can be used to angle the flow downward to create a direct vertical thrust component. In some cases, the thrust deflection or flap setting may be changed during the take-off or landing manoeuvre to achieve the best overall performance. However, the failure of such a system to operate could have an effect on performance of a similar magnitude to that of an engine failure. Active boundary layer control by blowing high-pressure air bled from the engines over the critical areas of the lift-producing surfaces is capable of substantially increasing the lift forces produced. The STOL aircraft depends on its powerplant to provide a proportion of the 'lift' force as powered lift. Should an engine fail during STOL flight, the possibility of recovery of the aircraft into controlled flight is likely to be marginal and engine failure accountability at all times may not be possible.

(iv) VTOL – Vertical Take-Off and Landing

This is the extreme case of STOL in which the total weight of the aircraft is supported by the engine thrust during the terminal manoeuvres, which can now be performed without the need for a runway as such. In VTOL operations, the failure of an engine will not only cause loss of propulsive thrust but also loss of the 'lift' force, and accountability for an engine failure during VTOL flight is unlikely to be possible.

(v) STOVL – Short Take-Off, Vertical Landing

This is an extended case of VTOL operation. The aircraft has the capability of VTOL but uses a short take-off run to reduce the take-off fuel demand and increase its range, or it may be required to take-off at a weight above its maximum VTOL weight. STOVL operations can be considered to include the 'ski-jump' take-off. Here, the aircraft is launched off a flight deck into flight with the assistance of the upwardly swept deck, following a short acceleration run that may be made with partially vectored thrust.

Although these definitions of the terminal phase performance separate the aircraft into broad categories, the take-off and landing performance, for all except the VTOL and STOVL aircraft, can be analysed in a broadly similar manner.

6.2 Take-off performance

In the conventional take-off manoeuvre, the aircraft is accelerated along the runway until it reaches a speed at which it can generate sufficient aerodynamic lift to overcome its weight. It can then lift off the runway and start its climb. During the take-off, consideration is given to the need to ensure that the aircraft can be controlled safely and the distances required for the manoeuvres do not exceed those available. The take-off manoeuvre is shown in Fig. 6.1 and some of the principal speeds and events described.

The aircraft starts the take-off at rest on the runway, take-off thrust is set and the brakes released. The excess thrust accelerates the aircraft along the runway and, initially, the directional control needed to maintain heading along the runway would be provided by the nose-wheel steering. This is because the rudder cannot provide sufficient aerodynamic yawing moment to give directional control at very low airspeeds. As the airspeed increases the rudder will gain effectiveness and will take over directional control from the nose-wheel steering. However, should an engine fail during the take-off run the yawing moment produced by the asymmetric loss of thrust will have to be opposed by a yawing moment produced by the rudder. There will be an airspeed below which the rudder will not be capable of producing a yawing moment large enough to provide directional control without assistance from either brakes or nose-wheel steering or a reduction in thrust on another engine. This airspeed is known as the *minimum control speed, ground*, V_{mcg}; if an engine failure occurs before this airspeed is reached, the take-off run must be abandoned.

During the ground run the nose wheel of the aircraft is held on the runway to keep the pitch attitude, and hence the angle of attack in the ground run, α_g, low. This will keep the lift produced by the wing to a small value so that the lift-dependent drag is minimized and the excess thrust available for acceleration is maximized. As the aircraft continues to accelerate, it will approach the *lift-off speed*, V_{LOF}, at which it can generate enough lift to become airborne. Just before the lift-off speed is reached, the aircraft is rotated into a nose-up attitude equal to the lift-off angle of attack. The

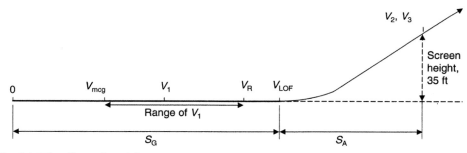

Fig. 6.1 Take-off speeds and distances.

rotation speed, V_R, must allow time for the aircraft to rotate into the lift-off attitude before the lift-off speed is achieved. As the aircraft continues to accelerate it reaches the lift-off airspeed and becomes airborne; this is the end of the *ground run distance*, S_G. The lift-off speed must allow a sufficient margin over the stalling speed to avoid an inadvertent stall, and a consequent loss of height. This may be caused by turbulence in the atmosphere or any loss of airspeed during the manoeuvring of the aircraft after the lift-off. The lift-off speed will usually be taken to be not less than $1.2V_{S1}$, where V_{S1} is the stalling speed of the aircraft in the take-off configuration. This will give a lift coefficient at lift-off of about $0.7C_{L\,max}$ and provide an adequate margin of safety over the stall. If the aircraft is over-rotated to a greater angle of attack at the rotation speed then lift-off can occur too soon and the aircraft start the climb at too low an airspeed. This can occur if, for example, the elevator trim control is set incorrectly or turbulence produces an unexpected nose-up pitching moment. The minimum speed at which the aircraft can become airborne is known as the *minimum unstick speed*, V_{MU}. It occurs when extreme overrotation pitches the aircraft up to the geometry limited angle of attack with the tail of the aircraft in contact with the runway. Tests are usually required to measure the take-off performance in this condition.

During the take-off run, should an engine fail between the minimum control speed (ground) and the rotation speed, the decision either to abandon or continue the take-off will have to be made. This decision is based on the distances required either to stop the aircraft or to continue to accelerate to the lift-off speed with one engine inoperative. There will be a point during the acceleration along the runway at which the distances required by the two options are equal. This point is recognized by the indicated airspeed of the aircraft and is known as the *decision speed*, V_1. The decision speed also determines the minimum safe length of runway from which the aircraft can take off. If an engine fails before the decision speed is reached then the take off is abandoned, otherwise the take-off must be continued. This topic will be discussed fully in Chapter 9.

Once the lift-off has been achieved the aircraft must be accelerated to the *take-off safety speed* (V_2, one engine inoperative, V_3, all engines operating). This is the airspeed at which both a safe climb gradient and directional control can be achieved in the case of an engine failure in the airborne state; this phase of the take-off path is known as the *transition*. The ability to maintain directional control in the climb is determined by the *minimum control speed, airborne*, V_{mca}. The minimum control speed, airborne, will be greater than the minimum control speed, ground, V_{mcg}, since the aircraft is not restrained in roll by the contact between the landing gear and the runway. In the event of an engine failure in the climb, the aircraft will depart in yaw, which will cause the aircraft to roll and enter a spiral dive if the yaw cannot be controlled. The take-off is complete when the lowest part of the aircraft clears a screen height of 35 ft (or 10.5 m) above the extended take-off surface. The distance between the lift-off point and the point at which the screen height is cleared is known as the *airborne distance*, S_A.

The total take-off distance required will be the sum of the ground run distance, S_G, and the airborne distance, S_A. To ensure that the take-off is performed safely, the take-off distances will be suitably factored to allow for statistical variation in the take-off performance of the individual aircraft and in the ambient conditions.

This outline of the events, which take place during the take-off, applies to CTOL and RTOL operations and, broadly, to STOL operations. The relationships between the various speeds referred to above are set out in the airworthiness requirements and depend on the classification of the aircraft and the particular code of practice under which it is certificated. This is discussed further in Chapter 9.

6.2.1 Estimation of take-off distances

At the design stage of the aircraft, estimates of the take-off distances will be needed to determine the minimum size of airfield from which the aircraft will be able to operate. Any such estimation will depend on the ability to estimate the thrust of the power-plant at very low forward speeds and the drag of the aircraft at low speeds and in ground effect. In addition, there will be other forces acting on the aircraft, for example, wheel spin-up, runway friction and side-wind loads, some of which are very difficult to quantify. In practice, although reasonable estimates of the take-off distances can be made, it will be necessary to determine the actual take-off performance by measurement.

There are several methods for the estimation of the ground run and airborne take-off distances; the methods given here are among the simplest but give reasonable results for conventional aircraft with moderate thrust.

The equations of motion of the aircraft are given in Appendix B, eqn (B.19). Using the simplifying assumptions and including the runway reaction force, the equations of motion during the take-off run can be written,

$$\left. \begin{aligned} F_N - D - W \sin \gamma_R - \mu_R R = m\dot{V} \\ R + L = W \cos \gamma_R \end{aligned} \right\} \qquad (6.1)$$

where γ_R = the runway slope, R = wheel load or runway reaction force, and μ_R = runway coefficient of rolling friction.

The ground-run distance, S_G

The ground run distance can be estimated from the time integral of the ground speed of the aircraft,

$$S_G = \int_0^t V \, dt = \int_0^{V_{LOF}} \frac{mV}{F} \, dV \qquad (6.2)$$

where F = the accelerating force acting on the aircraft.

Now, from eqn (6.1), the accelerating force, F, can be written in the form,

$$\frac{F}{W} = \left\{ \frac{F_N}{W} - \mu_R - \sin \gamma_R \right\} - \frac{L}{W} \left(\frac{C_D}{C_L} - \mu_R \right) \qquad (6.3)$$

In eqn (6.3), the curly bracket { } represents the net propulsive thrust–weight ratio, taking into account the runway rolling friction and the runway slope. The thrust force produced by the powerplant is unlikely to vary substantially during the take-off and, for the purposes of a simple estimation of the ground run, the net propulsive thrust–weight ratio can be assumed to be a constant, A. The round bracket ( ) is predominantly the aerodynamic drag–lift ratio at the ground run angle of attack, α_g; it also includes a runway friction term since any lift generated by the aircraft will alleviate the total runway friction force. The lift force, L, is the lift force that occurs during the take-off run and will be produced by the angle of attack, α_g, determined by the ground run pitch attitude of the aircraft. Since the aircraft pitch attitude is constant up to the point of rotation, the angle of attack is constant, which implies that the lift and drag coefficients will be constant during the ground run. Therefore, the aerodynamic drag term can be evaluated as,

$$\frac{\frac{1}{2}\rho S C_L}{W}\left(\frac{C_D}{C_L} - \mu_R\right) V^2 = BV^2 \tag{6.4}$$

where B is a quasi-constant determined by the aircraft configuration and ground run attitude.

Equation (6.3) can now be expressed in the form,

$$\frac{F}{W} = A - BV^2 \tag{6.5}$$

and eqn (6.2) becomes

$$S_G = \int_0^{V_{LOF}} \frac{V\,\mathrm{d}V}{g[A - BV^2]} \tag{6.6}$$

This can be integrated to give,

$$S_G = \frac{1}{2Bg}\left[\log_e(A - BV_{LOF}^2) - \log_e A\right] \tag{6.7}$$

If $A \gg BV^2$ a further simplifying approximation can be made by evaluating eqn (6.5) at $0.7V_{LOF}$ and integrating eqn (6.6) on the assumption that $[A - BV^2]$ is a constant, this gives,

$$S_G = \frac{V_{LOF}^2}{2g(A - BV^2)_{0.7V_{LOF}}} \tag{6.8}$$

Equations (6.7) or (6.8) can be used to estimate the ground run of a CTOL or RTOL aircraft up to the lift-off point. However, it may be necessary to divide the ground run into two parts. First, the ground run to the rotation, during which the ground attitude is constant. Secondly, the ground run during and after rotation in which the angle of attack increases to the lift-off angle of attack and the lift dependent drag will become significantly larger.

The airborne distance, S_A

After lift-off, the aircraft is accelerated to the safe climbing speed as it is rotated into the climb. The take-off is complete when the aircraft clears a screen height of 35 ft above the extended take-off surface at a speed not less than the take-off safety speed. The flight path in the airborne phase of the take-off is not, therefore, a simple path but combines both acceleration and climb. An approximation to the airborne distance can be made by considering the change of energy of the aircraft during the airborne phase. This form of approximation does not presuppose the flight path to be a smooth curve and allows for a variation in take-off technique. The energy change, ΔE, is given by,

$$\Delta E = \text{excess thrust} \times \text{distance moved}$$

giving,

$$\Delta E = \int (F_N - D)\, dS \approx (F_N - D)_{av} S_A \qquad (6.9)$$

Now, the energy change between the lift-off and the end of the airborne distance, at which the airspeed has been increased to the take-off safety speed, V_2 or V_3 as appropriate, and the potential energy has been increased by 35 ft, is,

$$\Delta E = [KE + PE]_{35'} - [KE + PE]_{LOF} \qquad (6.10)$$

so that, from eqns (6.9) and (6.10), the airborne distance is given by

$$S_A = \frac{W}{(F_N - D)_{av}} \left\{ \frac{V_2^2 - V_{LOF}^2}{2g} + 35 \right\} \qquad (6.11)$$

assuming that the horizontal airborne distance travelled is very much greater than 35 ft.

The difference between the lift-off speed and the take-off safety speed should be as small as possible to minimize the time during which the aircraft may be unable to meet the requirement for directional control following an engine failure. In most cases, the increment required in the kinetic energy is about equal to the change in potential energy.

The operation of an aircraft usually requires the take-off distances to be as short as possible, this applies to the ground run, the airborne distance and to their sum as the overall distance. For an aircraft with a given airframe–engine combination, any such optimization process will depend on the aerodynamic characteristics of the airframe, since the maximum available thrust of the engines is fixed. The ground run can be minimized by selecting the combination of high-lift devices that will produce the lowest lift-off speed together with a high lift–drag ratio. This will need to take into account the drag in the phase between rotation and lift-off and the performance with one engine inoperative. Similar criteria can be applied to the airborne distance and the overall distance, but will need to include the difference between the lift-off speed and the take-off safety speed. Clearly, such a problem cannot be solved by analytical means alone and, in practice, the final combination of high lift devices will be optimized by flight trials.

6.2.2 The effect on the take-off distances of the flight variables

The take-off distances will be affected by the weight of the aircraft, the state of the atmosphere and the airfield conditions. Equations (6.8) and (6.11) enable the approximate effect of variation of the flight variables to be discussed, and the influence of these effects on the performance data in the aircraft performance manual will be considered in Chapter 10.

Aircraft weight

If eqn (6.8) is expressed for a take-off on a level runway in still air then the approximation for the ground run distance is,

$$S_G = \frac{WV_{LOF}^2}{2g[F_N - D - \mu(W - L)]_{0.7V_{LOF}}} \tag{6.12}$$

An increase in aircraft weight can be seen to have two direct effects on the ground run distance.

First, the ground run distance is directly proportional to aircraft weight, so that the ground run will increase in proportion to the weight increase.

Secondly, the ground run distance is directly proportional to the square of the lift-off speed, V_{LOF}. Now, the lift-off speed is proportional to the stalling speed, V_{S1}, which, in turn, is proportional to the square root of the weight of the aircraft. Therefore, the take-off ground run distance will be increased, again in proportion to the aircraft weight.

In addition to the direct effects, the increased weight will increase the runway friction force acting on the aircraft, but the effect of this on the ground run will be relatively small compared with the direct effects of weight.

Summing the individual effects of a weight increase, it can be expected that increasing the aircraft weight by 10% will increase the take-off ground run distance by at least 20%

The airborne distance will be similarly affected.

First, from eqn (6.11) it can be seen that the airborne distance is directly proportional to the aircraft weight.

$$S_A = \frac{W}{(F_N - D)_{av}} \left\{ \frac{V_2^2 - V_{LOF}^2}{2g} + 35 \right\} \tag{6.11}$$

Secondly, both the lift-off speed and the take-off safety speed will be increased by the increase in aircraft weight. This will produce a proportional increment in the kinetic energy related term since the airborne distance is a function of the square of the airspeeds, which are proportional to the square root of the weight. However, as the potential energy and kinetic energy terms in the expression for the airborne distance are roughly equal, the effect of a weight increase of 10% on the kinetic energy term is to increase the airborne distance by about 5%.

Also, the increased weight will increase the airframe drag and reduce the excess thrust available. The magnitude of this effect will depend on the excess thrust and on the lift-dependent part of the drag characteristic of the airframe. For a transport aircraft, it will probably equate to a reduction in the excess thrust of the order of 2% for a weight increase of 10%.

By summing these effects, it can be expected that increasing the weight of the aircraft by 10% will increase the airborne distance by between 15% and 20%.

Atmosphere state effects

The state of the atmosphere has two basic effects on the take-off distances.

The take-off is performed with reference to indicated airspeed (IAS) displayed by the airspeed indicator. In the absence of Pitot-static errors, the IAS will be the same as the equivalent airspeed, EAS (since the take-off is made at low speed and low altitude and the scale-altitude correction will be negligible). However, the take-off distances are functions of true airspeed (TAS), since they are determined by the motion of the aircraft in Earth axes. Since the ground run distance is proportional to the TAS^2 it will be proportional to the inverse of the relative density, $1/\sigma$, or θ/δ. This implies that hot or high (low relative pressure) conditions will increase the take-off ground run since the TAS will be increased for a given EAS determined by the aircraft weight. The airborne distance will also be increased but, because only the kinetic energy term is affected, the effect will be about halved.

The output of the powerplant is roughly proportional to the relative density since the engine output is dependent on its air mass flow. Therefore, the net thrust will decrease in hot or high conditions increasing the take-off distances; the magnitude of the effect will depend on the characteristics of the powerplant.

Headwind

The effect of the headwind, V_w, is to change the datum speed of the take-off; the aircraft now only needs to be accelerated to a ground speed of $V_{LOF} - V_w$. From eqn (6.12), it can be seen that the effect of a headwind equal to 10% of the lift-off speed is to decrease the ground run by almost 20%. Conversely, the same tailwind would increase the ground run by a little over 20%.

The effect of the headwind on the airborne distance is approximately half as severe as it is on the ground run distance since it only affects the kinetic energy term.

Runway conditions

The effects of a runway slope (uphill) and the runway friction coefficient on the ground run distance can each be accounted for by considering them as equivalent to a decrease in the take-off thrust-to-weight ratio (see eqn (6.3)). There is, of course, no effect on the airborne distance.

It will be seen later (in Chapter 10 on the application of aircraft performance), that account is taken of the effect of the flight variables on the take-off performance data in their presentation in the aircraft performance manual. The effects of the weight, wind and runway slopes are each accounted for by using the corrections developed above to factor the datum performance measured at a known atmosphere state.

6.3 Landing performance

In the landing phase of the flight, the aircraft is on a descending flight path towards the runway. As it approaches the runway, the airspeed and the rate of descent are reduced in the flare so that a touchdown is achieved at a low rate of descent. After touchdown, the nose is lowered onto the runway and the aircraft brought to a halt. During the landing, consideration is given to the need to ensure that the aircraft can be controlled safely and that the distances required for the manoeuvres do not exceed those available. The landing manoeuvre is described in Fig. 6.2.

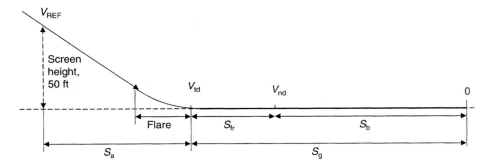

Fig. 6.2 Landing speeds and distances.

The approach path to the runway is protected by a safe approach sector, sloping downwards to the threshold of the runway, which is required to be clear of obstructions. The gradient of the approach sector is determined by the classification of the airfield and is a minimum of 2% for large airfields. The gradient of the approach path of the aircraft must be steeper than the minimum approach sector gradient and will usually be about 3° or 5% gradient for CTOL aircraft. However, it may be steeper for RTOL and STOL aircraft; a STOL aircraft may be able to make an approach at a gradient of 7° or more. The landing distance commences when the aircraft just clears a screen height of 50 ft above the extended landing surface.

The approach to the landing is made with the flaps and other high lift devices set to a high lift (and high drag) setting to enable the approach to be made at a low airspeed. The approach airspeed will usually be not less than $1.3V_{S0}$ to provide a safe margin over the stalling speed in the landing configuration. To maintain flight path stability, in which an increase in airspeed will produce an increase in drag, which will tend to restore the airspeed to its former value, the aircraft must be flown on the forward side of the drag curve (see Chapters 3 and 5). Since the approach speed is relatively low, it may be necessary to increase the zero-lift drag of the aircraft so that the minimum drag speed is less than the approach speed. Extending the landing gear will increase the zero-lift drag (in some cases to twice the value with the landing gear retracted), and usually flaps will be lowered to a high-lift, high-drag setting. If additional drag is needed, airbrakes and spoilers can be used. In the high drag state, the engine thrust required to hold the approach gradient would be relatively high. This is a further advantage since, if the approach has to be abandoned and a go-around has to be initiated, the engines are already producing high thrust and the time required for them to accelerate to maximum thrust will be shortened. By retracting the landing gear and high drag devices, the drag can be reduced quickly. This enables the maximum excess thrust for the go-around to be achieved quicker than by the acceleration of the engines from idle thrust to maximum thrust with the aircraft in a low drag configuration.

As the aircraft passes the screen height, its combined airspeed and gradient of descent will produce a rate of descent that would be too high for an acceptable touch-down. It would exceed the capability of the landing gear to absorb the kinetic energy of the descent. A manoeuvre, *the flare*, is needed to reduce the rate of descent for the touchdown. In the flare, the thrust is reduced to flight idle and the nose of the aircraft is steadily raised to increase progressively the angle of attack. This will allow the airspeed to decrease to a safe touchdown speed and the rate of descent to be reduced, ideally to almost zero, as the aircraft touches down. The pitch attitude of the aircraft at touchdown will be equal to the angle of attack, since the flight path gradient is zero, and the touchdown is made in a nose up attitude. The height at which the flare commences is judged by the pilot or, in auto-land systems, determined by reference to a measurement of true height above the runway from a radio altimeter. Since the aircraft is in a high drag configuration there will be a large decelerating force acting after the thrust is reduced. This will cause the airspeed to bleed away quickly enough to avoid a long float during the flare as the airspeed decreases. The distance between the 50 ft screen height and the touchdown point is known as the *airborne landing distance*, S_a.

At the touchdown, the aircraft will be in a nose up attitude with the nose wheel clear of the runway at the *touchdown speed*, V_{td}. In this attitude, the large angle of attack will produce a high lift-dependent drag that will help to decelerate the aircraft. As the airspeed decreases there will come a point at which there will be insufficient pitching moment produced by the elevator control to hold the nose-up attitude. The aircraft will then pitch down on to the nose wheel at the *nose-down speed*, V_{nd}. (In practice, the speed at which the nose wheel is lowered onto the ground will be controlled by the pilot.) The ground run distance with the nose wheel clear of the ground is known as the *free roll distance*, S_{fr}, since it is not generally possible to use any retarding system, other than the aerodynamic drag of the aircraft in the landing configuration, to assist the deceleration. The reason for this is the possibility of producing a pitching moment that could not be controlled by the elevator. This could result in a pitch up, which might result in the aircraft becoming airborne again, or a pitch down which would result in the nose wheel slamming down onto the runway.

Once the nose wheel is on the ground, the aircraft can be decelerated to a halt in the *braking distance*, S_b. The deceleration in this phase can be assisted by various devices. Reverse thrust from the engines and the wheel brakes provide direct retarding forces. Flaps can be deployed to a high drag, 'lift-dump', setting in which they are deflected to a very large angle to provide a high zero-lift drag. In this setting they also act as spoilers to decrease the lift produced by the wing, thus assisting the wheel braking by increasing the load on the wheels. Spoilers, airbrakes or drag parachutes can be employed to increase the aerodynamic retarding force; in some cases the aerodynamic retarding systems and wheel brakes can be armed in flight to activate automatically as the nose wheel contacts the ground. The landing run is complete when the aircraft has been brought to a halt on the runway.

The *landing ground run*, S_g, is the sum of the free-roll distance and the braking distance and can be minimized by selection of the best nose-down speed. If the aircraft produces a large lift-dependent drag force in the free-roll phase, it might be advantageous to maintain the nose up attitude as long as possible. The nose can then be allowed to descend onto the runway and braking used to bring the aircraft finally to a halt. If the lift-dependent drag is small then the nose should be lowered onto the runway as soon as possible and the deceleration assisted by any other means available; this is the more usual case for transport aircraft.

The *landing distance*, S_{ldg}, is the sum of the airborne landing distance and the landing ground run.

6.3.1 The estimation of the landing distances

The landing distances can be estimated in a very similar manner to the take-off distances.

The airborne distance, S_a

This is the distance required for the aircraft to clear the 50 ft screen height at the landing reference speed, V_{REF}, and to decelerate and descend to the touchdown. The distance can be estimated in the same way as the airborne take-off distance by considering the energy difference between the screen height and the touchdown point, this gives,

$$S_a = \frac{W}{(F_N - D)_{av}} \left\{ \frac{V_{REF}^2 - V_{td}^2}{2g} + 50 \, \text{ft} \right\} \tag{6.13}$$

In applying this method of estimation, it should be noted that during the flare there would be a thrust reduction and a drag increase that may need to be taken into account. It should also be noted that in landing a large aircraft having a high kinetic energy, it might be necessary to start the flare before reaching the 50 ft screen height. A small aircraft with less kinetic energy may not need to flare until well below 50 ft.

The ground-run distance

The landing ground run distance can be estimated by using the same technique that was employed for the take-off run by considering the integral,

$$S_g = \int \frac{WV}{g\Re} \, dV \tag{6.14}$$

where $\Re$ is the retarding force acting on the aircraft given by,

$$\Re = [D + \mu_R(W - L) + W \sin \gamma_R - F_N] \tag{6.15}$$

in which the drag, D, represents the aerodynamic drag force of the airframe and of any drag-producing retardation devices. Equation (6.15) can be expressed in the form,

$$\frac{\Re}{W} = \left[\frac{\rho S C_L}{2W} \left(\frac{C_D}{C_L} - \mu_R \right) V^2 + \left\{ \mu_R + \sin \gamma_R - \frac{F_N}{W} \right\} \right] = BV^2 + A \tag{6.16}$$

(NB The terms A and B are not the same as those used in eqn (6.5) relating to the take-off distances.)

Substituting eqn (6.16) into eqn (6.14) and integrating leads to a general expression for the ground run distance,

$$S_g = \frac{1}{2gB} \log_e (BV^2 + A) \tag{6.17}$$

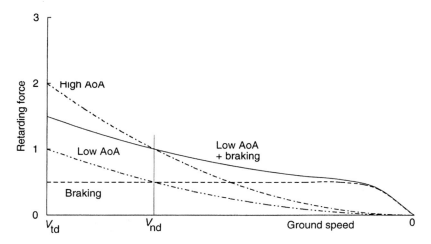

Fig 6.3 Retarding forces acting during the landing ground run.

This will need to be evaluated between the touchdown speed and the nose-down speed, and the nose-down speed and zero, to give the free-roll and the braked distances respectively. The evaluation will depend on the determination of the nose-down speed from the relative magnitudes of the terms A and B, which will be different in the free-roll and braked ground run.

In the free-roll, the retarding force will be dominated by the aerodynamic drag of the airframe in its landing configuration and in ground effect. The rolling coefficient of friction, μ_R, will be small and the angle of attack, and hence the lift coefficient, C_L, will be large. This will lead to a high drag-to-lift ratio, C_D/C_L, for the aircraft in its landing configuration, which may be further increased by being in ground effect; the thrust–weight ratio at flight idle thrust will be small. In this case the term A will be very small compared with BV^2.

In the braked ground run the retarding force will be dominated by the non-aerodynamic retarding devices, the brakes and reverse thrust. The braking coefficient of friction, μ_R, will be large, probably as high as 0.5. The angle of attack will be small, and hence the coefficient of lift will be small, but the aerodynamic drag coefficient, which includes the effects of any high drag devices, could be large. This would lead to a large drag-to-lift ratio at the higher airspeeds. If reverse thrust is used, F_N/W will be large and negative. In this case A will be of similar magnitude to, or larger than, BV^2 and will dominate at low speeds. The relative retarding forces are shown in Fig. 6.3.

Comparison between the two cases will determine the optimum nose-down speed for the aircraft, see Fig. 6.3. At high airspeeds, the aerodynamic drag of the airframe will tend to be dominant and favour the free-roll technique, whereas at low airspeeds braking will be more effective. However, the capacity of the brakes to absorb the kinetic energy of the aircraft as heat must be taken into account. If the aircraft is equipped with thrust reversers it is likely that there will be no benefit in using the free-roll retardation. The nose will be lowered onto the runway immediately after touchdown so that mechanical retardation and thrust reversal can be employed. It is not unusual to pre-arm thrust reversers prior to landing so that they operate automatically when the nose wheel touches the runway.

6.3.2 The effect on the landing distances of the flight variables

The landing distances will be affected by the weight of the aircraft, the state of the atmosphere and the airfield conditions, in a broadly similar manner to the take-off distances.

Aircraft weight

In the approach phase of the landing the thrust produced by the engines will be less than the maximum thrust available. Therefore, the gradient of the approach flight path can be controlled by the engine thrust and will not be affected by the weight of the aircraft. In the flare, the thrust is reduced to idle and the aircraft decelerates as the angle of attack is steadily increased towards the touchdown angle of attack to arrest the rate of descent. Since the angle of attack on the approach and at touchdown are usually determined by the stalling speed of the aircraft in the landing configuration, the lift–drag ratio during the flare will be practically independent of the aircraft weight. This implies that, unlike the airborne take-off distance, the only significant effect of the weight on the airborne landing distance will be due to the kinetic energy term in the flare. Because the approach is made at about $1.3V_{S0}$, the kinetic energy loss during the flare will be about twice as much as the potential energy loss. Therefore, a weight increase of 10% would be expected to increase the distance in the flare by about 7%.

The effect of the weight on the landing ground run distance is similar to the take-off case; it increases the ground speed at touchdown through its effect on the stalling speed. The direct effect of the increased weight on the kinetic energy will increase the ground run distance by about 20% for a weight increase of 10%. This affects both the free roll distance and the braked ground run.

Atmosphere state effects

The state of the atmosphere will affect the landing distances through its effect on the TAS. There will be no significant effect through the engine thrust since the thrust is either being controlled to maintain the approach gradient or will be at idle. As in the take-off case, the ground run distances will be increased in proportion to the inverse of the air density, σ, and the flare by about $2/3\sigma$.

Headwind

The effect of the headwind on the ground run distance is to change the datum speed of the landing since the aircraft now only needs to be decelerated from an effective touch-down speed of $(V_{td} - V_w)$. The effect of a headwind equal to 10% of the touch-down speed would be to reduce the landing ground run by about 20%; conversely the same tailwind will increase the ground run by 20%.

The effect of the headwind on the airborne distance is only felt in the flare since the approach gradient is relative to the ground and is controlled by engine thrust or power. A headwind of 10% of the touchdown speed will reduce the kinetic energy loss during the flare by 20% and the distance in the flare by about 14%.

Runway conditions

The effects of a runway slope (uphill) and the runway friction coefficient on the ground run distance can each be accounted for by considering them as equivalent to an increase in the braking force. There is, of course, no effect on the airborne distance.

6.4 STOL and VTOL considerations

So far, the take-off and landing performance analysis has been considered in terms of CTOL and RTOL aircraft. Aircraft designed specifically for STOL or VTOL operations need to be considered separately since their airfield performance depends on the contribution to the lift generated by the powerplant. It should be noted that an aircraft operated under STOL or VTOL conditions may use techniques and speeds that would depend on its STOL design features and which would not be considered suitable for CTOL aircraft. The performance of such aircraft cannot be analysed in such a general manner; each aircraft would need to be considered separately because of its particular STOL features.

As an example of the analysis of the take-off distance of a STOL aircraft, an aircraft with a power-induced lift contribution, can be considered.

The aerodynamic lift force required by a STOL aircraft during a STOL manoeuvre is relieved by power-induced wing lift augmentation and by deflected, or partially vectored, thrust. Thus, the vertical force equation, given in eqn (6.1), becomes,

$$R + L_a + L_p + T\sin(\alpha + \tau_1) = W\cos\gamma_2 \qquad (6.18)$$

where L_a is the aerodynamic lift due to the angle of attack, α, and dynamic pressure, q; L_p is the power-induced wing lift augmentation due to the slipstream velocity; and, $T\sin(\alpha + \tau_1)$ is the deflected or vectored gross thrust component, eqn (B.19) in Appendix B.

When substituted in eqn (6.3), and assuming the flight path gradient to be small, this leads to the expression for the net propulsive force,

$$\frac{F}{W} = \left\{\frac{F_N}{W} - \mu_R\left(1 - \frac{T}{W}\sin(\alpha + \tau_1)\right) - \sin\gamma_2\right\} - \frac{L_a}{W}\left(\frac{C_D}{C_L} - \mu_R\left(1 + \frac{L_p}{L_a}\right)\right) \qquad (6.19)$$

This separates the deflected, or vectored, gross thrust into the propulsive thrust term and the power-induced lift augmentation into the aerodynamic drag term.

Equation (6.18) can be expressed in the form,

$$\frac{R}{W} + \frac{L_a}{W}\left(1 + \frac{L_p}{L_a}\right) + \frac{T}{W}\sin(\alpha + \tau_1) = \cos\gamma_2 \approx 1 \qquad (6.20)$$

from which the reduction in airspeed required to lift-off can be deduced.

A simplified example can be used to demonstrate the effect of the STOL features of an aircraft on its take-off distances.

At lift-off, the runway reaction force is zero. It is assumed that the power-induced wing lift is equivalent to 20% of the total aerodynamic lift at the lift-off angle of attack of 10°, at which point the lift–drag ratio is 10 in the take-off configuration. The gross thrust-to-weight ratio of 0.6 is deflected through 20° throughout the take-off run and the lift off occurs as the aircraft reaches 10° angle of attack on rotation.

At lift off, the aerodynamic lift, L_a/W, is given by,

$$\frac{L_a}{W}\left(1+\frac{0.2}{0.8}\right) + 0.6\sin(10° + 20°) = 1$$

giving $L_a/W = 0.56$ and the lift-off speed, therefore, would be reduced to 0.748 of the speed required without the STOL features.

From eqn (6.19), and eqn (B.19), the gross thrust component during the take-off run would be reduced by cos 20° and the lift–drag ratio related term would be reduced by a factor of 0.56. Assuming the lift–drag ratio to be unaffected by the STOL features and the runway friction coefficient to be small enough to have negligible effect, the net propulsive force, F/W, would be practically unchanged.

The overall effect of the STOL features would be to reduce the ground run distance by a factor of 0.56. By similar reasoning, the airborne distance would be reduced by a factor of about 0.72. Whilst these factors have been estimated on gross simplifications, they do serve to indicate the potential effect of the STOL features.

A similar analysis can be applied to the landing performance, which would show the reduction in the airborne and ground run distances that could be achieved by the lower approach speeds possible with lift augmentation.

The airfield performance of VTOL aircraft raises special problems. During the lift-off and transition to forward flight or in the final approach and hover before landing, the aircraft is fully dependent on the engine for its 'lift' force. Should an engine fail during these phases of the flight the manoeuvre could not be continued and, at best, a forced landing would be inevitable. For this reason, VTOL civil transport aircraft cannot be considered as a realistic proposition since the current airworthiness requirements call for full engine failure accountability. In some military operations, the tactical benefits of VTOL outweigh the risks involved and aircraft can be operated without full engine failure accountability.

It should be noted that helicopters are able to operate in the VTOL mode, but with limited engine failure accountability. A helicopter generates its lift force through the rotor that is driven by the engine. The rotor must be turning at the required speed before the collective pitch is increased and take-off lift is produced. The rotor speed needs to be maintained throughout the flight so that the rotor always has a high level of angular momentum. In the case of an engine failure, the collective pitch of the rotor is decreased and the rotor auto-rotates, that is, it is driven by the airflow passing through it upwards. The auto-rotation maintains the angular velocity of the rotor and creates a high drag force that controls the rate of descent of the helicopter. On approaching the ground, the collective pitch is increased and the angular momentum of the rotor is converted into lift force to arrest the rate of descent and this allows the helicopter to touch down safely. The procedure is not foolproof, however, and there are combinations of speed and height, which need to be avoided, from which it is not possible to achieve a safe recovery from a sudden loss of engine power.

6.5 Conclusions

The ability to estimate the airfield performance of an aircraft is of considerable interest in the design phase of that aircraft. The minimum length of the airfield that the aircraft can take off from, or land on, needs to be estimated early in the design procedure since this could limit the routes on which the aircraft would be capable of operating and could significantly affect its marketability.

The methods of estimation of the airfield performance described here are not the only methods available. However, they are relatively straightforward and will give reasonable estimates of the take-off and landing distances for conventional aircraft if the basic ground rules around which the assumptions were made are followed. Aircraft of unconventional design, or having unusual powerplant features, may need adaptations of these estimation methods, or alternative methods, if the assumptions made in the methods developed in this chapter cannot be fully justified.

Bibliography

Airworthiness Authorities Steering Committee. Joint Airworthiness Requirements. JAR-1, *Definitions and Abbreviations*; JAR-25, *Large Aeroplanes* (CAA Printing and Publishing Services).

Engineering sciences data unit items
Performance Series, Vols 1–14.
Frictional and Retarding Forces on Aircraft Tyres: Part I, Introduction. Vol 5, No 71025, 1995 (Am D).
Frictional and Retarding Forces on Aircraft Tyres: Part II, Estimation of Braking Force. Vol 5, No 71026, 1995 (Am D).
Calculation of Ground Performance in Take-off and Landing. Vol 6, No 85029, 1985.
Force and Moment Components for Take-off and Landing Calculations. Vol 6, No 85030, 1985.
First Approximation to Landing Field Length for Civil Transport Aeroplanes. (50 ft, 15.2 m screen). Vol 6, No 84040, 1984.
Estimation of Airborne Performance in Landing. Vol 6, No 91032, 1991.
Energy Method for Analysis of Measured Airspeed Change in Landing Airborne Manoeuvre. Vol 6, No 92020, 1996 (Am A).
Estimation of Ground Run During Landing. Vol 6, EG6/4, 1992 (Am B).
Estimation of Take-off Distance. Vol 7, EG5/1, 1972 (Am B).
First Approximation to Take-off Field Length of Multi-engined Transport Aeroplanes. Vol 7, No 76011, 1985 (Am A).
First Approximation to Take-off Distance to 50 ft (15.2 m) for Light and General Aviation Aeroplanes. Vol 7, No 82033, 1982.
Examples of Take-off Field Length Calculations for a Civil Transport Aeroplane. Vol 7, No 87018, 1987.
Statistical Methods Applicable to Analysis of Aircraft Performance Data. Vol 10, No 91017, 1997 (Am D).

7

Aircraft manoeuvre performance

Manoeuvring is an essential part of all mission profiles; an aircraft needs to turn, pitch and to change its airspeed to carry out its mission. Whilst these manoeuvres may be of low significance in transport aircraft operations, they form the essential basis of combat aircraft missions and combat aircraft are designed around a need to manoeuvre aggressively.

7.1 Introduction

An aircraft can be said to be in manoeuvring flight when its flight path is in a continuous change of state and in which there is an inertial force due to acceleration. In Appendix B, it is shown that the inertial forces acting on the aircraft give rise to the statement of the accelerations acting in a general manoeuvre (eqn (B6)). Usually the rate of change of aircraft mass can be neglected and the statement of the inertial forces can be expressed as,

$$
\begin{bmatrix} F_x \\ F_y \\ F_z \end{bmatrix}_v = m \begin{bmatrix} \dot{V} \\ V\dot{\gamma}_3 \cos \gamma_2 \\ -V\dot{\gamma}_2 \end{bmatrix}_v \tag{7.1}
$$

which describes the three linear accelerations that occur in manoeuvring flight. These can be summarized as follows.

The linear acceleration, $\dot{V}$, arises from an imbalance of the forces in the direction of flight; this may be due to an excess of thrust or drag, or due to a component of weight in non-level flight. The linear acceleration is employed to control the airspeed in which thrust is increased or decreased to provide the necessary thrust–drag balance to achieve, or maintain, the required airspeed. When the aircraft is climbing or descending, the component of weight in the direction of flight will contribute to the accelerating force; thus, a rate of climb or descent can also be used to control the airspeed in non-level flight.

The lateral acceleration arises from the rate of turn, or rate of change of heading, $\dot{\gamma}_3$, which produces the centrifugal force in a turning manoeuvre. The balancing centripetal force is provided by a component of the lift force by banking the aircraft into the turn. The effect of the lateral acceleration will be perceived as a normal force, or 'g' force, acting on the aircraft during a turning manoeuvre.

The normal acceleration arises from the pitch rate of the aircraft, $\dot{\gamma}_2$, which produces the 'g' force experienced in a symmetric pull-up, or looping, manoeuvre.

The general manoeuvre produces a combination of these accelerations and the equations of motion for performance, developed in Appendix B, eqn (B19), describing the general manoeuvre, can be expressed as,

$$\left.\begin{array}{r} F_N - D - W\sin\gamma_2 = m\dot{V} \\ Y\cos\gamma_1 + L\sin\gamma_1 + T\{-\cos(\alpha+\tau_1)\sin\beta\cos\gamma_1 + \sin(\alpha+\tau_1)\sin\gamma_1\} = mV\dot{\gamma}_3\cos\gamma_2 \\ Y\sin\gamma_1 - \cos\gamma_1 \\ +T\{-\cos(\alpha+\tau_1)\sin\beta\sin\gamma_1 - \sin(\alpha+\tau_1)\cos\gamma_1\} + W\cos\gamma_2 = -mV\dot{\gamma}_2 \end{array}\right\}$$

$$(7.2)$$

in which the term for rate of change of aircraft mass as fuel is consumed has been omitted. The powerplant thrust is expressed as net thrust, F_N, and the total gross thrust component, T (Appendix B, eqn (B14)).

In a coordinated manoeuvre the sideforce, Y, and the sideslip angle, β, are both zero and it is assumed that the manoeuvres are coordinated. The equations of motion can then be reduced to the form,

$$\left.\begin{array}{r} F_N - D - W\sin\gamma_2 = m\dot{V} \\ \{L + T\sin(\alpha+\tau_1)\}\sin\gamma_1 = mV\dot{\gamma}_3\cos\gamma_2 \\ -\{L + T\sin(\alpha+\tau_1)\}\cos\gamma_1 + W\cos\gamma_2 = -mV\dot{\gamma}_2 \end{array}\right\} \quad (7.3)$$

where the term $\{L + T\sin(\alpha+\tau_1)\}$ represents the total normal force as the sum of the lift and the normal component of gross thrust from the engines.

The load factor, n, which characterizes the 'g' force is defined as the ratio of the overall normal force produced by the aircraft to the aircraft weight, thus

$$n = \frac{L + T\sin(\alpha+\tau_1)}{W} \quad (7.4)$$

This definition of the load factor allows the normal component of the gross thrust from the powerplant to be taken into account during a manoeuvre. This enables the equations of motion to be used to analyse the performance of vectored thrust aircraft and aircraft manoeuvring at a very high angle of attack. However, in the case of conventional aircraft, with lift–drag ratios of 10 or more, operating at angles of attack up to about 10°, and having little or no downward thrust deflection, the thrust component is small enough to be neglected. The load factor can then be taken to be the ratio of the aerodynamic lift force to the aircraft weight for all practical purposes, thus the approximation,

$$n = L/W \quad (7.4a)$$

can be used.

Substituting eqn (7.4) into eqn (7.3) gives the equations of motion for coordinated flight,

$$\left.\begin{aligned}
[F_N - D]\frac{1}{W} - \sin\gamma_2 &= \frac{\dot{V}}{g} \\
n\sin\gamma_1 &= \frac{V}{g}\dot{\gamma}_3\cos\gamma_2 \\
-n\cos\gamma_1 + \cos\gamma_2 &= -\frac{V}{g}\dot{\gamma}_2
\end{aligned}\right\} \tag{7.5}$$

In this form, the equations can be used to develop the basic expressions for co-ordinated manoeuvres.

7.2 The manoeuvre envelope

The manoeuvre performance of the aircraft will be limited by the structural strength of the airframe; there are two basic reasons for this.

- The pressure loading produced by the dynamic pressure of the airflow increases with the square of the airspeed and with it the air loads on structural components of the aircraft. This is particularly obvious in the case of the deflection of devices such as flaps or landing gear into the airstream. The design maximum dynamic pressure loading on the structure of the aircraft will determine the maximum equivalent airspeed, EAS.
- The normal acceleration associated with manoeuvring flight produces structural loads in the airframe. The maximum allowable load factor in a manoeuvre is determined by the load bearing capability of the airframe structure.

These limitations do not always apply universally. Since the configuration of the airframe may be changed for certain parts of the flight, for example by the lowering of the landing gear or the deflection of flaps, the structural loading and strength limits may be affected. Therefore, different maximum airspeed and load factor limitations may exist for each configuration. The design *manoeuvre envelope*, or *n–V diagram*, describes the design limitations on airspeed and load factor.

The structural strength of the airframe must be capable of sustaining the structural loading generated by flight manoeuvres and gusts at a given aircraft weight and by the dynamic pressure of the airflow at the maximum permissible airspeed. The load factor limits and maximum airspeeds depend on the role of the aircraft.

- A military combat aircraft requires a high level of manoeuvrability and, therefore, the capability of sustaining a high load factor. In addition, it will be called upon to perform the manoeuvres over a wide range of airspeeds. These requirements lead to the need for a strong airframe to withstand the combined aerodynamic and manoeuvre loading; this implies that the airframe will be relatively heavily constructed.
- A large, subsonic, transport aircraft needs only low manoeuvrability and can meet the entire manoeuvre and gust loading requirements with a relatively low load

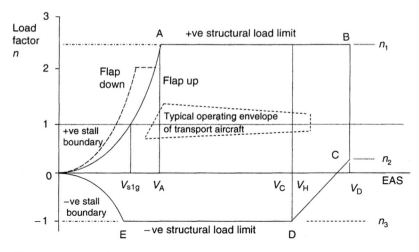

Fig. 7.1 The manoeuvre envelope.

factor. The airspeed is likely to be limited by a subsonic Mach number, which may enable a relatively low maximum EAS to be scheduled. These considerations lead to a much lower structural strength requirement and, consequently, a lighter airframe construction.

Figure 7.1 shows the main elements of a typical manoeuvre envelope. Although the manoeuvre envelopes for civil and military aircraft differ in detail, the definitions of the boundaries and the principal airspeeds are similar enough to be generalized.

7.2.1 The structural boundaries

The load factor limits, n_1 and n_3, are the maximum positive and negative normal acceleration loadings respectively.

The positive load factor, n_1 (boundary A–B in Fig. 7.1), is defined by requirements for the aircraft and the minimum load factor limit is laid down by the airworthiness codes of practice. The requirement for a high structural strength of the airframe increases the weight of the aircraft. Therefore, the maximum load factor needs to be kept as low as possible, commensurate with the ability of the aircraft to manoeuvre and to be able to sustain the loads induced by gusts. The positive load factor limit, n_1, is defined in FAR/JAR as a function of aircraft weight and is given by the expression,

$$n_1 = 2.1 + \frac{24000}{0.225W + 10000}$$

where W is the weight of the aircraft in Newtons. However, n_1 need not exceed the range $2.5 < n_1 < 3.8$ for normal category aircraft, which includes large transport aircraft.

For aircraft in the utility category, n_1 is taken to be $+4.4$, and for aerobatic aircraft, n_1 is taken to be $+6.0$. The maximum n_1 for military aircraft is not defined generally, but tends to be determined by the design role requirements, which may call for manoeuvres that would demand greater load factors than those required by any civil aircraft.

The negative load factor for civil aircraft, n_3, E–D in Fig. 7.1, is taken to be $-0.4n_1$ in the case of normal or utility category aircraft, or $-0.5n_1$ for aerobatic aircraft. In the case of large transport aircraft, n_3 is usually taken to be not less than -1.0 at speeds up to V_C decreasing to zero at V_D. For military aircraft the value of n_3 is taken to be $-0.6(n_1 - 1)$ up to V_H decreasing to $n_2 = 1 - 0.3n_1$ at V_D.

7.2.2 The airspeed boundaries

The airspeed limits, which are defined as equivalent airspeeds, are determined by the stall boundaries, O–A and O–E, and by the design diving speed, V_D, B–C.

The stalling speed, or the minimum airspeed at which the aircraft can maintain steady flight in a specified configuration, forms the low-speed boundary of the manoeuvre envelope. Since the lift force is a function of the load factor, the stalling speed is defined under steady, level, straight flight conditions to be V_{S1} or V_{S1g}, the $1g$ stalling speed. The minimum, practical, airspeed may be set by the stall buffet, which is caused by the initial separation of the airflow as the stalling speed is approached. Stall boundaries will usually be shown for flaps up and flaps down cases.

The high-speed boundary is determined by the maximum structural dynamic pressure loading, $q = \frac{1}{2}\rho_0 V_e^2$, which sets the design diving speed of the aircraft, V_D. Since V_D is defined in terms of equivalent airspeed, the Mach number associated with it will increase with altitude and may further limit the manoeuvre envelope. The high-speed boundary may be quoted as the lower of either V_D or M_D.

Other notable speeds within the manoeuvre envelope are as follows.

V_A, the design manoeuvring speed, where $V_A = V_{S1g}\sqrt{n_1}$, this is the minimum airspeed at which the aircraft can achieve the maximum positive load factor in a steady manoeuvre.

V_C/M_C, the design cruising speed or Mach number (civil transport aircraft), this is the normal cruising speed at which the structure can sustain the maximum load factors, V_C defines the corner D of the civil aircraft manoeuvre envelope.

V_H, the maximum speed in level flight with maximum continuous power, this airspeed defines the corner D of the military aircraft manoeuvre envelope.

(The relationships between these speeds depend on the size and purpose of the aircraft and are defined in more detail in the airworthiness code of practice under which the aircraft is to be certificated.)

In normal, civil, aircraft operations, a large transport aircraft uses only a portion of the available manoeuvre envelope, as indicated in Fig. 7.1. Operational speeds are usually limited by a safety margin over the stalling speed and by the cruise Mach number; turns do not generally exceed 30° bank angle so that load factors due to manoeuvre will rarely exceed $1.2g$.

7.3 Aircraft manoeuvres

The equations of motion for manoeuvring flight (eqn (7.5)) show that there are three components of inertial force, acting on the aircraft in the direction of the velocity axes. These are the longitudinal force, the lateral force and the normal force. Whilst the general manoeuvre of the aircraft may involve all three force components, the effect of each can be analysed individually.

7.3.1 The longitudinal manoeuvre

The longitudinal manoeuvre is the result of an imbalance of thrust and drag, which results in either a linear acceleration or a steady rate of climb, or in a combination of both acceleration and climb, in the direction of flight. It does not involve directly the accelerations that result from rates of pitch or turn, although those manoeuvres may produce increases in the drag force, which will have an indirect effect on the longitudinal force balance.

By expressing the gradient of climb in terms of the true rate of climb and true airspeed the longitudinal equation of motion for manoeuvring flight can be written as,

$$[F_N - D]\frac{V}{W} = \frac{d}{dt}\left\{H + \frac{V^2}{2g}\right\} \tag{7.6}$$

The term $\{H + V^2/2g\}$ is the *specific energy*, E_S of the aircraft. This is the sum of the potential energy and the kinetic energy of the aircraft per unit weight. It is also known as the *energy-height* since it represents the height the aircraft would attain if all the kinetic energy were to be converted into potential energy.

The term $[F_N - D]V/W$, the product of the excess thrust and the true airspeed per unit weight, is known as the *specific excess power* (SEP), of the aircraft and determines the rate of change of the specific energy. The excess power can be used to increase potential energy (climb), or to increase kinetic energy (acceleration) of the aircraft. It also can be used to increase the potential and kinetic energies in combination to achieve the maximum rate of change of total energy, the sum of the PE and KE, to minimize the time required to climb and accelerate the aircraft to its operating height and Mach number. This principle is employed by high performance aircraft in the optimization of their climb profile through the transonic flight region where the excess power is reduced by the increase in drag. This is also discussed in Chapter 5 under the climb performance of aircraft with a high excess thrust.

Any change in the specific excess power arising from an increment in either the thrust or the drag will produce either a rate of climb or an acceleration of the aircraft. If height is maintained constant then the airspeed will vary or, conversely, if the (true) airspeed is maintained constant the height will vary. This principle is important in the consideration of the overall effect of a manoeuvre on the flight path of the aircraft.

7.3.2 The lateral manoeuvre, or the level turn

In a level, constant airspeed, coordinated turn, the rate of climb, γ_2, the rate of pitch, $\dot{\gamma}_2$, and the rate of change of true airspeed, $\dot{V}$, are all zero and the equations for manoeuvre performance become,

$$\left.\begin{array}{c} [F_N - D]\dfrac{1}{W} = 0 \\[2mm] n\sin\gamma_1 = \dfrac{V}{g}\dot{\gamma}_3 \\[2mm] -n\cos\gamma_1 + 1 = 0 \end{array}\right\} \tag{7.7}$$

It should be noted that the bank angle used in the equations of motion is γ_1, the aircraft body axes relative to velocity axes, and not the Euler angle ϕ, the body axes relative to Earth axes. This enables the analysis to be applied in the general manoeuvring case in which the aircraft may be in a manoeuvre combining both turning and pitching motions.

The turn is shown diagrammatically in Fig. 7.2. From the normal force equation the load factor, n, is seen to be a function of the bank angle of the aircraft and is given by,

$$n = \frac{L}{W} = \frac{1}{\cos\gamma_1} \tag{7.8}$$

From the lateral force equation the rate of turn is given by,

$$\dot{\gamma}_3 = \frac{V}{R} = \frac{ng\sin\gamma_1}{V} = \frac{g\tan\gamma_1}{V} \tag{7.9}$$

where R is the radius of the turn which, using eqn (7.8), can be expressed as,

$$R = \frac{V^2}{g\tan\gamma_1} \tag{7.10}$$

From these expressions, it can be seen that both the rate and radius of the turn are functions of true airspeed and bank angle only and are independent of the weight of the aircraft.

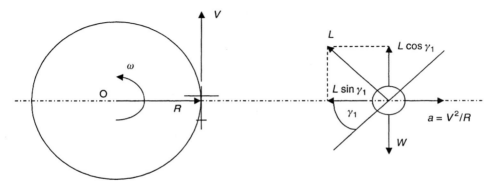

Fig. 7.2 The turning manoeuvre.

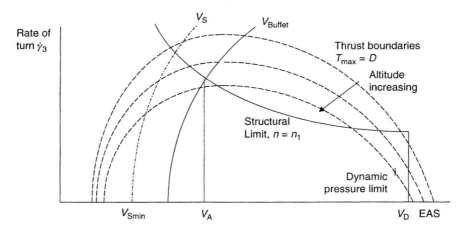

Fig. 7.3 The manoeuvre boundaries.

The turning performance will be constrained by the aerodynamic and structural limitations of the aircraft, together with the limit imposed on the turning manoeuvre by the excess thrust available. These are defined by the manoeuvre envelope, and are shown in Fig. 7.3 in terms of the rate of turn, $\dot{\gamma}_3$, and EAS.

From eqn (7.8) the maximum bank angle in the steady, level turn is determined by the limit load factor, n_1. Therefore, from eqn (7.9), the maximum rate of turn permitted by the structural strength of the aircraft is seen to be inversely proportional to the true airspeed. This forms the manoeuvre boundary set by the structural strength of the aircraft. The boundary will tend to move downwards as the air density decreases with increasing altitude and temperature.

The maximum angle of attack at which the aircraft can be flown in steady flight is limited by the stall (or stall buffet) boundary, this will determine the minimum airspeed in the steady, level turn. From the lift equation in turning flight, the airspeed and load factor are related by the expression,

$$\frac{V_e^2}{n} = \frac{2W}{\rho_0 S C_{Ls}} = \text{constant} \tag{7.11}$$

where C_{Ls} is the maximum steady, level-flight, lift coefficient.

The rate of turn is a function of the airspeed and load factor, eqn (7.9), so that there will be a minimum airspeed boundary set by the value of V_e^2/n determined from eqn (7.11). The maximum airspeed boundary is set by the maximum design speed, V_D, which is a structural strength limitation determined from the maximum dynamic pressure loading. However, there may be further restrictions to the high-speed boundary caused by the maximum design Mach number, M_D, and set by the aerodynamic limitations associated with changes in airflow characteristics leading to stability and handling qualities issues.

The maximum excess thrust power limitation is found from eqns (7.6) and (7.16) in turning flight. The maximum non-manoeuvring excess thrust-power is given from eqn (7.6) and is reduced by the increment in drag-power resulting from the rate of turn, eqn (7.16). The curves in Fig. 7.3 represent the maximum level flight manoeuvre boundaries for increasing WAT limits.

Figure 7.3 shows the limit of manoeuvre of the aircraft under these constraints. The maximum rate of turn occurs at V_A, the design manoeuvre speed. However, it should be noted that in steady, level flight the limiting manoeuvre is only possible if there is sufficient propulsive thrust to overcome the aerodynamic drag. The aircraft will be further limited in the manoeuvre by the thrust available if it is climbing. In the design of a combat aircraft the thrust and drag characteristics need to be developed to provide the maximum excess thrust-power as close as possible to the design manoeuvre speed. This is to avoid the thrust available limiting the turning performance at the most critical manoeuvre point.

The increase in the drag force in manoeuvring flight can be found from the drag characteristic of the aircraft. The drag polar of a conventional, subsonic, aircraft is assumed parabolic so that the drag coefficient can be expressed in the form,

$$C_D = C_{Dz} + KC_L^2 \tag{7.12}$$

and the drag force in level flight is given by,

$$D = \tfrac{1}{2}\rho S C_{Dz} V^2 + \frac{KW^2}{\tfrac{1}{2}\rho S} \frac{1}{V^2} \tag{7.13}$$

In the turn, the load factor of the aircraft is increased so that the lift generated by the wing will have to be increased to balance the forces. If the engine thrust component in the normal axis is negligible, the drag force becomes,

$$D = \tfrac{1}{2}\rho S C_{Dz} V^2 + \frac{Kn^2 W^2}{\tfrac{1}{2}\rho S} \frac{1}{V^2} \tag{7.14}$$

Thus, the increment in drag, ΔD, due to the turn is the difference between eqns (7.14) and (7.13), and

$$\Delta D = \left[\frac{KW^2}{\tfrac{1}{2}\rho S}\right] \frac{n^2 - 1}{V^2} \tag{7.15}$$

which, from eqns (7.8) and (7.9), can be written in the form,

$$\Delta D = \left[\frac{Km^2}{\tfrac{1}{2}\rho S}\right] \dot{\gamma}_3^2 \tag{7.16}$$

Here, the square bracket is a quasi-constant for the aircraft in the turn at a given weight and height. The increment in drag is thus seen to be proportional to the square of the rate of turn, $\dot{\gamma}_3$. Now, in the turn, the drag increases and, if the thrust is not increased to compensate, the drag increment ΔD will lead to either a rate of descent or a deceleration relative to the non-turning flight state. This can be deduced from eqn (7.6),

$$-\Delta D \frac{V}{W} = \Delta \frac{\mathrm{d}}{\mathrm{d}t}\left\{H + \frac{V^2}{2g}\right\} \tag{7.17}$$

In turning flight, the increase in the lift dependent drag coefficient leads to a modified expression for the minimum drag ratio, this is now given by,

$$\frac{D}{D_{\min}} = \frac{1}{2}\left(u^2 + \frac{n^2}{u^2}\right) \tag{7.18}$$

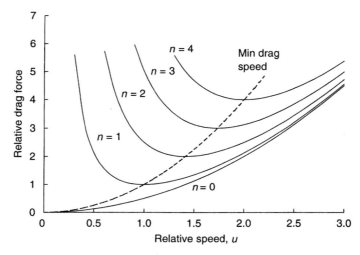

Fig. 7.4 The effect of load factor on the drag characteristic.

and the minimum drag speed in the turn, $V_{md(turn)}$ is given by,

$$V_{md(turn)} = \left(\frac{2W}{\rho S C_{Lmd}}\right)^{\frac{1}{2}} n^{\frac{1}{2}} = V_{md} n^{\frac{1}{2}} \tag{7.19}$$

so that the minimum drag speed increases as a function of the rate of turn. Figure 7.4 shows the increase in drag, and in the minimum drag speed, with load factor as the rate of turn is increased.

Unless the thrust is increased to compensate for the increased drag then the turn will cause the specific energy of the aircraft to decrease. If airspeed is maintained, then a rate of descent will occur or, if height is maintained, then the aircraft will decelerate. If the turn is initiated at an airspeed sufficiently above the minimum drag speed the airspeed will decrease, reducing the drag until the force equation is re-balanced and the level turn will continue at the lower airspeed. However, if the initial airspeed is close to or below the minimum drag speed, then any decrease in airspeed will lead to a further increase in drag and a consequent increase in the rate of loss of airspeed. If the thrust available is limited then the maximum airspeed in the level turn will decrease as the turn is tightened until the aircraft is at its minimum drag speed with maximum available thrust. At that point, the aircraft is performing its tightest, constant speed, level turn.

These effects can be very important in climbing turns with very little excess thrust available, for example, the after-take-off climb with one engine inoperative (Chapters 5 and 9). In such cases, the additional drag due to a turn can reduce the climb gradient to an unacceptably low level or even to a descent.

7.3.3 The pull-up manoeuvre or the loop

The pull-up manoeuvre is a coordinated manoeuvre in the vertical, or pitching, plane with no rate of turn or sideslip so that, $\gamma_1 = 0$, $\dot{\gamma}_3 = 0$, $Y = 0$ and $\beta = 0$. The equations

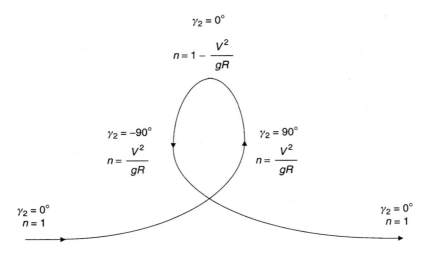

Fig. 7.5 The load factor in the pull-up manoeuvre.

of motion thus become,

$$
\left.
\begin{aligned}
F_N - D - W \sin\gamma_2 &= m\dot{V} \\
Y &= 0 \\
-L - T\sin(\alpha + \tau_1) + W\cos\gamma_2 &= -mV\dot{\gamma}_2
\end{aligned}
\right\}
\tag{7.20}
$$

Using the definition of the load factor, eqn (7.4), and writing the instantaneous rate of pitch as, $\dot{\gamma}_2 = V/R$, where R is the radius of the loop, leads to the load factor in the pull-up manoeuvre given by,

$$
n = \cos\gamma_2 + \frac{V^2}{gR}
\tag{7.21}
$$

and the radius of the manoeuvre given by,

$$
R = \frac{V^2}{g(n - \cos\gamma_2)}
\tag{7.22}
$$

The pull-up manoeuvre is shown in Fig. 7.5.

The load factor in the loop is not uniform and will vary with airspeed and flight path angle as the aircraft progresses around the loop, Fig. 7.5. In practice, the variation is complex since the increased load factor increases the drag force, eqn (7.14), which, together with the weight component, affects the balance of the longitudinal forces acting on the aircraft. This causes a continuous change in airspeed throughout the manoeuvre. To control the airspeed within acceptable limits the engine thrust must be increased in the upward segment of the loop and reduced in the downward segment, thus the loop cannot be regarded as a steady manoeuvre. Similarly, the radius of the loop is not uniform but tends to decrease to a minimum at the top of the manoeuvre and increase again on the descending path. Beyond aerobatic flight and some military aircraft combat manoeuvres, there are few practical applications of the extended pull-up, or looping, manoeuvre.

An extreme case of the pull-up manoeuvre is the 'Cobra' which is a *post-stall manoeuvre* involving a rapid pitch-up to increase in angle of attack to a state far beyond the stalling angle of attack. In this state, the lift force becomes small, since the aircraft is in the stalled condition, but the drag increases to a very large value and acts in the wind-axis direction. Since the lift force is no longer significant, the aircraft will not enter a looping manoeuvre but will tend to continue in its original direction of flight together with a rapid deceleration. When an aircraft is engaged in combat with another aircraft of similar performance, they may become locked into a circular tail-chase, each turning at maximum rate. Neither will be able to tighten the turn to bring its adversary into line of sight to fire its weapons. The aircraft will be burning fuel at a high rate, particularly if afterburners are being used. Unless the stalemate can be broken one aircraft will have to break off the combat because its fuel state is becoming critical – that aircraft is then at risk from its adversary. If the aircraft has suitable aerodynamic characteristics, a means of breaking the stalemate is for the aircraft to be pitched up to a very large angle of attack, far beyond the stalling angle of attack, in the 'Cobra' manoeuvre. This enables it to bring its adversary briefly into line of sight for the weapons to be fired. The angle of attack in the Cobra may be greater than 70° and although the aircraft is described as 'flying at post-stall angle of attack' this is an overstatement. The aircraft may be stable in this condition (inasmuch as it does not have a tendency to rotate uncontrollably about any of its axes), its engines may continue to produce thrust and it may possess some degree of controllability. However, it will not be producing aerodynamic lift and so cannot be described properly as 'flying'. It will produce a very large drag force, which will cause the aircraft to lose energy at a very high rate. This may assist it to escape retaliatory attack since its adversary is likely to over-fly the rapidly decelerating aircraft before it can aim and fire its weapons. Once the Cobra has been performed, the loss of total energy may have left the aircraft unable to resume normal flight without a great loss of height. It is an extreme manoeuvre that normally would only be used where no alternative escape is available.

In combat missions, turning manoeuvres are often combined with an extended pull-up and most manoeuvres, whether attacking or evasive, will contain rates of turn and pitch as well as longitudinal acceleration. If the manoeuvre is coordinated, eqn (7.3) can be used to analyse the combined manoeuvres including those of aircraft with vectored thrust. These aircraft have the ability to rotate the gross thrust vector in their Oxz plane so that the normal force consists of the aerodynamic lift and a large component of the engine gross thrust. If this facility is used in flight, the thrust component forms part of the normal force vector and some of the assumptions made in the development of the expressions for turning flight no longer apply. The rates and radii of turn will no longer be restricted by the maximum lift coefficient.

Some aircraft with thrust vectoring capability are able to perform a similar manoeuvre to the Cobra to avoid attack from behind by vectoring thrust in forward flight, known as 'Viffing'. In this case, the engine nozzles are rotated downwards together with some pitch-up of the aircraft. The result is a rapid deceleration that cannot be matched by the attacking aircraft, which is likely to overtake the decelerating aircraft before it can aim and fire its weapons. Since the vectored thrust aircraft may not need to have been pitched up to an angle of attack beyond the stall, it may be able to recover normal flight relatively quickly without great loss of height. This is an example of the way in which tactical combat manoeuvres can be developed to take advantage of the performance characteristics of a combat aircraft. By using the aerodynamic and propulsive characteristics of the aircraft type, offensive and evasive manoeuvres peculiar to that type can be developed from the equations of motion for manoeuvre and used to give that aircraft an area of air superiority in combat.

7.4 Transport aircraft manoeuvre performance

The effect of manoeuvres on the flight path performance of civil transport aircraft is generally not very significant. They spend only a very low proportion of their time in turning flight and, since the maximum bank angle used in the turn is typically 20°, the turns are generally of very low rate. The effect of such turns on the overall performance is minimal. En-route turns, which are generally associated with heading changes, are usually made through angles of less than 90°. If the aircraft is stacked in the hold during the descent to landing it is required to fly an oval flight path, with a 180° turn at each end, known as the holding pattern. The holding pattern turn is flown at Rate 1, or 3°/s, and takes one minute. This is probably the most sustained turn that the aircraft will be required to carry out in normal operations and will be flown at a speed commensurate with flight safety and air traffic control considerations. This speed should be close to the maximum endurance speed if possible.

The rate and radius of the turns made by aircraft with differing airspeeds can be used to advantage in control of air traffic. The smaller 'regional' aircraft, often turbo-props, can fly at lower speeds than the big jets and can use bank angles up to 30° compared with the normal maximum of 20° in the case of the big jets. This enables the regional aircraft to turn at a higher rate and with a smaller radius. These differences in performance can be used to aid the separation and flow of traffic in airport terminal areas.

The only case in which the effect of turning would be significant to a transport aircraft, in terms of its effect on flight safety, is a turn made with marginal excess thrust available. For example, during the after-take-off climb with one engine inoperative, the additional drag due to the turn may decrease the already small gradient of climb. Similarly, the sustained pull-up is not a manoeuvre associated with civil transport operations. The most significant pitching manoeuvres, other than the transition at take-off and the flare on landing, will be transient manoeuvres between the climb and cruising states or between cruising and descending flight.

7.5 Military aircraft manoeuvre performance

Military transport aircraft may be required to manoeuvre more aggressively than civil aircraft and sustained turning performance may be significant in the operational profile of the mission. Although pure pull-ups are still unlikely to be more than transient manoeuvres in pitch, they may extend, in the case of tactical transport aircraft, to 'pop-up' manoeuvres from low-level flight that would demand a higher rate of pitch than the simple gradient changing manoeuvre. The increased fuel burned due to the turning manoeuvres may need to be accounted for in the analysis of the fuel requirement for the mission.

The combat aircraft (or the aerobatic aircraft) spends a significant proportion of its mission time in manoeuvring flight and the design criterion for the aircraft is that it will be capable of sustaining high-*g* manoeuvres with high rates of turn. In the design phase of the development of the aircraft, estimates will be needed of the thrust required to achieve the design target manoeuvres, particularly in terms of rates of turn and acceleration. Complex manoeuvres combining turning with pull-ups and acceleration will also need to be considered to establish the limits of the aircraft in air-to-air combat situations. The performance in these manoeuvres will be part of the design specification of the aircraft. However, in service, combat manoeuvres will be developed from combinations of the three basic manoeuvres to take advantage of the performance qualities peculiar to the aircraft. It is unlikely that such manoeuvres could be foreseen at the design stage, but their development in service through experience of the aircraft, is a necessary phase of the maintenance of the aircraft air superiority.

7.6 Conclusions

The equations of motion given in this chapter have been developed to enable the most general manoeuvre cases to be analysed. Equation (7.2) gives the full equations of motion that cover all manoeuvres, including those involving asymmetric cases, whilst the simpler versions, eqns (7.3) and (7.5), assume coordinated flight. Using the expressions developed in this chapter, any manoeuvre can be analysed to provide the effect on the performance of the aircraft in terms of the additional thrust, and hence fuel, required and time taken. The broad effects of manoeuvring on the performance of aircraft engaged in various missions can be briefly summarized.

In the case of civil transport operations, and most military transport operations, the manoeuvres required are low rate and have only a marginal effect on the mission. Usually, any effect on the fuel required will be able to be taken into account through fuel contingency allowances. The only time when the effect of a manoeuvre is likely to be critical is when the excess thrust is marginal and the aircraft is unable to maintain height in turning flight. This case is considered in regulated performance (Chapter 9).

Some military transport and supply aircraft are required to manoeuvre with greater freedom than civil transport aircraft. This category could include some patrol and surveillance aircraft and aircraft engaged in activities, such as anti-submarine operations, which may be carried out by large aircraft. These manoeuvres call for a higher rate of turn, and more sustained turns, than transport aircraft, and the need for sufficient excess thrust is an important consideration in their design and operation. An activity such as in-flight re-fuelling also calls for a higher level of manoeuvre, particularly from the receiver aircraft. Although the aircraft may require an inter-mediate level of manoeuvre performance, the task is more dependent on the handling qualities and control of the aircraft than on the flight path manoeuvres associated with this treatment of performance.

Aerobatic aircraft, both civil aircraft and military combat aircraft, are designed around the ability to perform high rate, aggressive, manoeuvres and combinations of manoeuvres. They are optimized, both in terms of their structural design and their power or thrust available, to be able to achieve sustained 'g' levels and rates of rotation, which enable them to perform a specific set of basic manoeuvres. The operational manoeuvres are developed from this basic set.

Bibliography

Miele, A. (1962) *Flight Mechanics, Vol 1, Theory of Flight Paths* (Pergamon Press).

Engineering sciences data unit items
Performance Series, Vols 1–14.
Flight Path Optimization; Estimation (Manoeuvres). Performance, Vol 9.
Introduction to Data Items on Flight Path Optimization. Vol 9, No 89015, 1989.
Energy–Height Methods for Flight Path Optimization. Vol 9, No 90012, 1990.
Estimation of Turning Performance. Vol 9, EG8/1, 1992 (Am A).

8

Aircraft performance measurement and data handling

Following the estimation of the performance of the aircraft during its design process, flight trials will be required to measure the achieved performance. Performance data measured in flight need to be converted from the arbitrary weight, altitude and temperature conditions of measurement to the standard conditions used in the design process. They can then be used for verification of the design process and for the preparation of the aircraft performance manual.

8.1 Introduction

In Chapters 4, 5 and 6 the basic elements of the flight path were analysed and expressions were developed for the estimation of the performance of the aircraft. The expressions, which addressed cruise, climb and descent, and take-off and landing, were based on the assumption of a parabolic drag polar and a simple fuel flow law. In Chapter 3, it was shown that the real drag polar may not be exactly parabolic, and that it is unlikely that the specific fuel consumption would be a simple constant. Therefore, the actual performance achieved by the aircraft in any of the basic elements of the flight path will be likely to differ from the performance predicted by the theoretical expressions. In the design process, the performance of the aircraft is estimated on the basis of the best available assumptions of its aerodynamic characteristics and of the predicted output of the powerplant. These assumptions are drawn from theoretical analysis and empirical data, which may have been gathered over a period of time from many sources. It has been used to form a database which, whilst it may represent the best available information, is imperfect; any estimation of the performance of the aircraft based on this data will, therefore, be inaccurate. The estimates of performance made in the aircraft design process can be used to predict that the aircraft will meet its design specification. However, they could not be used to provide the data required for the aircraft *performance manual* or *Operating Data Manual (ODM)*, which contains all the performance data necessary for the flight plan to be prepared. (The flight plan will determine, for example, the fuel required for a flight and the maximum permissible take-off weight of the aircraft.) Neither would the estimated data be acceptable for the granting of airworthiness certification since the safety related performance must be demonstrated to the airworthiness authority; only data based on flight measurement can be accepted for this purpose.

The performance of an aircraft needs to be measured in flight for three principal purposes,

- To verify the estimates of performance made at the design stage. A comparison between the flight-measured performance and the estimated performance of the aircraft will enable the design database and performance prediction methods to be verified and improved.
- To obtain evidence of the performance of the aircraft for airworthiness certification. Before an aircraft can be granted a certificate of airworthiness, there must be a demonstration to the airworthiness authority of its ability to meet certain performance criteria associated with flight safety. This evidence must be obtained by measurement of the performance of the aircraft in flight.
- To obtain performance data for the performance manual of the aircraft. The performance manual contains information concerning the performance of the aircraft that is necessary for flight planning. These data must be obtained from flight measurement and converted into a form that can readily be used for the construction of the performance charts for the aircraft in terms of standard Weight, Altitude and Temperature (WAT) states.

Flight tests may be performed for purposes other than the aircraft development and certification programme, for example, research or the testing of equipment. The general principles of certification-related testing apply equally to the data gathering and data handling for any flight-test programme.

When the performance of the aircraft is estimated during the design process, the state of the atmosphere is taken to be either the ISA model or a design atmosphere model (Chapter 2). In addition, the weight of the aircraft will need to be assumed and, usually, a conveniently rounded figure will be taken as the datum. When the performance is measured in flight to verify the design estimates, the state of the atmosphere is arbitrary and will not, in general, correspond to any particular model. Neither will the weight correspond exactly to that used in the estimates. Any flight-measured data will need to be processed so that they correspond to the same flight conditions as those assumed in the estimation of the aircraft performance in order for comparison to be made between them. This will then enable the verification of the estimation process and validation of the database from which the performance was estimated.

During the measurement of performance in flight, the weight of the aircraft will be decreasing as fuel is burned, Fig. 8.1(a). This means that the weight can only be at the required, standard value, W_s, at one instant during a flight; at all other times the weight at test, W_t, will be either greater than or less than the standard. Therefore, at all other points in the test there will be a difference, ΔW_t, between the preferred standard, W_s, and the test weight, W_t

$$W_s = W_t + \Delta W_t$$

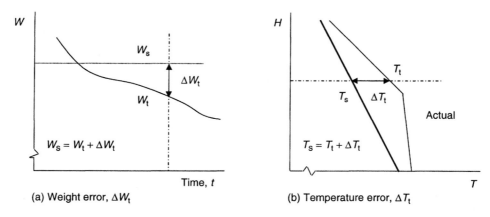

Fig. 8.1 Performance variables – weight and temperature. (a) Weight error from standard weight. (b) Temperature error from standard atmosphere.

A correction to the measured performance will be needed to account for the difference between the performance at the standard weight and at the weight at test. Weight, therefore, is a partially controllable variable. Although it can be arranged for the flight trial to be performed within a limited range of weights, the actual weight at any time is varying as fuel is consumed.

The actual state of the atmosphere at the time of test, defined by its pressure and temperature, will depend on the local meteorology. Although the local atmospheric pressure datum may vary this is not significant. This is because any measurement of altitude by an altimeter will be based on a standard atmosphere datum pressure setting of 1013 mb and so the pressure is a known function of height. Pressure is a controllable variable through altitude. The temperature–height profile in the actual atmosphere is unknown and will vary with time; there is no control over the temperature of the atmosphere. Temperature is an uncontrollable performance variable. Any flight test will take place in an arbitrary temperature–height profile and a correction to account for the difference from a standard temperature–height profile will be necessary, Fig. 8.1(b). The actual temperature at the test altitude, T_t, may differ from the standard temperature, T_s, the difference being given by ΔT_t, where,

$$T_s = T_t + \Delta T_t$$

A correction to the measured performance will be needed to account for the difference between the performance at the standard temperature and at the actual temperature at the test height.

Any method of performance measurement must enable data measured under arbitrary test conditions of Weight, Altitude (pressure) and Temperature (WAT), to be corrected to correspond to the preferred, standard WAT state. This will require some form of data handling process that will depend on the type of powerplant installed in the aircraft.

The measurement of the performance of the aircraft considered in this chapter will be centred on the basic elements of the flight path. The cruise performance will be concerned mainly with the measurement of the specific air range and specific endurance so that the fuel required and time required for any payload–range combination can be determined. The climb, and descent, performance will be concerned with rates and gradients of climb and with the fuel required to climb to cruising altitude using the recommended climb thrust setting. The take-off and landing performance will be concerned with the ground run and airborne distances required and with the critical speeds during the manoeuvres. Other aspects of performance may need to be measured but the same basic principles can be applied to any performance-related measurements.

A number of methods are available to convert, or correct, data measured in arbitrary WAT conditions in flight to preferred standard WAT conditions; each method has its particular application. In the following sections the principles of three methods of performance data handling are considered and their application to the basic elements of the flight path are discussed.

8.2 Parametric performance data analysis

This is a dimensionless method based on the well-established non-dimensional coefficient form of the aerodynamic forces acting on the aircraft. The method is particularly well suited to aircraft with thrust-producing engines.

The principle of dimensional analysis, Buckingham's Π theorem, states that, 'If an equation is dimensionally homogeneous, then the form of the equation will not depend on the fundamental units of measurement. Thus, if the performance characteristic 'X' is a function of variables a, b, c...etc, then all these variables and X have the same dimensions'. It is important that all the significant variables in any particular aspect of the performance are included, so that the method of dimensional analysis is only applicable where the phenomenon is well understood. Since the performance equation of an aircraft is dimensionally homogeneous, the variables can be reformed into groups, each group having the same overall dimensions. If the groups have zero dimensions, they will have the form of a performance characteristic.

A dimensional analysis of the forces acting on a body moving through a fluid leads to the well-known group of non-dimensional expressions based on either the velocity or the Mach number of the flow. (The alternative expressions of the coefficient form result from the relationship between true airspeed and Mach number given in eqn (2.26)). These are,

$$\left.\begin{array}{c} C_X = \dfrac{X}{\frac{1}{2}\rho V^2 S} = \dfrac{X}{\frac{1}{2}\gamma p M^2 S} \\[2mm] M = \dfrac{V}{a} \\[2mm] R_e = \dfrac{\rho V l}{\mu} \end{array}\right\} \qquad (8.1)$$

where C_X is the non-dimensional force coefficient (X being a general aerodynamic force), M is the flight Mach number, and R_e is the Reynolds number of the flow.

The set of equations, (8.1), can be expressed as a functional relationship,

$$C_X = f(M, R_e) \tag{8.2}$$

which shows the performance variable, X, in its coefficient form, as a function of the Mach number and the Reynolds number of the flow. It should be noted that, if the Reynolds number were constant, then the force coefficient would be a function of Mach number only; this condition will be used as the basis of the application of the method.

Since the performance variables involve the state of the atmosphere, both pressure (altitude) and temperature are implicit in the coefficient form of the aerodynamic force. Therefore, the force expressed in dimensionless coefficient form is valid for all values of the performance variables. This process enables the forces measured under one set of pressure and temperature conditions to be converted into coefficient form. They can then be re-converted to an alternative set of pressure and temperature conditions, providing that the Mach number, M, and the Reynolds number, R_e, are similar in each of the sets of conditions. In this way, wind-tunnel data from a scale model can be converted into estimations of the forces acting on the full-scale aircraft flying in the real atmosphere.

The measurement of the performance of the aircraft in flight takes place in an arbitrary state of the atmosphere, and at an arbitrary aircraft weight. These then need to be converted to a preferred standard atmosphere state and to a standard weight. The non-dimensional coefficient method can be used for this purpose. However, since the aircraft is at full-scale both before and after the conversion of the data, the scaling factor, usually the gross wing area, S, can be omitted, along with the other constant terms in the equation, and only the variable terms need be considered. Furthermore, since the arbitrary and preferred standard conditions are unlikely to differ by a large amount, any difference in the Reynolds number is not likely to be significant. It can be treated as a constant and the performance variable can then be considered a function of Mach number only. These considerations enable a simplified form of the non-dimensional expressions to be used in the analysis of the full-scale flight data.

Substituting for density in relative form, $\sigma = \delta/\theta$, in eqn (8.1), the aerodynamic force, X, can be expressed in terms of the force coefficient, the airspeed or Mach number and the atmosphere variables as,

$$\frac{X}{\delta} = \tfrac{1}{2}\rho_0 S \left(\frac{V}{\theta^{\frac{1}{2}}}\right)^2 C_X \quad \text{or} \quad \tfrac{1}{2}\gamma p_0 M^2 C_X \tag{8.3}$$

By omitting the constant terms, eqn (8.3) reduces to the functional relationship,

$$\frac{X}{\delta} = f\left[\frac{V}{\theta^{\frac{1}{2}}}, C_X\right] \quad \text{or} \quad f[M, C_X] \tag{8.4}$$

which now represents the aerodynamic force in *parametric form* in which the atmosphere variables are an integral part of the *parametric group*. In this form, the parametric groups appear to be dimensional and will usually be referred to in terms of the units of the performance variable. However, they are derived from a non-dimensional form in which the scaling constants do not appear and they behave in exactly the same way as the full coefficient form of the variable. The aerodynamic coefficient C_X is an aerodynamic characteristic of the aircraft and may itself be a function of the flight variables. Either of the alternative airspeed terms, Mach number, M, or parametric airspeed, $V/\theta^{\frac{1}{2}}$, can be used as the variable relating to the airspeed, the choice will usually depend on the aircraft and its range of operating airspeed. In the following discussion, Mach number will be used in the derivation of the methodology.

8.2.1 The parametric form of the aircraft aerodynamic forces

From Chapter 3 it was seen that, in the case of the drag force of a conventional, subsonic, aircraft having a parabolic drag polar, the drag in steady, non-manoeuvring flight can be expressed as,

$$D = \tfrac{1}{2}\gamma p M^2 S C_{\mathrm{Dz}} + \frac{KW^2}{\tfrac{1}{2}\gamma p M^2 S} \tag{8.5}$$

which, in parametric form becomes,

$$\frac{D}{\delta} = f\left(\frac{W}{\delta}, M\right) \tag{8.6}$$

since C_{Dz} and K are constants for any given aircraft configuration. Figure 8.2 shows the parametric drag force of the aircraft as a function of Mach number and parametric weight; this is valid for all combinations of the performance variables, weight, altitude and temperature. In this form, the drag is generalized for a particular airframe configuration that determines the drag characteristic of the aircraft. Any other configuration that affects the values of the zero-lift drag coefficient, C_{Dz}, or the lift dependent drag factor, K, will produce its own generalized characteristic.

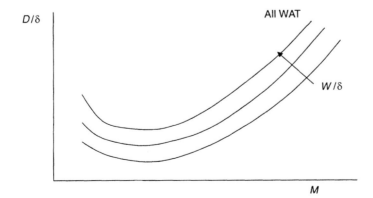

Fig. 8.2 Parametric form of drag force.

8.2.2 The parametric form of the thrust force

In Chapter 3, eqn (3.11), the net thrust of the thrust-producing powerplant, was shown to be given by,

$$F_N = \dot{m}V_j - \dot{m}V \tag{3.11}$$

where $\dot{m} = Q_a$ is the air mass flow entering the engine, and V is the true airspeed, which is related to Mach number, (eqn (2.26)).

The engine net thrust is seen, therefore, to be a function of the air mass flow, Q_a, the fuel mass flow (or energy added), Q_f, and Mach number so that,

$$F_N = f(Q_a, Q_f, M) \tag{8.7}$$

Now the engine is a volumetric device and the volume of air passed by the engine per unit time is a function of its rotational speed, N. The air mass flow, Q_a, which determines the thrust, is the product of the volume of air passed per unit time and the air density, ρ, and is governed by the engine rotational speed. Therefore, eqn (8.7) can be expressed in terms of the air mass flow governing parameter, N, the fuel flow rate, Q_f, the flight Mach number, M, and the atmosphere density that is determined by the atmospheric pressure, δ, and temperature, θ,

$$F_N = f(N, \delta, \theta, Q_f, M) \tag{8.8}$$

By applying dimensional analysis to the engine thermodynamic cycle, eqn (8.8) can be written in parametric form, with the scaling factors omitted. In this form, the thrust force from the powerplant is compatible with the parametric form of the drag force from the airframe,

$$\frac{F_N}{\delta} = f\left(\frac{N}{\theta^{\frac{1}{2}}}, \frac{Q_f}{\delta\theta^{\frac{1}{2}}}, M\right) \tag{8.9}$$

The form of the parametric thrust function is shown in Fig. 8.3.

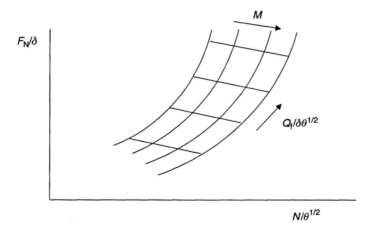

Fig. 8.3 Parametric form of thrust force.

In some cases, for example take-off thrust or climb thrust, the thrust of the engine is governed by limiting an engine parameter to prevent the engine from damage by over-stress. The parametric thrust may then be determined by one controlling parameter only since this now governs the other engine parameters. In these circumstances, a simplified expression for the parametric thrust can be accepted, for example,

$$\frac{F_N}{\delta} = f\left(\frac{N}{\theta^{\frac{1}{2}}}, M\right) \tag{8.9a}$$

Since the thrust and drag are now expressed in similar parametric form, they can be used together in the performance equations, whereas in the full coefficient form the dis-similar non-dimensionalizing factors would not permit their combination so readily.

(It should be noted that this derivation of thrust in parametric form is by no means exhaustive and is intended only to show the principle of the parametric method. A full analysis is necessary to establish all the conditions that would need to be met in the application of the method. These conditions would include, for example, Reynolds number, R_e, and the specific heats of the atmosphere, C_v and C_p. It should also be noted that turbo-fan and turbo-jet engines might have two or more shafts, each rotating independently, and the output of the engine may be a function of more than one parametric rotational speed. The full set of variables of the engine must always be considered in practice.)

8.2.3 Cruise performance measurement

In Chapter 4, the cruise performance was described in terms of the specific air range, SAR, and the specific endurance, SE, eqns (4.4) and (4.5), as

$$\text{SAR} = V/Q_f \qquad \text{and} \qquad \text{SE} = 1/Q_f$$

These are the basic cruise performance expressions, and the cruising range and endurance of the aircraft could be deduced by the integration of SAR and SE respectively over the change in aircraft weight as fuel is consumed during the cruise. The estimation of the aircraft cruise performance required the assumption of simplified forms of the airframe drag characteristic and of the powerplant fuel-flow law; neither of these are known exactly so that the estimated performance is also inexact. In practice, it is necessary to measure the cruise performance in flight to establish the achieved performance. This is required, first, for comparison with the estimated performance to validate the design estimation process and, secondly, to present the measured cruise data in terms of standard WAT conditions for the aircraft performance manual.

Since the cruise performance is measured in the real atmosphere, the data will have to be processed to present it in a preferred standard WAT state. This can be achieved at the data measurement stage by using the parametric form of the performance variables.

In cruising flight the aircraft is in a quasi-steady state so that,

$$F_N - D = 0 \tag{8.10}$$

which can now be expressed in parametric form, using eqns (8.6) and (8.9) as,

$$\frac{F_N}{\delta} - \frac{D}{\delta} = f\left(\frac{W}{\delta}, M, \frac{N}{\theta^{\frac{1}{2}}}, \frac{Q_f}{\delta\theta^{\frac{1}{2}}}\right) \tag{8.11}$$

This represents the performance of the aircraft under all combinations of weight, altitude and temperature. (The combination of the airframe drag and powerplant thrust forces can be made here because of the common parametric form of the performance variables.) The aircraft performance can be measured in such a way that the variables can be formed into the parametric groupings. A chart of performance in the parametric form can then be drawn to represent the generalized performance characteristic of the aircraft under all weight, altitude and temperature (WAT) conditions. By extracting data from the generalized performance characteristic, performance under the preferred WAT conditions can be produced. In this way, data measured under arbitrary WAT conditions can be converted to data representing alternative, specified WAT conditions for comparison with estimated data or for use in the performance manual.

The measurement of specific air range can be used as an example of the procedure. The specific air range, written in terms of the parametric variables and using the alternative expression for the Mach number, eqn (2.26), becomes,

$$\delta SAR = \frac{V}{\theta^{\frac{1}{2}}} \frac{\delta\theta^{\frac{1}{2}}}{Q_f} = f\left(\frac{W}{\delta}, \frac{N}{\theta^{\frac{1}{2}}}, M\right) \tag{8.12}$$

The flight measurement process involves keeping one parametric group constant whilst varying another and, at each test point, measuring the necessary data. In this case, the parametric weight, W/δ, can be held constant by adjusting altitude to compensate for the decrease in aircraft weight as fuel is burned. This is achieved by selecting a convenient value for the parametric weight for the test. The weight of the aircraft is determined from the aircraft zero fuel weight and the fuel quantity remaining in the tanks, and the pressure altitude at which the selected parametric weight will be achieved is then calculated. The aircraft is then flown at that altitude in level flight at a chosen Mach number or airspeed. When conditions are steady, the flight data are recorded, which will include:

- Airspeed, to give the EAS, V_e, or Mach number, M,
- Fuel flow, Q_f,
- Altitude, to give relative pressure, δ,
- Air temperature, to give relative temperature, θ,
- Engine rotational speed, N, and
- Aircraft weight, W, which will be derived from the total fuel burned and the initial aircraft weight.

The altitude can then be increased to allow for the decrease in weight as further fuel is burned, thus maintaining W/δ constant. The test is then repeated at another value of the Mach number, the process continuing through the Mach number range of the aircraft. The whole procedure is repeated for alternative values of parametric weight by using different combinations of aircraft weight and altitude.

From the flight measured data, the parametric groups can be calculated for each datum point at each parametric weight and plotted to give the generalized specific

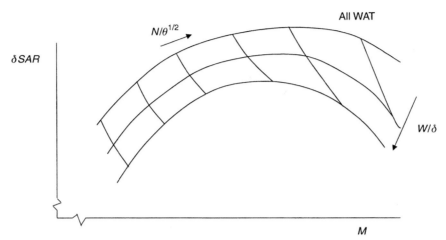

Fig. 8.4 Parametric form of specific air range.

air range performance for the aircraft in the form shown in Fig. 8.4. In the development of the generalized performance chart, it may be necessary to pre-plot the data, first to provide smoothing and, secondly, to enable integral values of the parametric groups to be extracted before the final chart can be produced.

Data at specific WAT combinations can now be extracted from the generalized SAR curves by the reverse process. For a specified weight and altitude the parametric weight can be calculated and interpolated on the general curves. From this curve, the parametric SAR and parametric engine rotational speed are read at intervals of Mach number. These are then scaled by the appropriate values of relative pressure and temperature to give the dimensional values of SAR and engine operating r.p.m. for the specified WAT conditions, Fig. 8.5. In this way, data measured under arbitrary WAT conditions have been processed through the generalized cruise performance chart into data relating to a specific combination of WAT. These can now be compared with estimated SAR performance data to verify the design estimation procedures or used in the construction of the performance manual for the aircraft.

(The specific endurance can be treated in a similar manner and the generalized SE characteristic can be constructed from the data used for the SAR characteristic.)

8.2.4 Climb performance measurement

The climb performance of an aircraft generally needs to be optimized for maximum climb gradient or for maximum climb rate since these criteria are used in the performance manual. In Chapter 5, expressions were developed to predict the optimum airspeeds for maximum rate and gradient of climb. These expressions assumed that the drag polar of the aircraft was parabolic and that the thrust from the powerplant was not a function of airspeed. In practice, these assumptions may not be fully valid and the actual airspeeds for best climb performance may differ from those predicted.

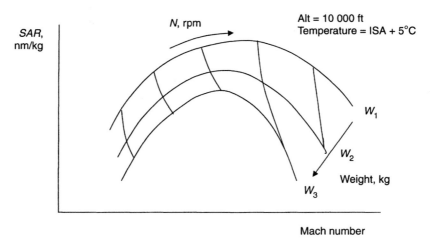

Fig. 8.5 Dimensional form of specific air range.

The first stage of the measurement of climb performance is to establish the actual airspeeds for best climb rate and gradient, and these can be found either by the partial climb technique or by the level acceleration technique.

Partial climbs are a series of short climbs, made at airspeeds above and below the predicted optimum climb speed, through a restricted height band. Each partial climb is flown at a different airspeed and the rate of climb is measured as the aircraft passes through the same datum height. The measured climb rate is corrected for the temperature deviation from standard atmosphere temperature, and the true rate of climb determined. The rate of climb, and the gradient of climb, can then be plotted as a function of airspeed or Mach number and the airspeeds for maximum rate of climb and gradient of climb found. During the partial climbs, fuel is being consumed and the weight of the aircraft will decrease. It will be necessary to correct for the change in weight between the climbs to obtain a consistent set of climb data. The correction can be made by parametric analysis or by one of the alternative methods given in Sections 8.3 and 8.4.

In the level acceleration technique, the aircraft is flown at minimum airspeed and climb thrust applied. The aircraft is allowed to accelerate as height is maintained constant and, during the acceleration, the airspeed is recorded. From the airspeed, the acceleration along the flight path can be calculated. The acceleration is proportional to the excess thrust and so the airspeed for maximum climb rate or gradient can be deduced. Again, account may need to be taken of the weight change due to the fuel consumed during the acceleration.

The airspeeds for best rate and for best gradient of climb, for an aircraft with thrust-producing engines, are determined by the angle of attack for minimum drag speed and will be functions of the aircraft weight. This was discussed in Chapter 5. Once the airspeeds for best rate of climb and best gradient of climb have been established for a particular aircraft weight, the airspeeds for best climb rate or gradient at any other weight can be deduced. This is done by finding the airspeed that gives the same angle of attack at the new weight.

In steady climbing flight, the performance equation is given by,

$$F_N - D = W \sin \gamma_2 = W \frac{dH/dt}{V} \qquad (8.13)$$

If the rate of climb is low enough to enable the acceleration due to the decrease in air density with height to be neglected, this can be expressed in parametric form as,

$$\frac{F_N}{\delta} - \frac{D}{\delta} = f\left(\frac{v_c}{\theta^{\frac{1}{2}}}, \frac{W}{\delta}, M, \frac{N}{\theta^{\frac{1}{2}}}, \frac{Q_f}{\delta\theta^{\frac{1}{2}}}\right) \qquad (8.14)$$

where v_c is the rate of climb, dH/dt, and $v_c/\theta^{\frac{1}{2}}$ is the rate of climb in parametric form.

Since there is an additional parameter in the parametric climb performance equation, the process of collecting the data to construct the generalized climb performance is not as straightforward as in the case of the cruise performance. However, the performance is normally required in terms of the best rate or best gradient of climb, for which the optimum airspeeds will have been established by the partial climb or level acceleration technique. Therefore, the airspeed used in the climb will be determined from the aircraft weight and is no longer an independent variable but a function of weight. This reduces the number of flight variables that have to be considered. Furthermore, the climb will be flown at a specified climb power setting, which will simplify the definition of the powerplant term since the flight Mach number is known and the engine control parameters are constant. These conditions enable the climb performance for best rate or best gradient to be presented in a generalized form if the powerplant output can be related to a single controlling variable.

In the measurement of the climb performance, both for best rate or for best gradient, the weight is determined and the aircraft is flown at the optimum equivalent airspeed, or Mach number. The engines are set to the climb thrust setting and the climb commences at the required airspeed. During the climb, data are collected at regular intervals of time and will include

- Altitude (relative pressure, δ).
- Air temperature (relative temperature, θ).
- Fuel flow (for aircraft weight, W, and flow rate, Q_f).
- Time (for rate of climb, v_c).
- Engine rotational speed, N, or other controlling parameters.

A number of climbs at different weights, and in differing atmosphere states, are needed to provide the necessary range of data to form the parametric performance characteristic. The measured climb rate data are converted into parametric form and used to construct a generalized climb performance chart, Fig. 8.6.

Dimensional data can then be extracted from the generalized characteristic, and are usually presented in a form similar to that shown in Fig. 8.7. The climb performance needs to cover all weight, altitude and temperature combinations and represent the performance of the aircraft flown at the recommended airspeed and engine setting for best climb rate or gradient. Because there are three variables, presentation of the climb data is usually split into two sub-charts to enable the data to be easily interpolated for any given WAT combination. The weight of the aircraft is interpolated on the weight axis and a vertical drawn to the appropriate temperature line. A horizontal is then drawn to the altitude line from which a vertical is dropped

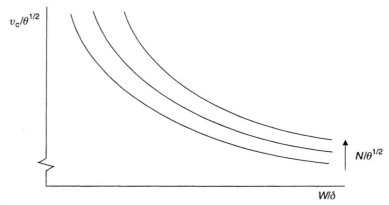

Fig. 8.6 Parametric form of climb performance.

onto the performance axis, and the climb performance (the rate, or gradient, of climb) can then be read from the axis. The chart can be used in reverse to give the maximum weight at which a required gradient of climb can be achieved in a given atmosphere state. This form of data presentation is commonly used in performance manuals and will be discussed further in Chapter 10.

This is a simplified summary of the process of collecting and handling the climb performance data. In practice, it may be necessary to incorporate further steps in the process to account for secondary effects of the atmosphere or flight Mach number on the powerplant characteristic.

8.2.5 Take-off and landing performance measurement

The theory of the take-off and landing manoeuvres, usually referred to as airfield performance, has been discussed in Chapter 6. In the measurement of the take-off

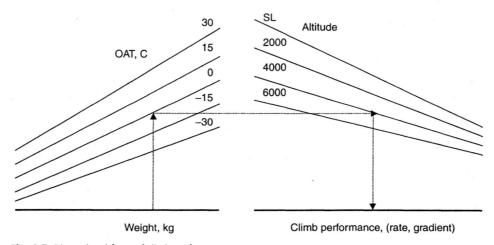

Fig. 8.7 Dimensional form of climb performance.

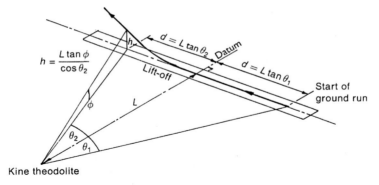

Fig. 8.8 Take-off performance measurement by tracking system.

and landing manoeuvres, the main concern is centred on the measurement of the distances required and the effect on those distances of the performance variables, weight, altitude and temperature. The speeds associated with the manoeuvres are determined by controllability or weight-related factors and these require monitoring during the measurement process to verify that the manoeuvre has been performed correctly.

Many ways of measuring airfield performance have been devised; some of which have involved the use of specialized equipment. Most of the traditional methods have relied on the reconstruction of the path of the aircraft from an optical tracking system, such as the kine-theodolite. This is a device that records the azimuth and altitude angles of the aircraft, relative to a datum in the horizontal plane, during the take-off or landing manoeuvre. From the angular data, and the geometry of the kine-theodolite installation, the ground and airborne distances can be determined – the principle of the system is shown in Fig. 8.8. With a reliable time base, velocities, and even acceleration, can also be determined. In some systems, tracking errors can be corrected and it is possible to achieve very high accuracy, particularly if multiple, synchronized, kine-theodolites are used. Events determining the manoeuvre distances, for example, rotation, lift-off, main-wheel touch-down and nose-wheel touch-down, are identified either by event-marking the data from visual cues or from a film recorded during the manoeuvre in synchronization with the data gathering system.

Alternative methods of flight path reconstruction are being developed and used. Combinations of Inertial Reference Systems, Satellite Navigation Systems and Radio Altimeters are capable of providing very good flight path reconstruction without the need for ground installations, and can even be used when optical observations would be impractical. In addition to the measurement of distances, these methods are capable of producing accurate acceleration, velocity and attitude data from which the manoeuvre events can be determined and the piloting technique monitored.

The data, measured under arbitrary WAT conditions, need to be presented in standard WAT conditions. Parametric analysis can be used for this purpose. From Chapter 6, eqn (6.12), the ground run distance on a level runway in still air is given as,

$$S_G = \frac{W V_{LO}^2}{2g[F_N - D - \mu(W - L)]_{0.7V_{LO}}} \qquad (6.12)$$

so that,

$$S_G = f(F_N, V, W, \sigma) \tag{8.15}$$

This can be expressed in parametric form as,

$$\frac{S_G}{\theta} = f\left(\frac{F_N}{\delta}, \frac{V}{\theta^{\frac{1}{2}}}, \frac{W}{\delta}\right) \tag{8.16}$$

The take-off and landing speeds are determined by the weight of the aircraft, as in the climb performance, so that parametric speed is a direct function of parametric weight. Therefore, it does not need to appear as a separate function in the expression for the ground-run distance. Since the engine will be set to the take-off thrust setting, the simplified thrust expression (eqn (8.9a)) can be used. The parametric ground-run can be reduced to the generalized form,

$$\frac{S_G}{\theta} = f\left(\frac{N}{\theta^{\frac{1}{2}}}, \frac{W}{\delta}\right) \tag{8.17}$$

Similarly, the parametric expression for the airborne distance reduces to the form,

$$\frac{S_A}{\theta} = f\left(\frac{N}{\theta^{\frac{1}{2}}}, \frac{W}{\delta}, \frac{h}{\theta}\right) \tag{8.18}$$

where h is the screen height at the end of the take-off flight path.

The dimensional data are extracted from the generalized take-off data to represent the take-off distances required, which can then be presented in a broadly similar form to the climb performance data, Fig. 8.9. The take-off distances are normally presented for 'level runway, zero wind' conditions and any runway slope and wind effects are accounted for separately. Since the main purpose of the take-off performance chart is to determine the maximum take-off weight of the aircraft, the starting point is a 'corrected take-off distance available'. This is measured data, which have been processed to take into account the runway slope, wind component and other factors that do not relate directly to the performance of the aircraft itself. This distance is

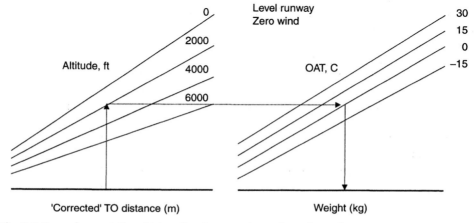

Fig. 8.9 Form of presentation of take-off performance (ground run distance).

then interpolated through altitude and temperature lines to give the maximum take-off weight. The general form of the presentation of the aircraft-related data is shown in Fig. 8.9 and will be considered further in Chapter 10.

The measured landing performance data can be treated in a manner broadly similar to the measured take-off data.

8.3 The equivalent-weight method

An alternative form of data analysis is the 'power-equivalent-weight, speed-equivalent-weight' method, PeW–VeW, which is more applicable to aircraft with power-producing engines driving propellers. If the engine power and propeller efficiency can be determined, so that the thrust-power can be evaluated, then the method can be used to form a generalized performance characteristic. This will be valid for all atmosphere states at a standard aircraft weight. Performance at specific WAT states can then be determined from the generalized performance.

The method is based on comparison of flight at constant lift coefficient with variation in aircraft weight. In the method, the performance equation is developed into a functional relationship that generalizes the equation in terms of weight and air density by using equivalent airspeed. The expression for the lift coefficient in terms of equivalent airspeed for the aircraft at a test weight W_t is,

$$C_L = \frac{W_t}{\frac{1}{2}\rho_0 V_e^2 S} \tag{8.19}$$

This can be written in terms of a preferred standard weight, W_s, as

$$C_L = \frac{W_s}{\frac{1}{2}\rho_0 S}\left[\frac{\{W_t/W_s\}^{\frac{1}{2}}}{V_e}\right]^2 \tag{8.20}$$

where $W_s/(\frac{1}{2}\rho_0 S)$ is constant and the function $V_e/\{W_t/W_s\}^{\frac{1}{2}}$ is known as the speed-equivalent-weight (VeW).

If, for a range of test weights, VeW is maintained constant by adjustment to the equivalent airspeed then the lift coefficient is also constant and compatibility of the performance of the aircraft between the test points is achieved.

Consider the performance equation for steady climbing flight in its power form, eqn (5.10),

$$\eta P = DV + WV \sin\gamma_2 = DV + W\frac{dH}{dt} \tag{5.10}$$

The drag can be expressed in terms of the equivalent airspeed,

$$D = YV_e^2 + \frac{ZW^2}{V_e^2}$$

where $Y = \frac{1}{2}\rho_0 S C_{Dz}$ and $Z = K/\frac{1}{2}\rho_0 S$,

Substituting into eqn (5.10) gives,

$$\eta P\sqrt{\sigma} = YV_e^3 + \frac{ZW_t^2}{V_e} + W_t\sqrt{\sigma}v_c \tag{8.21}$$

Since this expression is written in terms of equivalent airspeed, it is valid for all combinations of altitude and temperature but only for the test weight, W_t. This can be generalized to a standard weight, W_s, by expressing the equation in equivalent-weight terms by substituting VeW for the airspeed, giving

$$PeW = Y(VeW)^3 + ZW_S^2(VeW)^{-1} + W_sCeW \qquad (8.22)$$

where

$$PeW = \frac{\eta P \sqrt{\sigma}}{\{W_t/W_s\}^{\frac{3}{2}}} \quad \text{is the power-equivalent weight,}$$

and

$$CeW = \frac{v_c \sqrt{\sigma}}{\{W_t/W_s\}^{\frac{1}{2}}} \quad \text{is the climb-equivalent weight.}$$

The power-equivalent-weight may need to be expressed as a function of engine rotational speed, also in equivalent-weight terms, if the power and rotational speed can be varied independently. This is the case, for example, of some piston engines with constant speed propellers. In such powerplant systems, the engine rotational speed, N, is controlled by the variable pitch propeller but the inlet manifold boost pressure is controlled by the throttle; the power term will need statements of both variables in equivalent-weight form. The rotational speed, N, can be expressed in equivalent-weight form through the propeller advance ratio, J, given by,

$$J = \frac{V}{ND} \qquad (8.23)$$

where D is the propeller diameter.

By substituting the speed-equivalent-weight into eqn (8.23) the rotational speed-equivalent-weight, NeW, is seen to be,

$$NeW = \frac{N\sqrt{\sigma}}{\{W_t/W_s\}^{\frac{1}{2}}} \qquad (8.24)$$

The performance statements may require both PeW and NeW to be defined.

8.3.1 Cruise performance

In cruising flight, the rate of climb is zero and the thrust power required is equal to the drag power. From eqn (8.21) the power required for level, cruising, flight could be expressed as a function of the EAS at the aircraft test weight, W_t, see Fig. 8.10.

Figure 8.10 is valid for all altitude and temperature states but only for the test weight of the aircraft. If the data are expressed in PeW–VeW form, eqn (8.22), then it is valid for all WAT states based on the preferred standard weight, W_s. Figure 8.11 shows the generalized data in this form.

To construct the generalized PeW–VeW cruise performance characteristic the aircraft is flown at a number of steady cruising speeds using the standard density

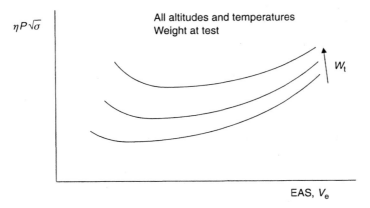

Fig. 8.10 Cruise power required as a function of EAS – at test weight.

altitude as the reference (Chapter 2, Section 2.4.6). During the cruise, airspeed, aircraft test weight, atmosphere data and engine data are recorded. From the data, the equivalent thrust power, $\eta P \sqrt{\sigma}$, is calculated and formed into PeW using the ratio of test weight to standard weight, which will be continuously varying as fuel is consumed. The VeW is calculated in a similar manner and the cruise performance curve plotted from the series of test points. If the engine rotational speed is independently variable it may be necessary to measure the cruise performance over a range of engine rotational speeds and to plot a set of curves for the range of NeW produced.

To construct cruise performance charts for specific WAT states, the process is reversed. Values of the VeW are used to define values of PeW and NeW from the generalized cruise performance data, and these values are then converted into data at the required WAT state. By this means, the flight data measured under arbitrary WAT conditions are first generalized into the PeW–VeW form. This is then used to form a set of cruise performance curves at selected WAT conditions from which data for the performance manual can be drawn.

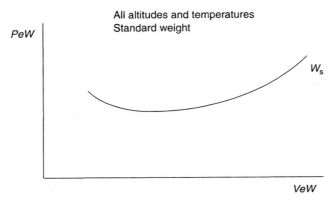

Fig. 8.11 Cruise power required – equivalent-weight form.

8.3.2 Climb performance

The measurement of the climb performance can be approached in a similar way to the cruise performance. However, in this case, there is an additional variable in the performance equation (eqn (8.22)) to be considered, leading to a different approach to the collection of the data during the test flights. The climb data are usually measured in a series of 'partial climbs'. These are short climbs, performed at a range of airspeeds, each climb passing through a datum altitude. During the climb, the airspeed, aircraft weight, atmosphere data, engine data and rate of climb are recorded. Each series of partial climbs is performed at a power setting that will enable suitable combinations of PeW and NeW to be interpolated from the data. The generalized climb performance chart can then be constructed in terms of the CeW and VeW at constant combinations of PeW and NeW (Fig. 8.12).

To construct climb performance charts from the generalized data, values of PeW and NeW are determined for the powerplant at the required climb power and WAT state, and are interpolated on the generalized climb performance chart. Values of VeW are then used to define values of CeW. These are then converted into the required WAT state to form a set of climb performance curves at a selected WAT state and power setting for the performance manual.

8.3.3 Take-off and landing performance

By considering the expressions for the take-off and landing distances defined in Chapter 6 (eqns (6.11) and (6.12) for the take-off, and (6.13) and (6.17) for the landing), the distances can be expressed in equivalent-weight form as,

$$S_GeW = \frac{\sigma S_G}{\{W_t/W_2\}}, \qquad S_AeW = \frac{\sigma S_A}{\{W_t/W_s\}} \quad \text{and} \quad heW = \frac{\sigma h}{\{W_t/W_s\}}$$

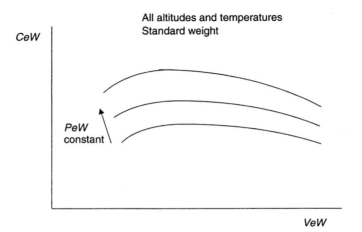

Fig. 8.12 Climb power required – equivalent-weight form.

where S_G, S_A and h are the take-off ground-run distance, airborne distance and screen height respectively; similar expressions can be deduced for the landing distances.

Following the same argument that was used in Section 8.2.5 for the parametric analysis the take-off ground run distance can be expressed in functional terms as,

$$S_G eW = f(\text{PeW}, \text{VeW}) \tag{8.25}$$

and the airborne distance as

$$S_A eW = f(\text{PeW}, \text{VeW}, \text{heW}) \tag{8.26}$$

and these are then used to form a generalized characteristic from which the take-off distances under a specified WAT state can be determined.

The landing distances can be treated in a similar manner.

8.4 Performance data reduction

If the aircraft is powered by piston engines driving propellers, the parametric analysis of its cruise performance is not appropriate and an alternative method of analysis, *performance data reduction*, may be employed. This differential technique depends on knowledge of the drag characteristic of the airframe and the ability to define a reasonable power law for the engine. In this method, the performance equation is differentiated to produce functions that will enable the effect of small changes in weight and air temperature on the performance to be deduced. In the flight tests, the point performance is measured in the arbitrary atmosphere at an aircraft weight close to the required standard weight. Corrections are then calculated for each datum point to account for the differences between the measured performance at the weight and temperature at the test point and the performance that would have been obtained at the preferred standard conditions. The corrections are then added to the measured point performance to give the point performance under the preferred standard conditions; corrections based on this method are generally accepted for weight differences of up to 5% and temperature differences of $\pm 5°C$. In this method, each individual performance point measured in steady, or quasi-steady, flight is corrected separately for the effects of the differences in weight and temperature.

This method of performance data handling was the principal method used in the UK before the turbo-jet engine was in common usage. It is cumbersome to use, and has limited application, but since it provides an acceptable means of correcting measured performance data to standard conditions, it needs to be considered.

The performance equation in its power form can be written as,

$$\eta P \sqrt{\sigma} = DV_e + W\sqrt{\sigma}v_c \tag{8.27}$$

where v_c is the true rate of climb, dH/dt, and ηP is the thrust power from the engine-propeller installation.

Since eqn (8.27) is written in terms of equivalent airspeed, the drag can be expressed in the form,

$$D = YV_e^2 + \frac{ZW^2}{V_e^2} \tag{8.28}$$

where $Y = \frac{1}{2}\rho_0 S C_{Dz}$ and $Z = K/\frac{1}{2}\rho_0 S$ from the parabolic drag characteristic assumed for the aircraft.

Equation (8.27) can now be expressed as,

$$\eta P \sqrt{\sigma} = Y V_e^3 + \frac{Z W^2}{V_e} + W \sqrt{\sigma} v_c \qquad (8.29)$$

Equation (8.29) can now be differentiated to provide corrections to measured performance to account for differences in weight and temperature between the test conditions and the preferred standard conditions. The performance is measured under arbitrary weight and temperature but at a known (and controllable) pressure height; the measured performance will be denoted by subscript t. The performance under the preferred standard WAT conditions will be denoted by subscript, s. Corrections to the measured performance for differences in weight and temperature are made separately.

In developing the corrections, reference will be made to the differences in weight and temperature between the standard and test conditions, these are defined in Section 8.1 as,

$$\Delta W_t = W_s - W_t$$

and

$$\Delta T_t = T_s - T_t$$

Corrections to the performance parameters are defined similarly.

8.4.1 Corrections to cruising speed for weight and temperature

Correction for weight
In steady, cruising flight the rate of climb is zero and the performance equation, in its power form, is given by,

$$\eta P \sqrt{\sigma} = D V_{et} = Y V_{et}^3 + \frac{Z W_t^2}{V_{et}} \qquad (8.30)$$

where W_t is the aircraft weight and V_{et} is the equivalent airspeed measured at the test point.

Since the airframe drag is a function of aircraft weight, any change in weight will affect the drag-power required for cruise; thus, the cruising speed for a given engine power will be a function of weight. The effect of a small variation in the aircraft weight on the cruising speed at a given power setting can be assessed by differentiating eqn (8.30) with respect to weight. In the differentiation, it is assumed that the thrust power is constant. This is reasonable since the correction to the EAS resulting from the weight difference is small enough to have a negligible effect on the engine power or the propeller efficiency. In addition, in cruising flight, the air density can be taken to be constant so that differentiation of eqn (8.24) leads to,

$$\frac{dV_{et}}{V_{et}} = -\left\{ \frac{2}{[3 Y V_{et}^4 / Z W_t^2] - 1} \right\} \frac{dW_t}{W_t} \qquad (8.31)$$

Now the minimum drag speed of the aircraft, V_{emP}, is given by,

$$V_{\text{emP}} = [ZW_t^2/3Y]^{\frac{1}{4}} \qquad (8.32)$$

and, letting $\varpi = [V_e/V_{\text{emP}}]_t$, eqn (8.32) becomes,

$$\frac{\Delta V_{\text{et}}}{V_{\text{et}}} = -\left\{\frac{2}{\varpi^4 - 1}\right\}\frac{\Delta W_t}{W_t} \qquad (8.33)$$

This determines the increment in level speed, ΔV_{et}, produced by a difference in aircraft weight, ΔW_t, between the preferred standard weight and the weight at test.

The correction for weight has to be applied to each measured datum point individually. For example, in a test to measure the specific air range, the airspeed is measured at the aircraft test weight for a series of engine power settings. The measured points will produce a specific air range curve at the test weight, W_t, which is varying throughout the test. At each datum point, the correction function, eqn (8.33), is used to calculate the correction, ΔV_{et}, to the airspeed to account for the error in weight, ΔW_t. The correction is added to the measured airspeed; the measured data point is then shifted to the new speed, $(V_{\text{et}} + \Delta V_{\text{et}})$. Secondly, the airspeed correction is converted to true airspeed, ΔV, and is used to correct the SAR; the separate corrections to the measured performance point are shown in Fig. 8.13. By correcting each measured data point in this manner, a new specific air range curve is produced that would have been measured if the aircraft had been at the standard weight, W_s, throughout the measurement process Fig. 8.13.

Correction for temperature

In this case, the aircraft weight will be considered constant at the test weight, W_t. Differentiating eqn (8.30) with respect to temperature gives,

$$\frac{\mathrm{d}(\eta P\sqrt{\sigma})}{\mathrm{d}T} = \left\{3YV_{\text{et}}^2 - \frac{ZW_t^2}{V_{\text{et}}^2}\right\}\frac{\mathrm{d}V_{\text{et}}}{\mathrm{d}T} \qquad (8.34)$$

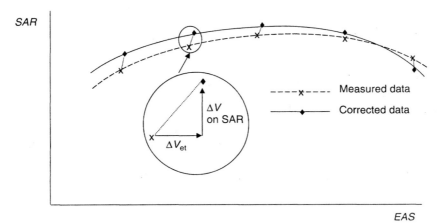

Fig. 8.13 Correction to cruise performance for aircraft weight.

and dividing by eqn (8.30) leads to,

$$\frac{1}{\eta P\sqrt{\sigma}}\frac{\mathrm{d}(\eta P\sqrt{\sigma})}{\mathrm{d}T} = \left\{ \frac{[3YV_{et}^4/ZW_t^2] - 1}{[YV_{et}^4/ZW_t^2] + 1} \right\}\frac{1}{V_{et}}\frac{\mathrm{d}V_{et}}{\mathrm{d}T} \tag{8.35}$$

Using eqn (8.32) in eqn (8.35) the correction to the level speed for temperature can be expressed as the product of two separate functions, one relating to the airframe drag characteristic and one to the engine power law,

$$\frac{\Delta V_{et}}{V_{et}} = \left[\frac{1}{\eta P\sqrt{\sigma}}\frac{\mathrm{d}(\eta P\sqrt{\sigma})}{\mathrm{d}T}\right]\left\{\frac{\frac{1}{3}\varpi^4 + 1}{\varpi^4 - 1}\right\}\Delta T_t \tag{8.36}$$

The function in eqn (8.36) relating to the engine power law needs further development.

Correction for temperature due to the engine power law

The engine power law correction function in eqn (8.36) can be expressed in the form,

$$\frac{1}{\eta P\sqrt{\sigma}}\frac{\mathrm{d}(\eta P\sqrt{\sigma})}{\mathrm{d}T} = \frac{1}{P}\frac{\mathrm{d}P}{\mathrm{d}T} + \frac{1}{\sqrt{\sigma}}\frac{\mathrm{d}\sqrt{\sigma}}{\mathrm{d}T} + \frac{1}{\eta}\frac{\mathrm{d}\eta}{\mathrm{d}T} \tag{8.37}$$

Since the propeller efficiency, η, of a constant speed propeller is unlikely to be affected by the small power changes associated with the effects of temperature, the term in $\mathrm{d}\eta/\mathrm{d}T$ can be taken to be zero and eliminated.

The power developed by the naturally aspirated piston engine at a constant rotational speed is proportional to the air mass flow through the engine and is, therefore, proportional to the density of the air entering the engine cylinders. The pressure of the air entering the cylinders is regulated by passing the intake air through a throttle, which reduces the pressure to the value needed to maintain the power at the required level. The simple throttling process provides a constant pressure ratio so that the power output of the engine will be proportional to ambient pressure. The temperature of the air entering the cylinders will have been modified by the throttling process and heated in the carburettor to ensure that the fuel is vaporized before combustion. If the temperature rise between the ambient atmosphere and the inlet valve is ΔT_c, then the cylinder entry temperature will be $(T_t + \Delta T_c)$. The simple power law can now be expressed in the form,

$$P = P_0\frac{p}{T + \Delta T_c} \tag{8.38}$$

where P is the power developed by the engine, and P_0 is the reference power, usually referred to ISA datum conditions.

Equation (8.38) represents the simplest form of the engine power law. It will be used to illustrate the principle of the correction to performance due to the engine power law. If the engine is supercharged, or the power is flat-rated to a full-throttle height, then a more complex law will be required.

Differentiating the power term in eqn (8.37), and using the atmosphere temperature–height relationship, leads to,

$$\frac{1}{P}\frac{\mathrm{d}P}{\mathrm{d}T} = \frac{1}{p}\frac{\mathrm{d}p}{\mathrm{d}H}\frac{\mathrm{d}H}{\mathrm{d}T} - \frac{1}{(T_t + \Delta T_c)} \tag{8.39}$$

giving,

$$\frac{1}{P}\frac{dP}{dT} = -\left[\frac{g_0}{RT_tL_0} + \frac{1}{T_t + \Delta T_c}\right] \tag{8.40}$$

Differentiating the density term in eqn (8.37) with respect to temperature leads to the result,

$$\frac{1}{\sqrt{\sigma}}\frac{d\sqrt{\sigma}}{dT} = -\frac{1}{2T_t} \tag{8.41}$$

Substituting eqns (8.40) and (8.41) in eqn (8.37), the engine power law correction now becomes,

$$\left[\frac{1}{\eta P\sqrt{\sigma}}\frac{d(\eta P\sqrt{\sigma})}{dT}\right] = -\left[\frac{g_0}{RT_tL_0} + \frac{1}{T_t + \Delta T_c} + \frac{1}{2T_t}\right] \tag{8.42}$$

and each of the terms in eqn (8.36) can be evaluated.

In the measurement of the cruise performance, the level speeds are measured at a number of engine power settings under the arbitrary temperature conditions at the time of test. The test temperature, T_t, will differ from the standard temperature, T_s, and the data will need to be corrected for the error, ΔT_t.

The correction for temperature, eqn (8.36), consists of the product of two functions. One is an expression involving the airframe drag characteristic; this is a function of the parameter ϖ and can be evaluated at the test point. The other is an expression involving the powerplant characteristic; this is a function of the test temperature T_t. The correction to the measured airspeed for the difference in test temperature can be calculated and added to the measured airspeed to give the equivalent airspeed that would have been attained at the preferred standard temperature.

The specific air range and specific endurance functions require the fuel flow, Q_f, to be measured in the cruise and this also needs to be corrected for the effect of temperature (obviously there will be no effect of aircraft weight). A reasonable approximation to a general fuel flow temperature law is that $Q_f \propto 1/\sqrt{T}$ which gives the simple correction,

$$\frac{1}{Q_f}\frac{dQ_f}{dT} = -\frac{1}{2T} \tag{8.43}$$

This is acceptable as a first-order correction to the fuel flow to account for small differences in temperature.

(It should be noted that the correction function to the fuel flow depends on the design of the carburettor, which may be temperature compensated or have a mixture control to maintain the required fuel–air ratio. If the engine is supercharged or flat rated, the throttle setting will be controlled by a boost control unit. These factors need to be considered before eqn (8.43) is used.)

The correction to the cruise performance for temperature is applied in a similar manner to the correction for weight and, as an example, its effect on the specific air range is shown in Fig. 8.14.

In this example, the temperature correction affects both the airspeed and the SAR through the correction to airspeed and there is a further correction for the effect of temperature on the fuel flow.

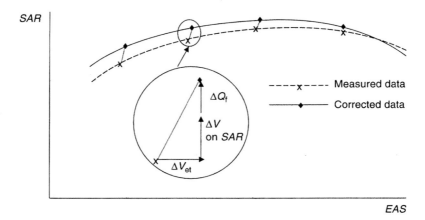

Fig. 8.14 Correction to cruise performance for temperature.

The overall correction to cruise performance

The combination of the two independent sets of corrections for weight, eqn (8.33), and for temperature, eqns (8.36) and (8.43), enable the cruising speed of the aircraft at standard WAT conditions to be deduced from the level speed performance measured under arbitrary WAT conditions. However, it should be noted that, at low airspeeds, approaching the minimum power speed, the value of ϖ tends towards unity and the correction functions fail; the method is not applicable under these conditions.

8.4.2 Correction to rate of climb for weight and temperature

The correction functions for rate of climb can be found by differentiation of eqn (8.29) with respect to weight and temperature respectively. This follows the same general procedure used to develop the correction functions for the effects of differences in weight and temperature on cruising performance. In the flight tests for climb performance, the equivalent airspeed is maintained constant and the effects of differences in weight or temperature affect only the measured rate of climb, v_{ct}.

The correction to the rate of climb for difference in weight is found to be,

$$\Delta(\sqrt{\sigma}v_{ct}) = -\left\{\sqrt{\sigma}v_{ct} + \frac{2ZW_t}{V_e}\right\}\frac{\Delta W_t}{W_t} \qquad (8.44)$$

and the correction to the rate of climb for difference in temperature is,

$$\Delta(\sqrt{\sigma}v_{ct}) = \left[\frac{1}{\eta P\sqrt{\sigma}}\frac{d(\eta P\sqrt{\sigma})}{dT}\right]\left\{\sqrt{\sigma}v_{ct} + V_e(\tfrac{1}{3}YZ)^{\frac{1}{2}}\left(\varpi^2 + \frac{3}{\varpi^2}\right)\right\}\Delta T_t \qquad (8.45)$$

in which the powerplant function is identical to that developed in eqn (8.42).

The speeds for best climb performance are measured by performing a series of partial climbs, each at a different equivalent airspeed, through a known height datum. The measured, pressure, rate of climb is first corrected to true rate of climb, v_{ct}, to account for the temperature effect on the altimeter calibration equation (eqn (2.9)). It is then multiplied by the square root of the air density to give the equivalent rate of climb, $\sqrt{\sigma}v_{ct}$. The corrections for weight and temperature, eqns (8.44) and (8.45), are calculated and added to each measured data point to give the equivalent rate of climb that would have occurred at the preferred, standard WAT state. Using the equivalent airspeed measured at the test point, the gradient of climb at the preferred WAT state can also be found. Climb tests will need to be carried out at altitudes up to the ceiling of the aircraft and at weights and atmosphere states that allow the required WAT envelope to be covered by the correction process.

These corrections can be applied at all airspeeds since the functions do not break down when ϖ is unity; this is important since the best gradient of climb will occur around the minimum power speed.

8.4.3 Correction to take-off and landing performance for weight and temperature

Since the differential method of data reduction is only applicable to 'point-performance' data in steady state, or quasi-steady, flight, it is not appropriate for the correction of the take-off and landing performance. This is because the 'flight path' of the aircraft is not steady during these manoeuvres and the data cannot be regarded as 'point-performance' data.

8.5 Conclusions

This chapter has concentrated on the handling of measured performance data to produce data under standard WAT conditions. The means of measuring the data in flight and the details of the measurement techniques have not been addressed fully. Only a minimal description of the flight test process has been given as a background to the data gathering and data handling process. To describe fully the flight test techniques used to obtain the data would be beyond the scope of this book. Test methods vary with the size and performance of the aircraft being tested, and the choice of test method and its execution is the preserve of flight test engineering.

Whatever method of performance data handling is employed, its purpose is to provide data at standard WAT states from data measured under arbitrary WAT states. Measured data can then be compared with estimated data, to validate the design process, and operating data can be constructed for the aircraft performance manual. The methods outlined in this chapter each have their specific applications in achieving this objective.

The parametric method of performance data analysis is the most commonly used method of performance data handling. It provides readily available data, directly from flight measurements, in the convenient form of a generalized performance chart from which the data at specific WAT states can be extracted. It is used almost exclusively for handling performance data from aircraft with turbo-jet or turbo-fan engines. It can be used for turbo-prop aircraft performance if care is taken to include the additional terms introduced by the propeller. In this case, the propulsive thrust is shared between the thrust-power from the propeller and the direct thrust from the engine exhaust, which adds further terms into the performance equation.

The PeW–VeW method is the most appropriate method to use for aircraft with power-producing engines. It is straightforward in its application and produces a generalized performance chart from which data at the required WAT state can be found. It requires more data manipulation than the parametric method, and the estimation of the engine power, but allows data at any WAT state to be extracted.

The performance data reduction method is very cumbersome to apply since every datum point needs to be corrected individually for both weight and temperature and the correction functions are complex. The corrections that are made are limited in magnitude and so the flight tests that produce the raw data need to cover the full WAT range. The corrections are simply used to 'move' the measured data points to the nearest standard WAT state. Although cumbersome, the method does enable individual data points to be corrected. If the drag characteristic of the aircraft is known to a reasonable accuracy, it can be used to reduce small quantities of data to standard weight or temperature relatively quickly.

Bibliography

Cameron, D. (1953) British performance reduction methods for modern aircraft. *Aeronautical Research Council*, R & M No 2447.

Durbin, E. J. and Perkins, C. D. (eds) (1962) *AGARD Flight Test Manual, Vol 1, Performance* (Pergamon Press).

Reed, A. C. (1941) Airplane performance testing at altitude. *Journal of Aeronautical Sciences*, **8**(4).

Ward, D. T. (1993) *Flight Test Engineering* (Elsevier).

Engineering sciences data unit items
Performance Series, Vols 1–14.

Non-dimensional Approach to Engine Thrust and Airframe Drag for the Analysis of Measured Performance Data: Aircraft with Turbo-jet and Turbo-fan Engines. Vol 13, No 70020, 1970.

Graphical Method for the Analysis of Measured Performance Data using Drag Determination: Aircraft with Turbo-jet and Turbo-fan Engines. Vol 13, No 70021, 1970.

Non-dimensional Graphical Method for the Analysis of Measurements of Steady Level Speed, Range and Endurance: Aircraft with Turbo-jet and Turbo-fan Engines. Vol 13, No 70022, 1973.

The Measurement and Analysis of Climb and Excess Power Performance. Vol 13, No 70023, 1982 (Am A).

Performance Analysis for Aircraft with Turbojets. 'Non-dimensional' Graphical Method. Rate of Climb. Vol 13, RJ1/2, 1971 (Am B).

Introductory Sheet on the Analytical Method of Performance Reduction for Aircraft with Turbo-jets. Vol 14, RJ2/0, 1970 (Am A).

Analytical Method of Performance Reduction for Aircraft with Turbojets. Rate of Climb. Vol 14, RJ2/1, 1948.

Reduction of Take-off and Landing Measurements to Standard Conditions. Vol 6, RG2/1, 1987 (Am A).

9

Scheduled performance

In this chapter, extensive reference is made to the airworthiness requirements, and relevant extracts from JAR 25, and other regulatory material, are paraphrased for use as the basis for examples within the text to illustrate the application of the requirements. Airworthiness requirements are amended from time to time and this chapter will not necessarily reflect the current requirements; if current information is required, the latest issue of the requirements must be consulted.

9.1 Introduction

An aircraft must be able to fly from its point of departure to its destination safely. This implies that it must have sufficient power, and carry enough fuel, for it to perform all the necessary manoeuvres required to achieve this objective. The objective is met by *performance planning* and *fuel planning*. Although this chapter is based on the requirements for civil transport aircraft, military aircraft are normally operated to similar standards and, generally, the same criteria are applied. However, reduced safety margins, known as Military Operating Standards, may be authorized in combat and combat support operations where situations of urgency arise. In such cases, it is accepted that the aircraft may not always have sufficient performance to ensure its safe operation at all points on its flight path but the additional risk can be justified operationally.

Performance planning is part of the Flight Plan and is concerned only with the safety of the aircraft. In the flight plan, the first essential is to ensure that there is sufficient space available at all times for the aircraft to perform the manoeuvres necessary for that part of the flight. The space required must never exceed the space available. This principle must be maintained in all events, whether the flight takes place without incident or whether an incident with a foreseeable result occurs. An example of such an incident is the failure of a power unit resulting in the loss of propulsive thrust. All safety critical systems on an aircraft are designed with functional redundancy so that, in the event of a failure of a system, there is an alternative available. However, in the case of an engine failure there is no such redundancy, the loss of thrust cannot be replaced and it must be shown that the aircraft can manoeuvre in the space available with one engine inoperative.

The space required by the aircraft to perform a particular manoeuvre increases with the weight of the aircraft. Performance planning involves the matching of the space required to the space available, and aircraft weight is, ultimately, the controllable variable. Therefore, the end product of performance planning is the maximum permissible weight at which the aircraft can take off and meet all the safety critical performance requirements throughout the flight.

Fuel planning ensures that the aircraft has sufficient fuel for the intended mission, together with allowances for a diversion, the effects of other contingencies and reasonable reserves. Since the fuel plan depends on the weight of the aircraft, and the maximum take-off weight is determined by the performance plan, it is evident that the performance planning and fuel planning are interdependent.

A *Certificate of Airworthiness (CA)* for an aircraft is only granted by the licensing authority when, in addition to other requirements relating to design and construction, it can be shown that the aircraft can satisfy the conditions laid down by the *Air Navigation (General) Regulations (AN(G)R)*, relating to the demonstration of certain flight safety criteria. Some of the most demanding of the criteria relate to the performance of the aircraft with one power unit inoperative. Because aircraft differ in size, number of power units and purpose, the levels of performance that can be achieved with a power unit inoperative differ. Allowances can be made for this in limitations on the operation of the aircraft, and the level of flight safety accepted, by giving aircraft *performance classifications*. The classifications are defined by the type of powerplant, the number of passenger seats and the *Maximum Take-off Weight Authorized (MTWA)*, of the aircraft and correspond to the airworthiness code under which they are designed. The performance grouping system was originally introduced in 1951 and was defined in the British Civil Aircraft Requirements (BCAR), which applied at that time. Since then, the groupings have been amended to take account of the development of aircraft in terms of their size and performance. The current performance classifications are fully defined in JAR-OPS 1, Commercial Air Transportation (Aeroplanes), and can be briefly summarized as follows.

Performance Class A: This applies to multi-engined aeroplanes powered by turbo-propeller engines with more than nine passenger seats and a maximum take-off weight exceeding 5700 kg, and all multi-engined turbojet powered aeroplanes.

These aircraft must have sufficient performance to enable them to,

1. account for reasonably expected adverse operating conditions such as take-off and landing on contaminated runways; and
2. take consideration of engine failure in all flight phases.

This is the most stringent of the performance classes and all large, turbine powered, transport aircraft, i.e. those covered by FAR/JAR 25, must be certificated to Performance Class A.

(This definition replaces a previous statement of Performance Group A derived from BCAR, Section D and FAR/JAR Part 25. This was defined as; 'An aircraft (having a MTWA greater than 5700 kg) with performance such that at whatever stage in the flight, including take-off, a power unit fails a forced landing should not be necessary'.)

Performance Class B: This applies to propeller driven aeroplanes with nine or less passenger seats and a maximum take-off weight of 5700 kg or less.

(This class applies to the smaller types of aircraft, covered by FAR/JAR Part 23. These may have less engine inoperative performance than their FAR/JAR 25 counterparts, the desired overall safety being achieved by restrictions on stalling speed, conditions of operation and by airworthiness design features. For example, single-engined aircraft, or two-engined aircraft, which do not meet the climb requirements, shall not be operated at night or in Instrument Meteorological Conditions (except under special visual flight rules).)

Performance Class C: This applies to aeroplanes powered by reciprocating engines with more than nine passenger seats or a take-off weight exceeding 5700 kg.

(The requirements for Class C are broadly similar to those in Performance Class A, but take into account the different characteristics of the pure power-producing engines.)

This treatment of scheduled performance will be concerned mainly with Performance Class A. It is important to note that the airworthiness regulations are subject to occasional revision in the light of experience and improved knowledge. However, it is not always possible to modify the aircraft or its performance to comply with such amendments. Therefore, aircraft are certificated to the airworthiness regulations in force at the time and retain that certification thereafter; the issue of the regulations to which it is certificated will be noted in the performance manual of the aircraft.

9.2 Flight planning

For the purpose of flight planning the flight is divided into four phases, in each of which the necessary level of flight safety must be demonstrated. These are as follows.

(i) *Take-off*. The take-off phase starts from the commencement of the take-off run, through the acceleration, rotation, lift-off and initial climb. It extends to the point at which the lowest part of the aircraft clears a *screen height* above the extended runway surface. The screen height is 35 ft in the case of a dry runway or 15 ft on a wet/contaminated runway, for aircraft in Performance Class A.

(ii) *Take-off flight path*. The take-off flight path extends from the take-off screen height to a height of 1500 ft above the airfield level.

(iii) *En route*. The en-route phase extends from the end of the take-off flight path (1500 ft), at the departure airfield to a height of 1500 ft above the landing airfield.

(iv) *Landing*. The landing phase extends from a height of 1500 ft above the landing airfield to the point where the aircraft is brought to a complete stop on the landing runway.

In each of the four phases of the flight the performance of the aircraft is measured to provide the data needed to demonstrate that the aircraft is capable of safe operation with respect to the prevailing conditions. The performance is measured in flight tests performed by an aircraft, or group of aircraft, and the data produced is referred to as *measured performance*. Due to the statistical variation between aircraft, it will usually be necessary to adjust the measured performance to *gross performance*. This is an average performance for the fleet of aircraft such that the performance of any aircraft of the type is at least as likely to exceed the gross performance as not. To allow for various operating contingencies, the need to manoeuvre, variation from the expected weight, altitude (pressure) and temperature or variations in piloting technique, the gross performance is diminished by a margin or factor to give the *net performance*. This is the performance used to show compliance with the airworthiness requirements by comparison with the performance published in the performance manual.

The flight-path related performance published in the performance manual is based on the powerplant manufacturer's declaration of the *guaranteed minimum thrust* or *power* produced by the engine. This is the minimum acceptable output of any engine of that type. It is determined by reducing the average thrust or power output of that type of engine by a statistical margin so that the probability of an engine having an output less than the guaranteed minimum is remote. Therefore, the performance published in the performance manual represents the lowest performance that can be expected from an aircraft. Thus, a further margin of safety is implicit in the published data in that the actual aircraft should always have better performance than that predicted by the performance manual. However, certain minimum control speeds, which may govern the minimum operating speeds of the aircraft, are based on maximum available thrust since this represents the most limiting case. The minimum control speeds are concerned with the recovery of the aircraft from the effect of a sudden failure of an engine.

The maximum permissible take-off weight for any flight is determined by considering the structural strength limitations of the aircraft, the airfields of departure and destination and the performance of the aircraft in each flight phase. The scheduled maximum take-off weight will be the lowest of the weights obtained after considering the following criteria.

(i) The maximum take-off weight, MTOW, and the maximum landing weight, MLW, specified in the Certificate of Airworthiness. These are structural strength limitations, based on the design of the airframe, which must not be exceeded.

(ii) The airfield lengths available at departure and destination airfields, together with the slopes and surface conditions of the runway. These are limitations on the space available for the take-off and landing manoeuvres.

(iii) The WAT limits for take-off and landing. This is a limitation based on the climbing capability of the aircraft.

(iv) The take-off net flight path. This is a performance limitation based on minimum safe performance criteria.

(v) The en-route terrain clearance with one (or more) power units inoperative. This limitation is related to the performance ceiling of the aircraft.

There may be further limiting factors relating to particular aircraft or airfields that would restrict the weight at which the take-off could be scheduled but are not directly a part of the performance of the aircraft. These include brake energy limitations, tyre speeds and pressures, crosswind components and runway pavement strengths.

9.3 Take-off performance

In Chapter 6, the take-off flight path was discussed. It was seen that the take-off consisted of two parts; a ground run in which the aircraft accelerated to the lift-off speed and an airborne phase in which the aircraft achieved its safety speed and cleared a screen height above the take-off surface to complete the take-off. Methods for the calculation of the take-off distances were postulated and would be used to estimate the airfield performance at the design stage of the aircraft. The expressions also showed the effect of the flight variables on the take-off distances. The actual take-off distances are measured by take-off trials performed by the aircraft; these establish the gross take-off distances as functions of weight, altitude and temperature, and take into account the wind component and runway conditions. The gross distances represent the average space necessary for the aircraft to complete the take-off manoeuvre under the specified conditions; in performance planning, the gross take-off performance will lead to a statement of the space required for the take-off manoeuvre. Since performance planning considers the *space available* and the *space required* for the take-off manoeuvre these terms need to be defined.

9.3.1 The space available

The space available for take-off is limited by the dimensions of the runway and the area beyond the runway in the take-off direction, see Fig. 9.1. The *runway* is defined as a rectangular area of ground suitably prepared for an aircraft to take-off or land. At the end of the runway, there may be *stopway*. This is an extension of the runway on which the aircraft can be stopped in the event of an abandoned take-off but which cannot be used for take-off. The *clearway* is an area of ground or water at the end of the runway, free from obstructions, over which an aircraft may make a portion of its initial climb to a specified (screen) height of 35 ft. The purpose of the screen height is to ensure that the aircraft clears the obstacle limitation surface, defined in Section 9.4, by the required margin.

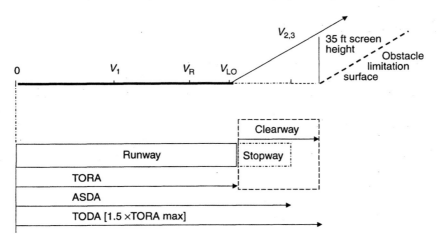

Fig. 9.1 Airfield distances available.

The *take-off run available, TORA,* is the length of runway declared available and suitable for the ground-run of an aircraft taking off. This is the distance available for the aircraft to accelerate from rest, achieve its lift-off speed and just become airborne; usually, this will be the physical length of the runway pavement. However, the TORA may be restricted by the take-off climb obstacle limitation surface if the obstacle clearance cannot be met by the required gradient from the end of the runway pavement. (This will be referred to in the *net flight path.*) In some cases, the runway length available may be limited by obstacle clearance limits in one direction but not limited in the opposite direction. This would lead to the TORA being different in reciprocal directions on the same runway; the restricted length being available for use as stopway.

The *take-off distance available, TODA,* is the length of take-off run available (TORA) plus the length of the clearway, if provided (but TODA must not exceed 1.5 times TORA). This is the distance available for the aircraft to complete its take-off run, lift off and achieve the screen height at a speed not less than the appropriate safety speed.

The *accelerate–stop distance available, ASDA* (also known as the *emergency distance available, EMDA*), is the length of the take-off run available (TORA) plus the length of the stopway, if provided. This is the distance available for the aircraft to accelerate on its take-off run, suffer the failure of a power unit, and be decelerated to a complete stop in an abandoned take-off.

In the simplest situation in which there is no stopway or clearway, TORA = TODA = ASDA.

These definitions, shown diagrammatically in Fig. 9.1, define the boundaries of the space available for take-off. All the distances are true lengths determined by the geometry of the airfield. There are no factors or margins applied to the space available. Lateral limitations to the minimum width of the runway and the clearway also exist but, as these do not have a direct effect on the definition of the take-off distances, they will not be considered further here. A full definition of the airfield dimensions and distances can be found in ICAO Annex 14.

9.3.2 The space required

In considering the space required for the take-off it is assumed that, although the aircraft has all power units operating normally at the start of the take-off run, one power unit will fail before the take-off is complete. In the event that a power unit does fail at some point during the take-off, the pilot must decide whether to abandon the take-off and bring the aircraft to a stop or to continue the take-off run and become airborne. In either case, there must be sufficient space available to carry out the manoeuvre. The first consideration must be to determine the balance between the distance required to abandon the take-off and the distance required to continue the take-off to ensure that the decision is always made objectively.

The distances required depend on the acceleration of the aircraft, assuming 'all power units operating' net performance, up to the point of engine failure and the deceleration, or continued acceleration, assuming 'one power unit inoperative' performance, thereafter. There will be a point, determined by the speed at which the engine fails, at which the distances required to abandon or to continue the take-off are equal, thus determining the minimum take-off distance required. The speed associated with that point is known as the *decision speed* for the take-off and is determined by the balance between the *maximum refusal speed* in the abandoned take-off and the *minimum continue speed* in the continued take-off. They are often referred to as ratios of the *rotation speed*, V_R, since this is a definable, and unique, function of the take-off weight of the aircraft. The definitions of these speeds are given below.

The rotation speed, V_R

Towards the end of the take-off run the aircraft is rotated from the nose-down attitude with the nose wheel on the runway to the nose-up, take-off, attitude. The rotation is made at a speed that will enable the aircraft to continue its acceleration and achieve the lift-off speed, V_{LOF}, in the shortest overall distance commensurate with the achievement of a safe climb. The lift coefficient at the lift-off must be sufficiently below the stalling lift coefficient to avoid an inadvertent stall. To achieve this, the maximum (stalling) lift coefficient is factored to give a lift coefficient at lift-off that will provide the necessary speed margin. To enable the lift-off to be performed smoothly and predictably, the rotation to the lift-off angle of attack will be made at a speed margin below the lift-off speed. The aircraft is allowed to accelerate to the lift-off speed, in the nose-up attitude, at which point it 'unsticks' and becomes airborne. The rotation will, therefore, be initiated at a lift coefficient, C_{LR}, related by the speed margin to that at lift-off. This implies that the equivalent airspeed at rotation will be a unique function of the weight of the aircraft given by,

$$V_{eR} = \left[\frac{2W}{\rho_0 S C_{LR}} \right]^{\frac{1}{2}} \tag{9.1}$$

thus $V_{eR} \propto W^{\frac{1}{2}}$ and increases as weight increases, see Fig. 9.2.

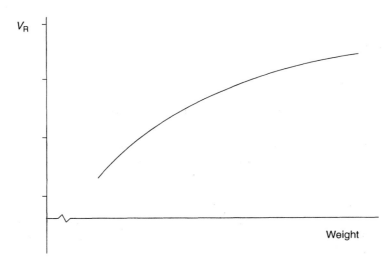

Fig. 9.2 Variation of rotation speed, V_R, with weight.

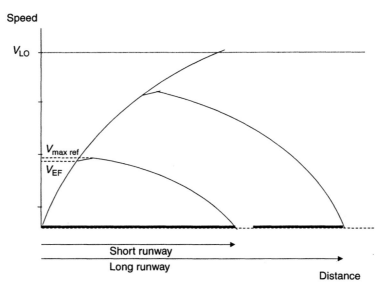

Fig. 9.3 Engine failure speed, V_{EF}, and refusal speed, $V_{max ref}$.

The maximum refusal speed, V$_{max ref}$

If it is decided to abandon the take-off then the aircraft must be brought to a halt within the runway length available, the distance required is known as the *accelerate–stop distance*. The accelerate–stop distance is the distance required to accelerate with all engines operating, to experience the failure of an engine at the engine failure speed, V_{EF}, to recognize the engine failure and identify the failed engine, to respond to the failure and to decelerate the aircraft to a complete halt. This is shown in Fig. 9.3. Since the aircraft will continue to accelerate under the excess thrust available following the engine failure, the speed will increase during the engine failure recognition and response time. The maximum speed that occurs before deceleration is initiated is known as the *maximum refusal speed, $V_{max ref}$*. If the engine failure occurs at a low speed, early in the take-off run, then the accelerate–stop distance will be short. As the speed at which the engine failure occurs increases so does the accelerate–stop distance. An engine failure speed from which the aircraft can just be brought to a halt in the stopping distance available defines the *maximum refusal speed, $V_{max ref}$*, for that distance available.

Since the accelerating force acting on the aircraft is independent of its weight, the all-engines operating acceleration will be inversely proportional to the aircraft mass. Therefore, the distance required to accelerate to the engine failure recognition speed will increase as the aircraft weight increases. In the deceleration from the engine failure recognition speed, the kinetic energy that has to be lost is proportional to the aircraft mass and therefore the deceleration distance will be increased. Thus, the accelerate–stop distance will increase as the aircraft weight increases and the refusal speed, which can be scheduled for a given take-off run available, will decrease as the aircraft weight increases, see Fig. 9.4.

Both the rotation speed and the refusal speed are functions of aircraft weight, making it convenient to express them as a ratio, $V_{max ref}/V_R$; the ratio is shown in Fig. 9.5 for a

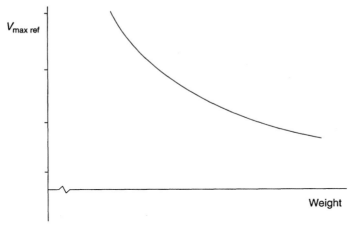

Fig. 9.4 Refusal speed, V_{maxref}, as a function of weight for a given ASDA.

given take-off run available. Since the take-off cannot be safely abandoned after rotation the refusal speed cannot exceed the rotation speed and V_{maxref}/V_R cannot exceed unity.

The minimum continue speed, $V_{min con}$

If, following the failure of an engine, it is decided to continue the take-off, then the aircraft must continue to accelerate to its lift-off speed on the thrust of the operating engines and complete the take-off in the distances available. If the engine failure speed is close to the lift-off speed then the increase in distance required to reach the lift-off speed under the reduced acceleration will be relatively small. As the speed at which the engine failure occurs decreases the *continued take-off distance* required to achieve the lift-off becomes greater, see Fig. 9.6. The lowest engine failure speed from which a take-off can be continued in the take-off run available defines the *minimum continue speed* $V_{min con}$, based on the take-off run available, TORA. The continued take-off distance to the screen height must also be considered and there will be a minimum

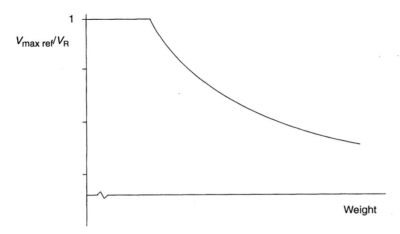

Fig. 9.5 V_{maxref}/V_R ratio for a given ASDA.

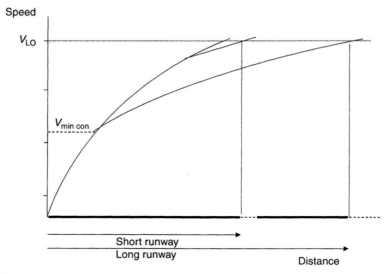

Fig. 9.6 Minimum continue speed, $V_{\text{min con}}$.

continue speed based on the take-off distance available, TODA; either of the minimum continue speeds may provide the limiting condition.

Since the one-engine inoperative accelerating force acting on the aircraft is constant, the acceleration of the aircraft will be inversely proportional to the aircraft mass and the continued take-off distance will increase as a function of the aircraft mass. Thus, the speed at which the continued take-off can be scheduled from a given take-off run available, or take-off distance available, will increase as the aircraft take-off weight increases, see Fig. 9.7.

The ratio $V_{\text{min con}}/V_{\text{R}}$, is shown in Fig. 9.8 for a given take-off run available. The ratio increases as the aircraft weight increases until either the all-engines operating take-off run required (TORR) is equal to the TORA, or take-off distance required (TODR) is equal to the TODA. This determines the maximum weight at which the

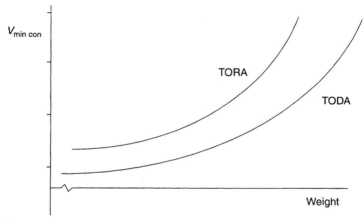

Fig. 9.7 Minimum continue speed, $V_{\text{min con}}$, as a function of weight for given TORA or TODA.

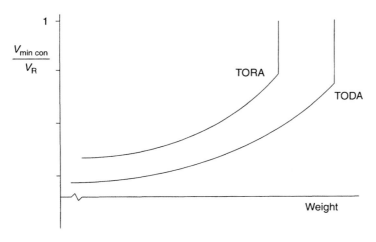

Fig. 9.8 $V_{min con}/V_R$ ratio for given TORA or TODA.

aircraft could be scheduled to take-off with all-engines operating on that length of runway, or within the take-off distance available, respectively.

The decision speed, V₁

For a safe take-off, the maximum refusal speed must not be less than the minimum continue speed. Otherwise, there will be a period during the take-off run during which, should an engine fail, there is insufficient distance available for the take-off either to be abandoned or to be continued. This can be seen by combining Figs 9.5 and 9.8 into Fig. 9.9. The point at which the curves meet determines the *maximum take-off weight, MTOW*, at which the aircraft could take-off in the distance available and comply with both requirements. The speed ratio at which the requirements are met produces the *decision speed, V_1*, which is the point on the take-off run beyond which the aircraft is committed to take-off even in the event of the failure of an engine. Up to the decision speed, the take-off must be abandoned. There are further

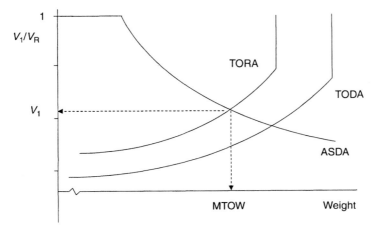

Fig. 9.9 V_1/V_R ratio for given ASDA and TORA or TODA.

practical limitations on the decision speed. It cannot be scheduled at a speed less than the minimum control speed, on or near the ground, V_{mcg}, since directional control of the aircraft cannot be maintained in the event of an engine failure. Neither can it be scheduled at a speed greater than the rotation speed, V_R, since braking cannot be applied safely once the nose wheel is lifted off the ground.

The decision speed is, in fact, used to determine the 'decision distance' from the start of the take-off run at which the decision to go or stop is made. A speed is used rather than a distance for convenience, since distance along the runway cannot be readily measured and displayed to the pilot whereas speed can. In using the decision speed, it is assumed that the aircraft is operating normally up to the point of engine failure and that the distance it has travelled when it reaches the decision speed is the 'decision distance'.

Although in Fig. 9.9 the TORA is shown to be the limiting distance that determines V_1, there will be cases in which the TODA may be the limiting distance. This will be the case, for example, if there is no clearway and TORA = TODA.

If the weight of the aircraft at take-off is less than the MTOW limited by the take-off distances available then there will be a range of V_1 / V_R available which can be used when the runway conditions are unfavourable. This leads to two further definitions of the maximum refusal speed and minimum continue speed associated with operation from wet runways. V_{Stop} is the highest decision speed from which the aircraft can stop within the accelerate–stop distance available and V_{Go} is the lowest decision speed from which a continued take-off is possible within the take-off distance available. V_1 can be scheduled at any speed between V_{Go} and V_{Stop}.

The balanced field length and critical field length

The decision speed defines, by implication, the minimum length of airfield from which a safe take-off can be made at a given take-off weight. Whilst the two terms are similar, they are based on slightly different criteria. The *balanced field length* refers to the balance between the distance required to continue the take-off to clear the screen height (TODR) and the distance required to bring the aircraft to a halt. This is usually the criterion assumed in the civil airworthiness codes. The *critical field length* refers to the balance between the length of runway required (TORR) either to continue the take-off or to bring the aircraft to a halt, this criterion is more often associated with military codes of practice.

Unbalanced field length

If the distance available for the continued take-off and the distance available to bring the aircraft to a halt are not equal, as in the case of a runway with stopway, then the field is referred to as an *unbalanced field*. In this case, the decision speed may be determined by separate consideration of the continued take-off and the abandoned take-off. However, for simplicity, a V_1 / V_R, based on the assumption of the shorter distance as a balanced field length, is often used and the small weight penalty involved accepted.

9.3.3 Take-off performance safety factors

The take-off run required (TORR) and take-off distance required (TODR) both depend on the acceleration of the aircraft, which is a function of the flight variables

(WAT) and on the runway conditions of slope, surface condition and wind component. The effect of these parameters on the estimated take-off performance was discussed in Chapter 6.

In practice, some of the variables will be subject to statistical variation that will affect the TORR and TODR.

- The aircraft take-off weight may not be known exactly since, for example, passengers are not usually weighed individually but taken at a 'standard weight' together with their hand baggage. The range of weights of individual human beings varies considerably so that the actual total weight of a given number of passengers making up a payload is subject to statistical variation about a mean.
- The forecast pressure and temperature for the take-off available at the time of flight planning may not be the same as the actual pressure and temperature at the time of the take-off. The air density, therefore, may not be as forecast and the assumed performance will not be achieved if the pressure or temperature variation leads to a lower than forecast air density.
- The forecast wind component may not occur at the time of take-off, and the wind strength and direction may vary during the take-off run.

In addition to variation in the airfield parameters, the aircraft airspeed indicator may have an error (within its permitted tolerance), and pilot techniques at the rotation may vary. Hence, the achieved rotation and lift-off speeds over a number of take-offs will be subject to a statistical variation about a mean.

To allow for these variations in the parameters affecting the take-off performance, factors and margins are applied to the measured take-off distances. These reduce the statistical probability of the aircraft exceeding the TORA or TODA to an acceptably low level; generally set at 1 in 10^6 events. As an example of such a factor, statistical analysis has shown that the combined effect of all the errors in the variables affecting the take-off distance required produces a standard deviation in the gross take-off performance of 3%. To achieve the necessary statistical probability of a take-off within the relevant distances available, a margin of 5 standard deviations is required and the gross performance is factored by 1.15, or 115%, in this case, to give the net performance.

In the simplest case, in which there is no clearway, only the TODR need be considered, since the aircraft must achieve the screen height before it reaches the end of the runway available, TORA. The requirements are paraphrased here from JAR/FAR 25.113 and associated references.

- The TODR is taken to be the greater of:

  (a) 115% of the horizontal distance along the take-off path, *with all engines operating*, from the start of the take-off to the point at which the aircraft is 35 ft above the take-off surface, and
  (b) the horizontal distance along the take-off path from the start of the take-off to the point at which the aircraft is 35 ft above the take-off surface *during which the critical engine has failed at the engine failure speed* V_{EF}.

If the take-off distance includes a clearway then the TORR also needs to be considered and can be paraphrased similarly as follows.

- The TORR is taken to be the greater of:

    (c) 115% of the horizontal distance along the take-off path, *with all engines operating*, from the start of the take-off to the point equidistant between the point at which V_{LOF} is reached and the point at which the aircraft is 35 ft above the take-off surface, and

    (d) the horizontal distance along the take-off path from the start of the take-off to the point equidistant between the point at which V_{LOF} is reached and the point at which the aircraft is 35 ft above the take-off surface *during which the critical engine has failed at the engine failure speed V_{EF}*.

It should be noted that in these statements the comparison is made between the 'net' performance in the case of the all-engines-operating take-off and 'gross' performance with one engine inoperative.

In the event of a wet runway the take-off distance required must allow for the effect of the reduced coefficient of friction on the accelerate–stop distance since this will increase the braking distance required. A runway is considered wet when there is sufficient moisture on the surface to cause it to appear reflective but there are no significant areas of standing water. In the wet condition, the surface friction is reduced and so the braking distances increase, adversely affecting the accelerate–stop distance and, consequently, reducing the decision speed. This leads to the need to consider the engine failure case on either a wet or a dry runway. On a wet runway the engine failure speed, V_{EF}, from which the aircraft can be brought to a halt in the distance remaining, is reduced. This sets a lower value of the decision speed V_1 than in the case of the dry runway. To determine the minimum take-off distance required by the aircraft, comparison is made between the take-off distances with and without an engine failure, and on a wet or dry runway as appropriate, taking into account the safety factors and margins. By comparing the distances required, the most limiting case determines the maximum weight at which the aircraft can be despatched from an airfield with given TORA and TODA. Alternatively, it determines the minimum TORA and TODA required by an aircraft at a given despatch weight.

(If the runway is contaminated beyond the state of being wet, for example by standing water, slush, snow or ice, there is not only a reduction in runway friction but also displacement drag, impingement drag and increased skin friction drag. (Displacement drag and impingement drag are additional drag forces due to displacement of the precipitation by the wheels and its impingement on the airframe.) This needs to be taken into account in the determination of a decision speed, and of a take-off weight, which will provide an acceptable level of engine failure accountability. The factors for the reduction of runway friction depend on the type of contamination, its depth and the aquaplaning speed; the displacement drag is a function of the width of the tyre, the depth of the contamination and its density. The contaminated runway is an extreme case of the 'wet' runway and will not be considered further in this study since the decision to take-off often requires consideration of factors – and operators' policies – that cannot be generalized.)

At this point, the conditions for a safe take-off have been complied with and the aircraft, having completed the take-off, is now commencing the climb phase of its flight from the take-off screen height.

9.4 Take-off net flight path

After take-off, the aircraft climbs away from the ground towards its intended cruising altitude. During the initial part of the climb, from the take-off to 1500 ft, the landing gear will be retracted, the flap setting changed from take-off to climb setting and thrust reduced from take-off thrust to climb thrust. This is known as the *after-take-off climb*. Unless there are obstructions that require a further extension of the after-take-off flight path above 1500 ft the aircraft is then considered to be 'en route'. In the after-take-off climb, with all engines operating, the aircraft should have sufficient excess thrust or power to achieve a gradient of climb steep enough to enable it to clear all known obstructions along its flight path. However, in a take-off during which an engine has failed just after the decision speed has been reached, the after-take-off climb gradient will be considerably reduced and, with it, the ability to clear obstructions. The problem of obstacle clearance now becomes an issue.

To guarantee acceptable handling qualities and the best climb performance available, the airspeed in the after-take-off climb must be not less than the appropriate take-off safety speed, V_2 or V_3. The take-off safety speed, all engines operating (aeo), V_3, would be the intended airspeed for the climb to 1500 ft following the take-off. This would be based on the airspeed for best climb *rate* commensurate with a margin over the stalling speed and not less than the minimum control speed. In the event of a take-off during which an engine has failed, the airspeed for the initial phase of the climb would be V_2, the take-off safety speed one engine inoperative (oei). V_2, which must not be greater than V_3, would be based on the airspeed for best *gradient* of climb, commensurate with a margin over the stalling speed and not less than the minimum control speed. For convenience, V_3 is normally expressed as V_2 plus a speed margin, e.g. $V_2 + 10$ kts, rather than as an independent speed. The critical take-off speeds, V_1, V_R, V_2 and V_3, are noted before take-off and set as 'bugs' on the airspeed indicator scale. They may also be recorded on a vital actions card, which is usually clipped to the control column for quick reference by the handling pilot.

During the after-take-off climb the Air Navigation (General) Regulations, AN(G)R, require that the aircraft must remain clear of any obstructions in its flight path by a vertical distance of at least 35 ft, in all conditions of visibility, up to a height of 1500 ft above the take-off airfield. This requirement can be met in one of two ways. First, by establishing an obstacle free 'safe zone' in which the take-off climb can be performed. Secondly, by providing climb performance data for the aircraft from which to calculate a take-off climb path which will clear any known obstacles.

The 'safe zone' is defined by specifying an obstacle limitation surface, which extends from the end of the take-off distance available, the space above which is maintained free of any obstruction. The provision of such a zone is part of the airfield licensing procedure. The definition of the obstacle limitation surface depends on the classification of the airfield and on the maximum size of the aircraft that are permitted to take off from it. Basically, for the larger airfields, the surface consists of a slope, of gradient 2%, extending from airfield ground level at the end of the TODA for a horizontal distance of 15 000 m in the direction of the runway. There are lateral dimensions that allow the aircraft some limited deviation from the extended runway centreline and the ability to make turning manoeuvres during the after-take-off climb. These do not have a direct

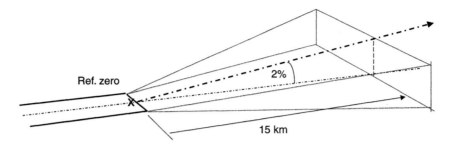

Fig. 9.10 Take-off obstacle limitation surface.

effect on the flight path gradient as far as this simplified treatment of the subject is concerned. The general form of the take-off obstacle limitation surface is shown in Fig. 9.10 for large airfields, and full definitions for all airfield classifications can be found in ICAO Annex 14. Since the origin of the obstacle limitation surface is the *reference-zero* at the end of the TODA, at which the aircraft must have achieved the screen height of 35 ft, the vertical separation between the aircraft and any obstacle required by the AN(G)R is guaranteed. This is the purpose of including the screen height in the take-off manoeuvre. Provided the aircraft remains above the obstacle limitation surface, by a margin of 35 ft, its flight path can be considered to be safe. This general form of protection can be provided by showing that the aircraft can achieve a climb gradient throughout the after-take-off climb that exceeds that of the obstacle limitation surface. This protection is limited to the area directly above the obstacle limitation surface.

Although the requirements for the licensing of airfields may specify obstacle limitation surfaces, it is not always possible to comply fully with them. Because of the local terrain or buildings and structures near the airfield, there are likely to be some obstructions that penetrate the boundaries of the safe zones. Charts will be published of each licensed airfield to show any obstacles that need to be taken into consideration during flight planning. In the flight planning procedure, a take-off climb flight path must be calculated that takes into account all known obstructions in the intended path of aircraft and must demonstrate that the aircraft can avoid those obstructions by the appropriate margins. This calls for a specific analysis of the take-off climb path that is intended to show that all known obstructions have been considered and taken into account. The take-off climb path consists of several climb segments, in each of which the aircraft will be in a different configuration with respect to its landing gear, flap setting and thrust or power setting. In the construction of the climb path, information will be needed on the performance of the aircraft in each segment of the climb and at the appropriate WAT state. This information is given in the generalized performance charts contained in the aircraft performance manual.

9.4.1 Climb performance WAT charts

As in the case of the take-off, the climb performance of the aircraft will be affected by a variation in the performance variables, WAT. It will also be affected by the sequential changes in configuration, power and speed that occur during the after-take-off climb. Statements of the climb performance, in the form of charts, will need to be produced that will enable the flight path to be calculated for all WAT states that are likely to occur. The climb performance WAT charts are constructed in terms of the measured, or gross, performance of the aircraft. They show, for example, the gradient of climb for given configurations of landing gear and flap settings for all combinations of weight, altitude and temperature (WAT). Thus, each segment of the after-take-off climb can be represented by a WAT chart that can be used as a building block for the construction of the take-off net flight path.

Take-off net climb performance

The gross performance is the average fleet performance and does not take into account any errors in indications of airspeed or assumed WAT states or any differences in airframe drag and powerplant output between individual aircraft and engines. In addition, in constructing the flight path from the WAT charts, account needs to be taken of the transition phases between the climb segments. The transition will tend to add time increments to the actual climb path and hence increase the distances required to complete the segment; these time increments may be affected by variation in individual piloting technique. To allow for the effect on the climb performance of any such statistical variations, the actual, or *gross, climb performance* is reduced by a margin to give the *net climb performance*. This provides the statistical performance margin necessary to ensure that the actual performance should not fall below the minimum performance needed to comply with the airworthiness requirements relating to obstacle clearance. The performance margins used to diminish the climb gradients are quoted in JAR 25.115 as,

0.8% for two-engined aircraft,

0.9% for three-engined aircraft, and

1.0% for four-engined aircraft.

The difference in the margins reflects the increasing statistical probability of one or more engines being below average performance. The net climb performance is used to construct the *take-off net flight path* for the demonstration of the obstacle clearance criteria.

The take-off net flight path

The take-off net flight path, which is measured from the *reference zero*, consists of, typically, four segments (but possibly more). These are based on the actions necessary if, during the take-off, failure of the critical engine has occurred after the decision speed has been reached. The *critical engine* is the engine that, in the event of its failure, produces the most severe effect on the handling or performance of the aircraft. In the case of an aircraft with power-producing engines, there is an asymmetry of flow into the propeller when flying at higher angles of attack. This will usually cause the failure of one of the engines to produce a greater yawing moment on the aircraft than the failure of its opposite counterpart. (This is particularly true if the propellers are not counter-rotating.) This has two effects. First, it produces a higher minimum control speed. This is the lowest speed at which the rudder can overcome the yawing

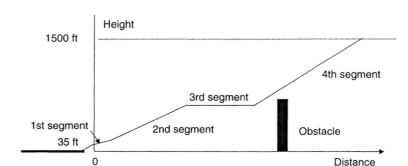

Fig. 9.11 Take-off net flight path.

moment at the point of engine failure; this is principally a handling consideration although it may set the climbing speed of the aircraft. Secondly, the rudder angle required to hold the aircraft on a steady heading is increased. Thus, the rudder trim drag is higher, reducing the excess thrust available for the climb. The engine that produces the higher minimum control speeds, and therefore the most limiting performance and handling criteria, is known as the critical engine. Aircraft with thrust-producing engines do not normally have critical engines other than the most outboard ones.

In the construction of the net take-off flight path it is assumed that the critical engine fails as the aircraft passes the decision speed. It continues to accelerate and lifts off to clear the screen height of 35 ft, at a speed not less than V_2, to complete the take-off manoeuvre. The climb segments that make up the after-take-off climb, one engine inoperative (oei), are shown in Fig. 9.11. The actions described in each segment are dependent on individual aircraft characteristics and operating procedures. The take-off net flight path will also be referred to in Chapter 10.

The reference zero
The reference zero is taken to be the point 35 ft vertically below the aircraft at the end of its required take-off distance. At this point, the aircraft must be flying at a speed not less than the take-off safety speed, V_2, in the take-off configuration with the landing gear extended.

The first segment climb
In the first segment, the landing gear is retracted to reduce the drag and, in the case of propeller-driven aircraft, the propeller of the failed engine is feathered unless this action has occurred automatically. The first segment extends from the screen height, 35 ft, to the point at which the landing gear is fully retracted and, where applicable, the propeller of the failed engine is feathered. In this segment the flaps remain at the take-off setting and the aircraft must be able to maintain a positive gradient of climb at an airspeed as close as possible to V_2. The distance flown in the first segment is given by the product of the airspeed and the time taken to retract the landing gear. The height gained is found from the gradient and the distance flown. The aircraft then climbs with landing gear retracted and take-off flaps set in the *second segment climb*.

The second segment climb

The second segment climb extends from the end of the first segment climb to the height selected for flap retraction. It is made in the take-off configuration at an airspeed as close as possible to the take-off safety speed V_2 with landing gear retracted and one engine inoperative. This is one of the most demanding climb conditions and will often determine the maximum take-off weight at which the aircraft can be dispatched. Normally, with all engines operating, the installed thrust of an aircraft will be sufficient to produce a gradient of climb that will enable it to clear all obstructions comfortably. However, should an engine fail, the excess of thrust over drag will be reduced, thus reducing the climb gradient. Since the engines are required to overcome the drag as well as to provide power for the climb the effect on the gradient of climb of a single engine failure is very significant and dependent on the number of engines. This can be illustrated by a simple example.

Consider the second segment climb gradients of an aircraft powered by four, three or two engines in the one-engine inoperative (oei) state. From eqn (5.5),

$$\sin \gamma_2 = \frac{F_N}{W} - \frac{D}{W} \tag{5.5}$$

Typically $D/W = 0.1$ for an aircraft in the take-off configuration with landing gear retracted. Assuming that $F_N/W = 0.24$ (a typical value for a transport aircraft), then, with all engines operating,

$$\sin \gamma_2 = 0.24 - 0.1 = 0.14; \quad \gamma_2 = 8.05°$$

$$\text{or gradient} = 14.1\% \text{ (gradient } \% = 100 \tan \gamma)$$

This represents the gross gradient of climb that would be shown in the WAT chart; this would need to be reduced to the net gradient before being used in the take-off net flight path calculation.

With one engine inoperative, the thrust-to-weight ratio will be reduced in proportion to the number of engines producing the thrust. Therefore, neglecting any additional drag that may occur due to the inoperative engine or asymmetry of flight, the gross gradients that would be achieved by an aircraft with two, three or four engines are given in Table 9.1. These are compared with the minimum second segment climb gradients required by the airworthiness requirements. This simplified example shows that the aircraft powered by two engines, and having an all engines operating thrust–weight ratio of 0.24, could not achieve the required minimum climb gradient with one engine inoperative. The aircraft with either three or four engines, and the same thrust–weight ratio, would comfortably exceed the minimum gradient requirement. Thus, to comply with the second segment climb gradient WAT limit requirements, two-engined aircraft of a given weight would need to have a greater installed take-off thrust-to-weight ratio than three- or four-engined aircraft of the same weight. Alternatively, if the installed thrust was the same in each case, the aircraft with two engines could only comply with the minimum gradient at a lower weight than the aircraft with three or four engines.

The second segment climb is one of the most critical areas of aircraft design. In fact, in the setting of the requirements for the second segment, the climb gradient was deliberately chosen to be an envelope case that would cover several other phases of

Table 9.1 Second-segment gradients of climb, oei, WAT, limits

All engines operating	One engine inoperative		
	4 engines	3 engines	2 engines
F_N/W 0.24	0.18	0.16	0.12
Gross gradient 14.1%	8%	6%	2%
Min gradient required, OEI [JAR 25.121(b)]	3.0%	2.7%	2.4%

flight with one engine inoperative. In this way, compliance with the second segment requirements ensures compliance with other gradient requirements without having to address them separately.

In the design estimation of the performance of the aircraft, it is usually the minimum gradient required in the second segment climb that determines the installed take-off thrust requirement. Operationally, the second segment climb will be one of the considerations that could limit the maximum weight at which the take-off is scheduled. If it is a limiting case, then the aircraft could not be despatched at that weight since there would be insufficient thrust to provide the climb gradient required to demonstrate obstacle clearance. By decreasing the weight of the aircraft, the thrust-to-weight ratio would be increased but there would be no effect on the maximum drag-to-weight ratio since this is a function of the drag characteristic. A weight could be found at which the aircraft would be able to meet the climb gradient criterion; this would define the maximum weight at which a take-off could be scheduled. The weight reduction is achieved primarily by reduction of the payload. The only other disposable load is the fuel load required for the trip, which can then be adjusted for the effect of the reduced payload. Thus, the final weight reduction is achieved by a combination of reduction in payload and in fuel. In addition, it should be remembered that the state of the atmosphere will affect the engine thrust. Therefore, if the example shown in Table 9.1 assumes a standard atmosphere, sea-level state, the gross gradients of climb shown will be reduced by temperatures above ISA or pressure altitudes above sea level. A further weight decrease would then be needed for the aircraft to meet the climb gradient limits.

The take-off flap setting used for take-off would have been selected to minimize the take-off distance required whilst enabling an acceptable second segment climb gradient to be achieved. This flap setting, however, may not produce the best climb gradient and the flaps may need to be retracted to a setting that will improve the climb gradient. Before this can be done, the aircraft must be accelerated to the flaps-up safety speed in the *third segment*.

The third segment

At a safe height, but not less than 400 ft above the ground, the excess thrust can be converted into acceleration in level flight. The airspeed is allowed to increase from the take-off safety speed, V_2, used in the second segment, to the flaps-up safety speed at which the flaps can be retracted safely. Since the excess thrust is small, the acceleration will be slow and the distance travelled during the third segment may be considerable. The benefits of flap retraction need to be balanced against the effect of the third segment on the overall, after-take-off, net climb gradient, and the height at which the third segment is initiated may need to take account of the distance flown during the acceleration.

The engines cannot produce take-off thrust or power indefinitely and there will be a time limit after which the thrust will need to be reduced to maximum continuous thrust or power. No reduction in thrust or power is permitted until a minimum height of 400 ft is reached and so the thrust or power reduction is usually made at the end of the third segment. With climb thrust or power set, the aircraft then continues its climb in the en-route configuration in the *fourth*, or *final*, *segment*.

The fourth segment

The *fourth segment*, or *final take-off, climb* extends from the end of the third segment to the point at which the aircraft reaches a height of 1500 ft above the level of the runway. The climb is made at the airspeed for best gradient of climb, but not less than the safety speed appropriate to the flap setting being used.

The overall climb gradient in the after-take-off climb is calculated from the distance flown and the height gained in each segment of the climb. Since the obstacle clearance data are normally quoted in relation to the brake-release point, the net flight path is constructed from brakes-off and includes the take-off distance required. The height gain and distance flown in each segment can then be added to form a plot of the net flight path. This can then be compared with the obstacle clearance heights to confirm that at no point in the climb does the aircraft infringe the clearance margins. The climb strategy, particularly with respect to the third segment, will need to consider the overall climb gradient; this will be discussed further in Chapter 10.

9.5 En-route performance

The en-route phase of the flight extends from the end of the after-take-off climb, 1500 ft above the runway level, to the beginning of the approach to landing, 1500 ft above the level of the landing runway. It includes the en-route climb, the cruise and the en-route descent phases of the flight. The Air Navigation Regulations state that, in the event of any one power unit becoming inoperative at any point on its en-route phase the aircraft must be capable of continuing its flight to its destination, or to a planned diversion, clearing by a vertical interval of at least 2000 ft obstacles within 10 nm either side of the intended track (or 5 nm if the navigation aids available can guarantee a track error margin of less than 5 nm from the intended track). It further requires that the net flight path gradient shall not be less than zero at 1500 ft above the landing airfield with any one engine inoperative. In the case of an aircraft with three or more engines, these requirements must be met with any two engines inoperative.

As an aircraft climbs, the engine output, either as thrust or as power, decreases due to the decreasing air density. A height will be reached at which the maximum available thrust is equal to the drag; this is the absolute ceiling of the aircraft. Usually the cruising height is such that the aircraft has a reserve of thrust and, with all engines operating, it can maintain height and has excess thrust available for manoeuvre. If an engine should fail, the thrust–weight ratio is reduced. Even with the operating engines set to maximum continuous thrust or power, and flying at the optimum airspeed, there may be insufficient thrust to enable the aircraft to maintain height and, if so, it will be committed to a descent. As the aircraft descends the engine output will increase and eventually the aircraft will reach, asymptotically, its new cruise ceiling, one engine inoperative. The descent is known as the *drift-down*.

The rate of drift-down will depend on the weight of the aircraft and on the state of the atmosphere. Flight tests will be performed to measure the rate of descent with one (or more) engines inoperative to determine the drift-down rate as a function of weight, altitude and temperature. This will form part of the en-route climb performance for the aircraft. Once this is known, the distance flown during the drift-down between two heights can be calculated and made available in the performance manual. Figure 9.12 shows the drift-down as a function of the aircraft weight for a given atmosphere state.

From the consideration of the en-route performance the drift down is assumed to start from either the cruising altitude or the maximum engine re-start altitude, whichever is the lowest. Knowing the height of any obstacle on track ahead of the aircraft, and allowing for the necessary clearance of 2000 ft, the drift-down distance can be assessed. This is then compared with the position of the aircraft relative to the obstacle, if clearance can be maintained the flight could continue. If not the aircraft will need to divert to avoid the obstacle or, possibly, to an alternate airfield. From a known cruising altitude, the obstacle clearance range from a known obstacle can be determined from the drift-down distance. For example, on a route over a mountainous ridge, a point can be predetermined from which, if an engine failure occurs, the obstruction can be cleared or, if the engine failure occurs before that point, a diversion will be necessary. Since the obstacle is ground based any headwind or tailwind will have to be accounted for in the assessment of the drift-down distance.

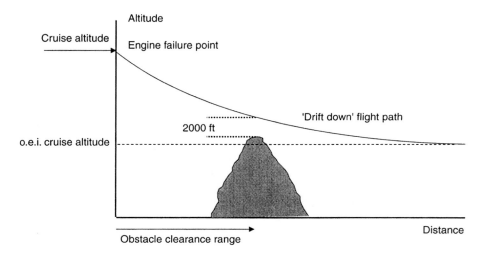

Fig. 9.12 Drift down and obstacle clearance range.

9.6 Landing performance

The requirements for landing commence when the aircraft is 1500 ft above the landing airfield and cover the approach and landing to bring the aircraft to a complete stop on the runway. The landing flight path was discussed in Chapter 6 and was seen to be in two parts, an airborne phase and a ground phase. Methods of calculation of the landing distances were determined for the estimation of the airfield performance at the design stage of the aircraft showing the effects of the flight variables.

As in the case of take-off, landing performance data are required for the performance manual. However, due to the nature of the landing manoeuvre, the provision of that data from measurement of the landing distances is not straightforward. In the measurement of the take-off distances, there was a clearly defined datum point at brake release from which the take-off flight path was measured. The landing manoeuvre is subject to very much greater statistical error than the take-off since it is not possible to position the aircraft with absolute accuracy at the screen height above the runway threshold on the approach. Neither is it possible to guarantee that the aircraft is flying at exactly the correct approach speed, V_{REF}, as it passes the screen height. Consequently, the datum point for the start of the manoeuvre is not well fixed, as it is for the take-off, and this leads to a statistical error in the touchdown point on the runway. The distance required to bring the aircraft to a halt following the touchdown depends on the level of braking and other methods of retardation applied. If a minimum distance landing were attempted, the severe braking required and the extreme use of retardation devices would not be acceptable for normal operations. It would compromise passenger comfort and safety and cause unacceptable wear and tear on the aircraft. Defining a 'normal' landing manoeuvre thus becomes a subjective problem. Due to the statistical errors implicit in the landing manoeuvre, large safety factors need to be applied to the landing distances required to ensure that the landing will not exceed the space available. For these reasons, the landing distances required for the aircraft flight manual are mainly determined by calculation rather than from measured data, but usually there will be some measured data to provide verification. In the landing case, calculation of performance data for the flight manual is acceptable because of the large factors applied to the calculated distances required. A parametric analysis will enable the effect of the flight variables to be assessed.

The *maximum landing weight* of the aircraft is the lowest of the weights necessary to comply with the limitations imposed by,

(i) the maximum design structural landing weight,
(ii) the landing distance available, and
(iii) the WAT limit set by the climb performance following a discontinued approach or a baulked landing.

The maximum design structural landing weight is an absolute limit that must not be exceeded. It is determined by the structural strength of the airframe needed to withstand the loads imposed by the vertical velocity of the aircraft at touchdown.

9.6.1 The space available

As in the case of take-off, the space available for landing is limited by the dimensions of the airfield and the approach to the runway in the landing direction, see Fig. 9.13.

The approach path is protected by an obstacle limitation surface in a similar manner to the take-off net flight path. The definition of the obstacle limitation surface depends on the classification of the runway. Figure 9.13 shows the limitations for a large airfield with a precision approach category; full definitions for all runway classifications can be found in ICAO Annex 14. As in the case of the take-off obstacle free zone, it may not be possible to comply fully with the requirements, and the flight planning will need to consider any known obstructions near the airfield.

The first section of the obstacle limitation surface starts from a point 60 m from the threshold of the runway and extends for a distance of 3000 m along the approach path at a gradient of 2%. The second section extends a further 3600 m at a gradient of 2.5%. The horizontal section extends a further 8400 m with a base height of 150 m above the runway surface. The total length of the approach obstacle limitation surface is 15 000 m. There are, of course, lateral dimensions associated with the obstacle limitation surface to enable the aircraft to make turning and positioning manoeuvres on the approach. As these do not affect the performance considerations addressed here, they need not be considered in detail.

In the event of a missed approach or a baulked landing, in which the aircraft abandons the landing and climbs away from the airfield, there needs to be an obstacle free area in the direction of the climb-out. The baulked-landing obstacle clearance surface extends from 1800 m beyond the threshold of the landing runway, or the end of the runway whichever is the lesser, at a gradient of 3.33%, to a distance of 4000 m.

The *landing distance available* (*LDA*) is defined as the length of runway that is declared available and suitable for the ground run of an aircraft landing.

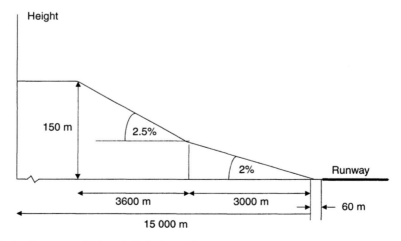

Fig. 9.13 Landing approach obstacle limitation surface.

9.6.2 The space required

The *landing distance required (LDR)* is the gross distance required to land on a level, smooth, hard-surfaced runway from a specified screen height at the runway threshold and to come to a complete stop, multiplied by a suitable safety factor.

The landing flight path assumes that the aircraft approaches the runway in a steady descent to a screen height of 50 ft at the runway threshold. The gradient of descent in the approach is usually 5% (or 3°), but may be steeper in some restricted airfield operations. The airspeed at the screen height should be not less than the greater of V_{MCL} or $1.3V_S$ in the landing configuration. Between the screen height and touchdown, the airspeed is reduced to a safe touchdown speed in the flare. This is done by progressively increasing the angle of attack, so that the touchdown is achieved at an acceptable vertical speed avoiding excessive vertical acceleration or any tendency to bounce. In the flare, no changes to the configuration, addition of thrust or depression of the nose should be required. After touchdown, brakes and other means of retardation, for which a satisfactory means of operation has been approved, can be used to decelerate the aircraft to a halt. If any retarding system depends on the operation of any engine, for example a thrust reverser, then it may be necessary to assess the landing distances with that engine inoperative to determine the most critical landing distance required.

To account for wet runway conditions, the dry runway distances are factored by a function of the coefficient of friction of the runway under wet and dry conditions.

The maximum landing weight based on the landing distance required is determined by the balance between the landing distance available and the greatest of the landing distances required. This takes into account the factors for the runway surface, slope and condition, the wind and an overall safety factor to account for the statistical variation in the parameters affecting the landing performance.

9.6.3 The discontinued landing

There are two cases to be considered in which the landing has to be abandoned. These are,

(i) the *discontinued approach* in which the aircraft terminates the approach before the thrust is reduced in the flare and continues the flight to a point from which a new approach can be made, and

(ii) the *baulked landing* in which the aircraft is required to go-around after the thrust has been reduced in the flare.

The discontinued approach occurs, for example, when the aircraft has reached the minimum decision height on the approach to landing and the runway is not in sight. At that point the approach must be discontinued and a climb initiated; it is assumed for the purposes of the requirements that the aircraft is flying with one engine inoperative. In a go-around from a discontinued approach, the aircraft must be capable of climbing away, with the critical engine inoperative and in the approach configuration, to a safe height from which it can make another approach or a diversion. The steady gradient of climb must not be less than 2.1% for two-engined aircraft, 2.4% for three-engined aircraft and 2.7% for four-engined aircraft; the different gradients reflect the effect of statistical variability in the thrust of the operating engines. (For Category II operations the gradient is 2.5% for all types.) The climb gradient must be demonstrated with the critical engine inoperative and the operating engines at take-off thrust, at maximum landing weight and with the landing gear retracted. The configuration for the discontinued approach must be such that the stalling speed is not greater than $1.1V_S$ for the all-engines operating landing configuration.

In the case of the baulked landing the aircraft must be capable of climbing at a gradient that will maintain clearance from all obstructions with all engines operating in the landing configuration and with the landing gear down. When the go-around is initiated, take-off thrust is selected and the aircraft is rotated into the climb. With all engines operating, the aircraft must be capable of achieving a gradient of climb of not less than 3.2% in the landing configuration. The engine thrust or power used to calculate the gradient of climb is that which is available 8 seconds after-take-off power is selected from the flight idle condition. The airspeed used for the climb is $1.2V_S$, but must not be less than V_{MCL} or more than the greater of V_{MCL} and $1.3V_S$.

The gradients of climb in the discontinued landing will determine the maximum weight at which the aircraft can safely terminate the approach, taking into account the WAT conditions. In most cases the limit of compliance for the discontinued approach and the baulked landing can be combined into a single WAT chart for the aircraft, this will be discussed further in Chapter 10. The maximum landing weight for the aircraft will be the lowest of the weights determined by consideration of the climb gradient in the discontinued landing, the space available and the maximum design landing weight.

9.7 Summary of performance planning

In performance planning, the critical areas of each phase of the flight have been examined, and the maximum aircraft weight to comply with the performance-related safety criteria in each phase determined. The most critical case is used to define the maximum scheduled take-off weight that enables the aircraft to be dispatched in the knowledge that it has the performance needed to maintain a safe flight path throughout its mission.

The next consideration is the fuel requirement for the mission.

9.8 Fuel planning

The flight plan for a transport aircraft operation states the point of departure of the mission, the intended route to the destination and an alternate destination in case of a diversion. The aircraft needs to be dispatched with sufficient fuel for the intended mission and the diversion, together with sufficient reserves of fuel for en-route contingencies and delays. A typical mission profile on which the fuel planning is based is shown in Fig. 9.14.

Military aircraft operations can be considered in a similar manner to civil transport operations but show some characteristic differences. Military flights often return to

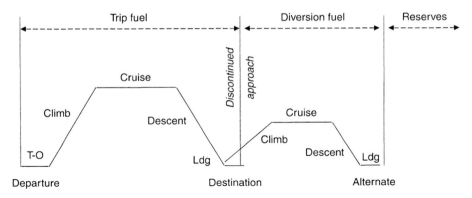

Fig. 9.14 Mission profile and fuel planning.

their point of departure without the opportunity to re-fuel and may not need to have a diversion, other than their base. However, they may be able to re-fuel in flight to extend their range or to permit a take-off from a short airfield at a reduced weight. In the absence of airborne re-fuelling, the general principles of fuel planning for civil transport operations can usually be extended to most military operations.

The weight of the aircraft is made up of three basic elements, the *aircraft prepared for service (APS) weight,* also known as the *operating empty weight (OEW),* the *payload* and the *fuel weight.* The APS weight consists of the weight of the aircraft, the crew and any non-disposable equipment it is required to carry for the purposes of the flight; this is a fixed weight that cannot be reduced. The remaining weight is made up of the *disposable load,* which is the sum of the payload and the fuel; the payload being the weight carried as the 'revenue-earning' part of the disposable load.

The total fuel load for the mission can be broken down into three parts,

(i) the *trip fuel* for the intended flight from the point of departure to the destination,
(ii) the *diversion fuel* for the diversion to the alternate airfield, and
(iii) the *reserves.*

Fuel planning strategies and regulations may differ between National Regulatory Authorities, aircraft operators and routes. The fuel planning can only be generalized here to illustrate the principles involved.

The trip fuel

The flight from the point of departure to the intended destination is made up of the basic elements of the flight path, take-off, climb, cruise, descent and landing. In addition, the aircraft will need to taxi from the ramp to the runway for take-off and from the runway to the ramp after landing.

The fuel required for each element of the intended flight is derived from flight-measured data. This is reduced to a form from which the effects of weight, altitude and temperature can be interpolated separately. This is discussed in more detail in Chapter 10. The data are presented in graphical or tabular form so that the fuel required for each element of the flight can be found for the particular WAT state and operational parameters. For example, the cruising segment will need a statement of the distance between the end of the climb and the beginning of the descent and the climbing segment will need the height increase between take-off and cruise altitude. In some cases, it may be sufficient to state an allowance of fuel for a manoeuvre or fuel burned per unit time. For example, in the case of smaller aircraft it is not unusual to quote a fixed quantity of fuel for the take-off and landing and for the taxiing fuel in units per minute. This is acceptable if the effect of variation of WAT on the fuel required for these manoeuvres is not significant.

Since the fuel consumed during the trip depends on the weight of the aircraft, a 'starting weight' is needed before the trip fuel can be calculated. The initial weight of the aircraft will depend on the quantity of fuel that will be burned during the trip. Hence, the only weight that can be determined is the weight at the end of the diversion at which time only the reserve fuel remains. Therefore, the calculation process is reversed, so that the 'starting weight' for the fuel calculation is the weight at the end of the flight, rather than that at the beginning. The starting weight for the calculation of the trip fuel is, therefore, the aircraft prepared for service (APS) weight of the aircraft plus its payload and the fuel remaining at the end of the trip. The fuel required for landing, descent, cruise, climb and take-off can be calculated in their reverse order and summed to give the trip fuel. This process estimates the minimum quantity of fuel required for the trip based on assumptions that include the route to be flown, cruise altitude, atmosphere state and forecast winds. In practice, however, the ideal flight plan is rarely achieved and the actual trip will differ from the trip assumed for flight planning; the reasons for this include the following.

- The flight planning will be completed some time before the flight and will use forecast weather (temperature profiles and winds), request the best cruise altitude and route and will be based on a specified departure time. The flight plan may not be confirmed until just before take-off, or even after-take-off, and may require changes to the route and cruise altitude to coordinate the flight with other traffic. Any such changes will usually increase the time and distance of the flight and extra fuel will be required.
- Delays in departure times may result in different air temperatures at take-off and en-route from those used for the flight plan. For example, if an early morning departure is delayed until midday, the increase in air temperature may increase the trip fuel required. Consequently, it could mean that the take-off WAT limit would be exceeded and the payload would then have to be reduced to comply with the scheduled performance requirements.
- On arrival at the destination, the aircraft may have to hold to await its turn to land. Flying in the hold will require additional fuel.

To account for the additional fuel requirement caused by environmental effects (temperature and wind), the trip fuel can be increased by a percentage of the calculated en-route fuel. In addition, 'contingency' allowances can be added to account for holding and unscheduled manoeuvres or route changes. The percentage increase of the en-route fuel may be based on a requirement set by the regulatory authorities or decided by the operator. The contingency fuel allowances are usually determined by the operator and based on knowledge and experience of the route.

The diversion fuel

The fuel for the diversion is calculated in the same manner as the trip fuel, but usually it assumes the same contingency fuel as the trip.

Reserves

Minimum reserves are usually set by the regulatory authorities although the operator may increase them at his or her discretion. An aircraft should not need to use fuel from its minimum reserves except in an emergency. The en-route percentage reserve and the contingency reserve are additional to the minimum reserve.

Tankering

In addition to the fuel required for the mission, the opportunity may be taken to carry extra fuel if it is economically advantageous to do so, this is known as tankering. Fuel can be tankered if the price at the departure point is sufficiently below the price at the destination to make it worth using any available surplus weight to carry the extra fuel. In this way, the cost of fuel uplift for the next flight can be reduced and the overall economy of the operation improved. Tankered fuel cannot be considered as fuel reserve for fuel planning purposes.

The fuel planning progression is shown in Fig. 9.15. The *payload weight* is added to the *aircraft prepared for service weight* to give the *zero fuel weight*, this is the basic weight of the aircraft for that mission. The *landing weight at the alternate airfield* is the sum of the zero fuel weight and the fuel reserves that have not been used, together with any tankered fuel being carried. The *landing weight at the destination* is the landing weight at the alternate plus the fuel for the diversion and, possibly, a part of the percentage en-route reserve if it has not been required. The greatest landing weight at the destination would occur if no reserves had been used; this case is shown in Fig. 9.15. The trip fuel added to the landing weight at the destination gives the *take-off weight*, which must not exceed the maximum scheduled take-off weight determined by the performance planning. Taxiing fuel may be carried to permit the aircraft to taxi to the take-off point and take-off at its limiting weight, this is added to the take-off weight to give the *ramp weight*. The fuel plan usually contains a simple self-check by adding the total fuel weight to the zero fuel weight to give the ramp weight directly. Comparing this with the ramp weight from the detailed plan helps to eliminate any arithmetic errors that may occur in the process.

Although the performance plan takes precedence over the fuel plan in flight planning, the fuel plan may need to be completed first in order to find the fuel required to transport the aircraft and payload to the intended destination. When this has been done, the take-off weight of the aircraft is known and the performance planning will be used to confirm that the aircraft can meet all the performance criteria along the intended route.

If the aircraft cannot meet any of the performance requirements then its weight must be reduced until it is able to do so. The APS weight is fixed and cannot be

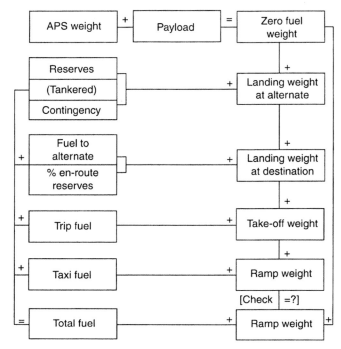

Fig. 9.15 Fuel planning.

reduced. The fuel load, including the reserves, required for the mission will also be fixed for that aircraft weight. Only the payload can be reduced, and to reduce the aircraft weight to comply with the performance criteria payload will have to be off-loaded. This is very serious economically since the reduction can only be made in the first instance in the revenue-earning part of the weight of the aircraft. However, since this will reduce the total aircraft weight the fuel required for the mission will also be reduced so that the final weight reduction will not be all payload but will include some fuel. A new weight schedule can then be made in which it may be possible to recover some of the payload reduction.

9.9 Conclusions

It has been seen that the purpose of performance scheduling is to ensure that aircraft are operated with safety as the primary objective.

Clearly, no aircraft operation can be absolutely safe and it is accepted that failures may occur in the aircraft or its engines. The first line of defence is that the design of the aircraft and of the engines is 'fail-safe'; that is, that any foreseeable failure should not lead to the immediate loss of the aircraft. The second line of defence is that the aircraft must be operated with the possibility of such failures in mind and that, if a failure occurred, it should be possible to continue the flight to a safe landing. In the case of the failure of an essential aircraft system, continued safe operation is normally provided by redundancy and the failed system could be isolated and its function taken over by an alternative system. In the event of an engine failure, there is no other engine that can replace the failed unit and it has to be accepted that the available propulsive thrust or power is reduced for the duration of the flight. This consideration leads to the regulations and requirements relating to performance scheduling.

The regulations and requirements ensure that proper margins of safety are included in the performance aspects of the planning of any flight. They reduce the risk of the aircraft being unable to manoeuvre within the space available to an acceptable low level, usually 10^{-6} or 10^{-7}. By statistical analysis of each element of performance, it is possible to determine factors and margins which, when applied to the operational performance of the aircraft, will reduce the risk of the aircraft being unable to meet the requirements. The flight plan, combining the performance plan and the fuel plan, ensures that the mission can be undertaken by the aircraft in the knowledge that, should an engine failure occur, the risk to the safety of the aircraft will be acceptably low.

The reader is reminded that this chapter is intended only to explain the broad principles of performance scheduling. In the explanation summaries of the requirements and regulations relating to performance scheduling, detail has been omitted in order to provide a simpler interpretation: under no circumstances should this chapter be considered to be definitive. In all cases in which the scheduled performance of an aircraft is being considered, reference must be made to the current issue of the relevant airworthiness requirements and regulations.

Bibliography

Airworthiness Authorities Steering Committee. *Joint Airworthiness Requirements, JAR-1, Definitions and Abbreviations; JAR-25, Large Aeroplanes; JAR-OPS 1, Commercial Air Transport (Aeroplanes)* (CAA Printing and Publishing Services).

Federal Aviation Administration. *Code of Federal Regulations; Title 14, Aeronautics and Space. Part 1, Definitions and Abbreviations; Part 25, Airworthiness Standards: Transport Category Airplanes* (Office of the Federal Register).

Grover, J. H. H. (1989) *Handbook of Aircraft Performance* (BSP Professional Books).

ICAO *Annex 14, Aerodromes Vol 1, Aerodrome Design and Operations.*

Wagenmakers, J. (1991) *Aircraft Performance Engineering* (Prentice Hall).

Engineering sciences data unit items
Performance Series, Vols 1–14.

Statistical Methods Applicable to Analysis of Aircraft Performance Data. Vol 10, No 91017, 1997 (Am D).

Variability of Standard Aircraft Performance Parameters. Vol 10, No 91020, 1991.

10

The application of performance

This chapter, which describes the use, or application, of performance data in practice, needs to cover a wide range of applications by both civil and military users of aircraft, often with very different operating criteria. It is not intended to explain in detail how each user would apply the aircraft performance to their particular purpose, but to give a broad insight into the ways in which the aircraft performance data are used in practice. It shows typical forms in which the performance of the aircraft needs to be presented.

10.1 Introduction

The performance of the aircraft, either estimated during the design stage or measured in flight, needs to be presented in various forms. These are required for application to such diverse purposes as the marketing of the aircraft, its certification, operational analysis and flight planning. The form in which the performance needs to be presented will differ according to the purpose for which the aircraft is to be used. In the case of a transport aircraft, it may need to be a general statement of the overall route performance or specific to a particular part of that route. It may be presented in terms of an aircraft weight and a standard atmosphere profile or may need to be interpolated for any WAT state. Military combat aircraft performance may need to address the ability of the aircraft to carry out a manoeuvre or task. Applied performance, therefore, requires a wide range of forms of presentation of the aircraft performance data. These range from the *performance summary* through to *operational performance*. The performance summary is a generalized statement of the aircraft performance used for a broad assessment of the ability of the aircraft to carry out a mission. Operational performance requires validated performance data, specific to the individual aircraft, for flight planning and other operational purposes. Some of the main topics that need statements of the performance of the aircraft are discussed below.

10.1.1 Marketing and fleet selection

At the design stage of the aircraft the manufacturer needs to supply a performance summary for the aircraft giving an estimation of its essential, overall, performance characteristics with respect to its design target criteria. The purpose of the performance summary is to enable the suitability of the aircraft for its intended role to be assessed by the potential operator. In civil transport operations, the aircraft must be seen to be capable of transporting the necessary payload over the required routes economically. In military operations, the aircraft must be shown to have air superiority and to be seen to have the potential to be able to carry out specified missions effectively. The performance summary, therefore, needs to enable the operator of the aircraft to assess its performance over a typical route or mission, in general terms. The aircraft can then be evaluated against operating criteria and design performance targets or it can be compared with other aircraft for cost-effectiveness studies. In the foregoing chapters, methods of estimating the performance of an aircraft were discussed. These allowed each segment of the flight to be analysed and the flight path related performance to be optimized with respect to some specified criterion. By adding the individual segments of the flight together, the overall performance of the aircraft, in terms of the fuel and time required to complete the mission, can be estimated and, from this, the operational effectiveness of the aircraft assessed. The performance summary used to market the aircraft will usually consist of statements of the fuel required and time taken for typical operational flight profiles; these are known as *block fuel* and *block time* diagrams. It is through the estimation of the performance summary that the customer would initiate the selection of the aircraft for his or her fleet and would consider placing options on the acquisition of a new type of aircraft. Similarly, in the case of a new aircraft, the manufacturer would consider the performance summary and assess the viability of the proposed aircraft as part of the decision to go ahead with the design and development of the project.

Another important use of the performance summary is in the assessment of the overall effect of a modification to the aircraft. During its life, the aircraft may be modified either to improve its payload capacity and performance or for reasons of flight safety relating to its airworthiness. Such modifications may be mandatory and affect the whole fleet. Examples of modifications to the airframe might be structural changes, such as a fuselage stretch, or involve the carriage of external equipment, both of which will affect the drag characteristic and aircraft weight, or it may involve a change in powerplant affecting the fuel consumption and thrust or power characteristics. By producing a performance summary for the aircraft before and after the modification, its effect on the overall performance can be evaluated through a cost-benefit analysis from which the economic and operational consequences of the modification can be assessed.

10.1.2 Certification

A civil aircraft must be granted a *Certificate of Airworthiness* before it can be used for any commercial purpose; similarly, a military aircraft must obtain the equivalent *Release to Service*. The purpose of the certification process is to demonstrate that the aircraft can be operated safely and to establish the limitations on the operation of the aircraft, which ensure safe flight. One element of the certification process, described fully in Chapter 9, is to ensure that the aircraft has sufficient performance to be able to carry out all necessary flight path manoeuvres in the space available. Generally, only measured data can be used for certification, since the estimated performance is derived from models of the characteristics of the aircraft and its powerplant, which are inevitably imperfect. It is essential for certification to determine the actual performance characteristics of the aircraft rather than to accept those that have been computed or predicted from an imperfect model. The measured performance is used to define the limits of aircraft weight, airspeed and environmental conditions within which the aircraft can comply with the airworthiness requirements and regulations.

10.1.3 Operational analysis

Before a civil transport aircraft enters service, or a new route is opened, the aircraft operator needs to analyse the aircraft performance in detail with respect to the intended route structure. The economic viability of the combination of the aircraft and route can then be fully assessed. Generally, this will need a greater depth of performance information than the block performance of the performance summary used for marketing purposes. It will be necessary to provide flight measured performance data that represents the average or *fleet mean performance*. This form of performance data may not need to be as detailed as that used for flight planning. However, it needs to be sufficient to enable the operation to be analysed with respect to a typical payload and either seasonal mean or specific atmosphere states. In practice, standard fuel reserves and diversion distances will normally be assumed in the operational analysis to simplify the process. Similarly, the military aircraft operator needs to be able to assess the viability of the military aircraft, either as a transport aircraft or as a combat aircraft, on typical operational mission profiles using standard WAT states.

To produce data for the operational analysis, flight tests are used to provide the actual performance data of the test aircraft or test fleet. The measurement procedures and data handling methods, described in Chapter 8, are used to produce data for the necessary range of WAT conditions to cover the expected operating states of the aircraft over the planned route structure. The measured performance is then used to provide fleet mean performance data for the aircraft type from which the route planning and operating costs can be established. The first stage of the process is to produce the operating data for each of the individual segments of the flight path and to reduce these data to standard weight, altitude and temperature (WAT) conditions. From the performance data in the individual segments of the flight path, the overall performance for the route can be assessed and the economic analysis completed. (The validated operational data may also be used to form a revised performance summary based on actual performance data, which can be used to replace the performance summary developed from the aircraft performance characteristics estimated in the earlier design stages.)

The operational analysis is only intended for the general economic and operational assessment of the aircraft on a given route or mission. *Flight planning* concerns the performance analysis of the aircraft on a particular route, at a particular time, under actual WAT conditions.

10.1.4 Flight planning

Flight planning concerns the analysis of the performance of the individual aircraft for a particular flight in which the payload and route are known and the probable operating conditions are well forecast. The principal objective of the flight planning, described in detail in Chapter 9, is to determine the maximum permissible take-off weight of the aircraft. It is also to verify that the aircraft is able to meet all the airworthiness requirements demanded by the performance scheduling. In this case, actual performance data for the individual aircraft concerned are needed since aircraft of the same nominal type may have minor differences that affect their individual performance.

From the flight measured data, the definitive performance data are used for the construction of the aircraft performance manual that contains all the performance data required for the flight planning process. The flight planning consists of performance calculations for the safety critical points on the route to be flown to demonstrate that the aircraft can meet all the necessary safety criteria at its take-off weight. It also covers the fuel planning, which calculates the quantity of fuel that will be required for the mission and the reserves of fuel for diversion and safety.

Each of these applications of performance require different forms and levels of data presentation and some of the typical forms of performance data presentation generally used for these purposes will be considered in the following sections.

10.2 The performance summary

There are two principal forms of the performance summary used to present the performance of the aircraft for a broad analysis of its capability, the *block performance* and the *payload–range*. Both are calculated on the assumption of an idealized route flown under optimum conditions so that they tend to show the best performance that the aircraft will be likely to achieve. Additionally, there may be simplified statements of other significant performance characteristics chosen to demonstrate the important design features of the aircraft with respect to its operation or marketability, for example, short take-off distances or performance with one engine inoperative. The ability of the aircraft to comply with any safety criteria relating to its airworthiness certification is normally taken to be implicit to any performance summary.

10.2.1 The block performance

A quick estimate of the fuel and time required to fly a transport mission is useful to the aircraft operator for a number of purposes. First, it enables the cost of operating the aircraft on a specified mission to be assessed so that its economic viability can be determined. This is an important preliminary stage in the selection of a suitable aircraft for a particular route system. Secondly, the process of flight planning can be considerably simplified if a quick estimate of the fuel and time required for a

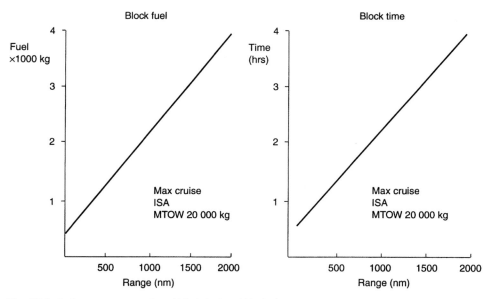

Fig. 10.1 Performance summaries of block fuel and block time.

route is available. This is particularly helpful if rapid flight planning is necessary, since some of the iterative calculations can be accelerated or avoided.

The block performance summarizes the time taken and the fuel required to fly a route over a given distance assuming a given set of WAT conditions; the summaries are known as the *block time* and the *block fuel*. It is a simplified calculation of the overall route performance in which allowances of fuel and time for a typical diversion, usually of a given distance, and standard fuel reserves are assumed. The atmosphere structure is taken to be ISA, or a parallel atmosphere, and, generally, the weight will be taken as the MTOW unless a wide range of take-off weights can be used. The block performance is calculated on the assumption of a hypothetical, but typical, flight profile flown under specific conditions, usually close to the optimum. The fuel and time required to fly from a departure point to a destination at a number of ranges is calculated and plotted as a function of the range, see Fig. 10.1.

In the early design stages of the development of the aircraft, the estimated performance would have to be used to calculate the block performance. Relatively simple assumptions would be used for the time taken and fuel used in the terminal phases of the mission and simple algorithms for climb, cruise and descent based on the estimated aircraft drag characteristic and engine performance. As the design progresses, so better information becomes available and the block performance estimates can be revised. Eventually, flight measured data will enable the block performance to be calculated with much greater confidence.

The block performance is not intended to be anything more than a guide to the fuel and time required to fly a typical, but hypothetical, mission. It is not a substitute for flight planning but simply a broad statement of the overall route performance of the aircraft.

10.2.2 The payload–range diagram

The payload–range diagram, Fig. 10.2, expresses the potential range of the aircraft based on its maximum disposable load. In this way it differs from the block time and block fuel, which are based on the time and fuel required for a specified range and take-off weight.

To explain the payload–range diagram it can be broken down into its basic elements.

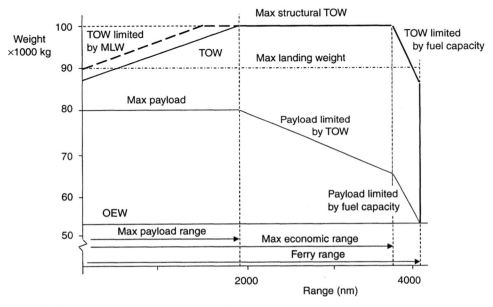

Fig. 10.2 Payload–range diagram – definition of limits.

Aircraft weight limits and weight breakdown

The diagram is based on the weight limitations of the aircraft, which are shown in Fig. 10.3. The weight datum is the *aircraft prepared for service weight (APS)*, or *operating empty weight (OEW)*, which is the weight of the aircraft prepared for flight but with no payload or fuel. The OEW includes the weight of the crew and their baggage, any special or essential equipment required on the aircraft, together with any unusable fuel that cannot reliably be pumped to the engines.

The aircraft is designed to a *maximum (structural) take-off weight (MTOW)*, which is the maximum weight at which the aircraft can be permitted to take off under any circumstances. The MTOW may be further limited by the considerations of the WAT limits and scheduled performance, but for the purposes of the payload–range diagram the MTOW is assumed the structural design limiting weight.

The OEW and the MTOW form the lower and upper weight boundaries respectively of the payload–range diagram. The *disposable load* consists of the sum of the payload and the fuel load and the difference between the MTOW and the OEW is the *maximum disposable load (MDL)*.

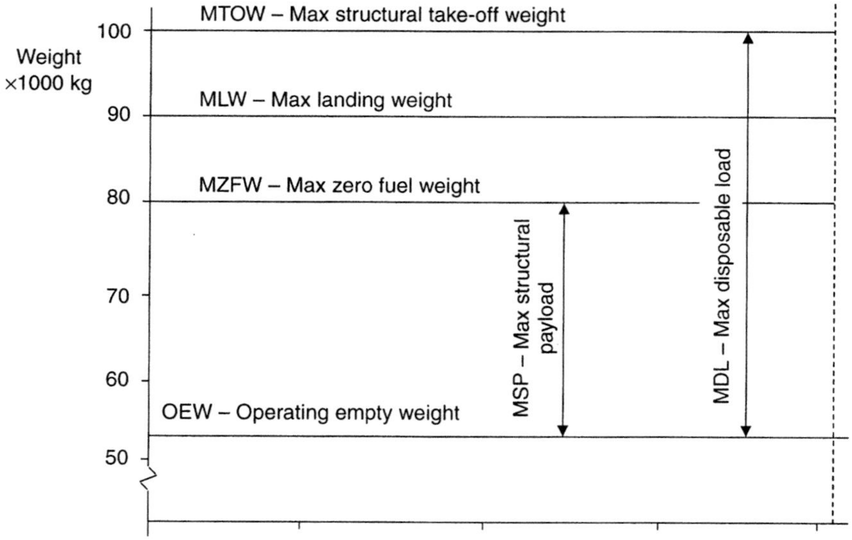

Fig. 10.3 Payload–range diagram – weight boundaries.

The *maximum structural payload (MSP)* is the maximum weight of payload that can be carried in the aircraft. In transport aircraft, the payload is usually carried in the fuselage and the maximum weight that can be carried will be limited by the structural design strength of the airframe, typically by the maximum wing-root bending moment. Adding the maximum structural payload to the operating empty weight gives the *maximum zero fuel weight (MZFW)* (note that $OEW + MSP = MZFW$).

The maximum load of fuel that can be carried is determined by the *maximum fuel capacity (MFC)* of the aircraft. In most cases, the sum of the MSP and the fuel load at MFC will exceed the MDL so that the *maximum fuel load (MFL)* that can be carried may be limited by the payload. This will be seen to be an important factor in the development of the payload–range diagram.

The *maximum landing weight (MLW)* of the aircraft is based on the energy that needs to be absorbed by the landing gear at the maximum design rate of descent at touchdown. It may be limited to a weight that is less than the maximum take-off weight. This is because it is assumed that the minimum foreseeable mission will be long enough to reduce the fuel load of the aircraft and bring the total weight down below the MLW. In such cases, the aircraft will usually be equipped with a fuel dumping capability to enable the weight to be reduced to a safe landing weight should an emergency occur soon after take-off that necessitates an immediate landing.

Fuel breakdown

The fuel required for the mission consists of the trip fuel and the reserves. If the normal fuel allowances are made, reserves will be carried for a reasonable diversion, company reserves and a percentage of the trip fuel for contingencies. The fuel reserves can then be treated as the trip fuel (factored for the contingency percentage) plus a constant weight representing the diversion and essential reserves. Figure 10.4 shows the fuel required by the aircraft as a function of the distance flown to the destination.

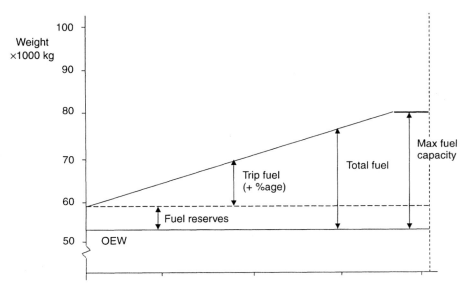

Fig. 10.4 Payload–range diagram – fuel definitions.

The maximum fuel capacity will determine the maximum weight of fuel that can be carried.

Payload–range diagram
The outline of the payload–range diagram is shown in Fig. 10.5. It is assumed that the aircraft is carrying its maximum structural payload so that the weight datum is the maximum zero fuel weight.

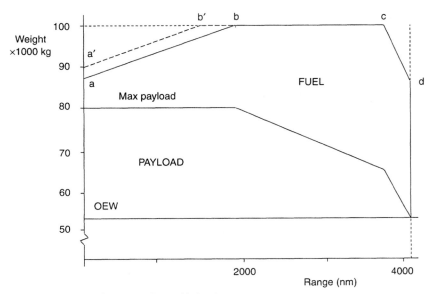

Fig. 10.5 Payload–range diagram – disposable load.

The weight of the fuel required for the mission is added to the maximum zero fuel weight to give the take-off weight (TOW) of the aircraft (a–b). As the mission range increases, the take-off weight increases until the MTOW is reached (b), this range is known as the *maximum-payload range* since it represents the maximum range over which the maximum structural payload can be transported. If a greater range is required, the additional fuel required can only be carried at the expense of payload (b–c). The exchange of fuel for payload can continue until the maximum fuel capacity is reached (c). Further range can only be gained by reducing the payload to reduce the weight, and hence the drag, of the aircraft (c–d). Operation of the aircraft at ranges greater than the maximum fuel capacity limit (c) is unlikely to be economical. The absolute maximum range, with zero payload, is the *ferry range* of the aircraft (d). A further limitation may be set by the maximum landing weight if very short-range missions are considered. If the aircraft were to be dispatched on a short-range flight with its maximum disposable load, it would reach its destination at a weight greater than the maximum landing weight. It would need to dump fuel or cruise until the weight was reduced to the maximum landing weight before it could land. Clearly this is not an economical situation and the maximum take-off weight would be limited by the maximum landing weight for short-range missions (a′–b′). The excess weight available between the TOW and the MTOW limited by the MLW is available for fuel tankering, as is any surplus payload weight at greater ranges. This enables excess fuel to be carried where it is economically beneficial to do so.

The payload–range diagram enables the operational efficiency of the aircraft to be assessed in, for example, payload tonne-miles or passenger seat-miles, as a function of cruising range, so that the maximum revenue earning potential of the aircraft can be estimated. It also indicates the effect on the revenue earning potential of cruising beyond the maximum payload–range point in terms of the payload reduction necessary. As a simple example, Fig. 10.6 shows the payload–range expressed in terms of seats available for a medium size transport aircraft.

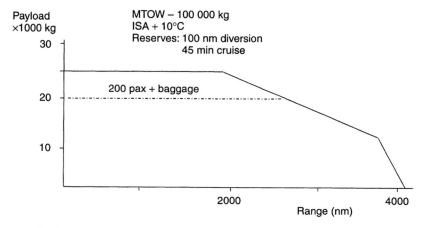

Fig. 10.6 Payload–range summary.

10.2.3 Additional performance summary data

Simplified statements of other performance characteristics, which affect the overall mission of the aircraft, may be included in the general performance summary for the aircraft. These statements may be expressed in the normal way as functions of aircraft weight or as secondary functions of other performance characteristics. Typical examples are as follows.

- *Take-off distances*, which are functions of aircraft take-off weight, can be expressed in terms of cruise range since this will determine the take-off weight. From such a summary, the range to the destination determines the minimum length of airfield from which the aircraft can operate, or, alternatively, the maximum airfield length available limits the range that the aircraft can fly from that airfield. Either case may be limiting to the operation of the aircraft.
- *The one-engine-inoperative cruise ceiling* is also a function of aircraft weight and will determine the ability of an aircraft to cross high ground or mountainous regions after the failure of an engine. A summary of the one (or more) engine-inoperative cruise ceiling and the drift down distance may be needed to assess the implications of an en-route engine failure on a flight over a mountainous region. The summary will usually show the one-engine-inoperative ceiling and drift-down distance from the point of engine failure to the new cruise ceiling.

10.3 Operational analysis

Operational analysis is a basic tool for optimizing the economics of a business system. In the case of aircraft operations, it involves the analysis of the actual routes that are to be flown, rather than hypothetical missions. Hence, the cost of any specific operation can be estimated, in terms of the fuel and time required, and planned so that the mission can be undertaken in the most fuel efficient or cost effective manner. Usually the primary mission – the flight to the intended destination – will be flown in the most cost-effective manner whilst the diversion will be flown in the most fuel-efficient manner. For the operational analysis, the operator needs to know the fleet-mean performance of the aircraft type that is proposed for the route so that the typical fuel and time required for the mission can be determined. In addition, it will be necessary to take into account the variable payloads and atmosphere conditions that will occur in actual operations. Therefore, the performance data will need to be presented in a form that will allow a route to be analysed over a range of WAT states. It should be noted that the atmosphere state may not be the same throughout the flight; for example, the departure may be made in arctic conditions and the arrival made in tropical conditions. It is normal practice to present the performance data for operational analysis, either graphically or in tabular form. It will be in terms of the International Standard Atmosphere and parallel atmosphere states and a range of weight increments so that each element of the mission can be estimated for the appropriate WAT state.

The principal performance parameters needed for the operational analysis are the fuel, time and distance required for each segment of the mission profile. This

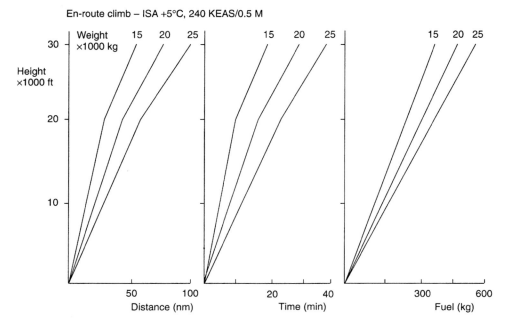

Fig. 10.7 En-route climb performance summary.

information needs to be supplied in an easily interpolated form and to be based on an initial aircraft weight, either for that segment or for the mission, and on a set of representative atmosphere states. The performance data are usually presented either as tables or charts that are constructed from flight measured data, corrected to a fleet mean, and reduced to a form from which the time, fuel and distance can be interpolated. Typical forms of data presentation for operational analysis are shown in Figs 10.7 to 10.9. In the data, certain aircraft conditions and states will need to be specified. These include the configuration of the airframe (e.g. settings of flaps, airbrakes, etc), and the status of any systems that affect the aircraft performance (e.g. pressurization and air conditioning, anti-icing, etc). The airspeed or Mach number used in the mission segment also needs to be specified, either as a fixed value or as a function of the initial weight and airframe configuration. The optimum combination of airspeed or Mach number and airframe configuration for each segment of the mission will have been determined from flight tests.

Figure 10.7 shows the distance, time and fuel used in a climb from sea level to cruise altitude in a given atmosphere state for a range of weights at the start of climb. The climb is made at maximum climb thrust using a specified schedule of airspeed and height in still air; the distance can be corrected for the effect of wind if necessary. The data for the climb segment of the mission, or diversion, can be read from the figure.

Figure 10.8 shows the cruise performance in terms of a chart of specific air range, SAR. Since the cruise segment of the mission is variable in terms of distance and

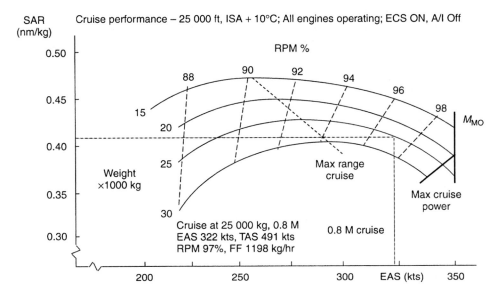

Fig. 10.8 Cruise performance summary – specific air range.

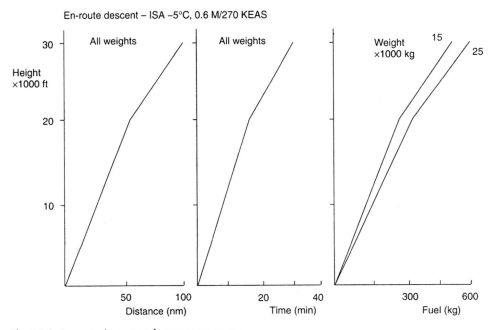

Fig. 10.9 En-route descent performance summary.

atmosphere state it is not always possible to reduce the cruise performance to a simple table or graph. SAR charts are drawn for specific cruise altitudes and atmosphere states. They are annotated with recommended cruising conditions for normal cruise (usually minimum time or maximum Mach number) or maximum economy (for diversion or holding) and enable the point cruise performance to be interpolated. This can be used to calculate the fuel used and distance flown in a unit of time, say one hour. The cruise sector can then be constructed of a number of hourly segments, adjusting the weight and atmosphere state as necessary. The recommended cruise conditions are often interpolated from the SAR charts and published in tabular form giving, for example, EAS, TAS, fuel flow and engine setting, as the essential cruise information at each cruise state.

Figure 10.9 shows the time distance and fuel used in a descent from cruising altitude to sea level; it is in a similar form to the climb performance. In this case, there may be different descent strategies to consider, for example, high rate descent or long range descent.

The performance data for operational analysis are usually optimized both for minimum cost and for minimum fuel consumption, each having its particular application to the mission. The minimum cost performance is usually flown at high speed to reduce the time element of the operation. Since time is a high unit-cost element of the overall cost of the mission, the operating airspeed is optimized so that the reduction in mission time outweighs the increased fuel consumption in the high-speed operation. The primary mission is flown to the minimum cost criteria to maximize the cost effectiveness of the revenue earning operation. The time constraints on the mission economy cease to apply when the mission is interrupted by an unscheduled event such as holding in a stack or a diversion to an alternate airfield. Where time is no longer the main criterion, the fuel cost becomes dominant. This is not only in unit cost terms but also in terms of the excess weight of fuel that needs to be carried as reserves for any non-scheduled manoeuvres such as holding and the diversion. These manoeuvres are flown to the minimum fuel criteria to minimize the weight of fuel that needs to be carried as reserves since this is non-revenue earning fuel and hence parasitic weight.

10.3.1 Operational analysis procedure

Once the route is defined and the conditions are set, the operational analysis can proceed. The process is iterative, calculating one phase of the flight from a known datum and then moving on to the next phase using the resultant revised datum as its starting point. Since fuel is burned during the flight, the weight of the aircraft will be decreasing with time and the only known weight datum is at the end of the diversion. Here, the weight is the sum of the operating empty weight of the aircraft, the payload and the fuel reserves. The actual calculation, therefore, progresses backwards from the end of the mission to the beginning.

First, an approximate analysis can be made to establish some of the principal criteria for the mission. Typically, a route will be divided into its segments and the most critical segment used as the base for the optimization process. In a normal transport mission, for example, the cruise segment of the primary mission

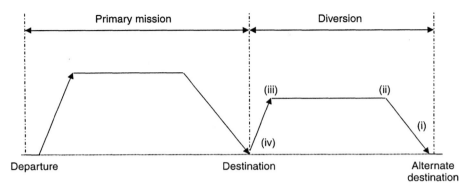

Fig. 10.10 Transport aircraft mission profile.

generally will be the critical segment. Here, the greatest quantity of fuel is used, and time spent, so that this is likely to be the segment that dominates the optimization process. The optimum cruising altitude can be determined on the assumption of the probable take-off weight of the aircraft and its cruise Mach number; this was seen in Chapter 4. The planned route will determine the actual cruise altitude since the system of vertical separation of flight paths assigns altitudes based on the aircraft heading. Once the cruise altitude is determined, the time, distance and fuel used in the descent, cruise and climb segments can be found from the performance planning data. The take-off and landing performance will provide the time and fuel used in the terminal segments of the flight as simple allowances or as functions of WAT. The overall mission performance can then be made up by adding together the time, fuel and distance calculated for the optimized segments, Fig. 10.10 outlines the process.

In the mission, there are two flight profiles to be considered:

1. the primary mission, or trip, from the point of departure to the planned destination; and
2. the diversion to an alternate destination.

Each flight profile consists of a climb, cruise and descent. There is, however, only one take-off (from the point of departure) and one landing (at either the planned destination or the alternate destination in the case of a diversion being necessary). The object of the operational analysis is to estimate the fuel and time requirements for the mission and to check for any operational restrictions imposed, for example, by airfield lengths available or high ground on the route. Since the performance of the aircraft is dependent on its weight, and the weight is decreasing throughout the mission, the diversion needs to be considered first followed by the primary mission.

A diversion would be required if the airfield at the planned destination were not available for landing due, for example, to the weather being below the landing minima for the airfield or the runway being blocked by an incident. If this is known well before the aircraft arrives at the planned destination the diversion can be initiated early, from the end of the descent phase 1500 ft above the destination airfield or even from the cruise. However, in case of marginal weather conditions the decision to divert may have to be taken at the decision height towards the end of the final approach to landing. If the runway is not seen at the decision height, typically 200 ft above the landing runway, the approach is discontinued and the climb to the *diversion cruise altitude (DCA)* commences. Whilst the probability of having to make a diversion is low, provision must be made for it in the flight plan on the assumption that it will be made from the decision height. To minimize the weight of fuel required for the diversion it is usual to adopt a minimum fuel cruise technique, rather than a minimum time cruise technique. It is usually better to cruise as high as possible, but not necessarily as high as the primary mission.

The fuel required for the diversion can now be calculated on the assumption that minimum-fuel flying techniques will be used throughout the diversion. The only known datum is the maximum landing weight at the diversion, point (i) in Fig. 10.10

The maximum landing weight at the end of the diversion would be the OEW + payload + company fuel reserves + contingency fuel.

The fuel used and distance travelled in the descent can now be calculated to give the weight at the end of cruise, point (ii). This may need to be an iterative calculation since the fuel burned in the descent will affect the weight of the aircraft during the descent.

Starting from the weight at the end of the cruise, the fuel required for the cruising distance in the diversion can be calculated to give the weight of the aircraft at the start of cruise, point (iii). Allowance will need to be made for the distance travelled in the climb so that the calculation will not be exact at this stage.

An iterative process can now be used to find the quantity of fuel used, and distance covered in the climb from the discontinued approach to the cruise altitude. This is based on the weight of the aircraft at the end of the climb, point (iii), and gives the weight at the start of diversion, point (iv), and hence the diversion fuel requirement. The cruise and climb phases may need to be iterated until the weight at the start of diversion converges.

Since the variable part of the aircraft weight at the end of the diversion contains only the payload and fuel reserves it may not vary greatly. Hence, it is possible to compute the fuel and time required for diversions of known distance for a set of standard WAT states. The results can be formed into a table or graph as a performance summary, which would enable the diversion fuel and time to be interpolated for a given diversion distance.

The primary mission is analysed in the same manner as the diversion. Starting with the landing weight at the destination (which is the weight at the start of diversion), the fuel and time required for the descent, cruise and climb phases can be calculated. In the calculation, account will need to be taken of the contingency fuel and en-route reserve, which is usually taken to be a percentage of the primary mission fuel. An iterative process will be needed to arrive at the take-off fuel load, and hence the take-off weight of the aircraft. Using this process the overall time and fuel requirements for the mission can be determined and the cost of the operation assessed.

The larger aircraft operators will usually have an operational analysis program that will estimate route performance in the manner outlined above. The charts and tables of the aircraft performance can be contained in a database from which the performance over any route and under any WAT state can be computed. That database will contain much of the performance data required for flight planning.

10.4 Flight planning

The purpose of flight planning is to show that all the performance related safety criteria can be met at all points throughout the actual mission that is to be flown and that sufficient fuel for the mission is carried. This was discussed fully in Chapter 9 under scheduled performance. This principal is satisfied by performance planning, which considers the safety related criteria, and fuel planning, which determines the fuel load that will be required.

The form of the data required for flight planning needs to provide more information concerning the conditions affecting the performance of the aircraft than is needed for operational analysis. The reason for this is that compliance with the safety critical performance requirements is taken to be implicit in the operational analysis. Therefore, it should not need to be considered separately in detail – that being the specific purpose of the flight planning process. Flight planning, which must be carried out before any flight, is intended to ensure that the aircraft is not despatched at a weight greater than that which will guarantee its safe performance under the prevailing conditions. The information that is needed for performance planning is contained in the performance manual of the aircraft in the form of charts or tables that cover all the essential parts of the mission. The charts and tables can be interpolated to give the performance under all WAT states and enable the critical parts of the mission to be checked against the airworthiness criteria. Although it is a requirement that a copy of the performance manual must be carried on the aircraft, most large aircraft operators will have a flight-planning program to compute the performance plan and fuel plan before the flight. It will also verify compliance with the airworthiness regulations. However, the captain of the aircraft is still ultimately responsible for the safety of the aircraft and may need to refer to the manual to confirm that the safety criteria are met. The flight management system (FMS) on board the aircraft may also contain a copy of the flight-planning program. In addition, it will contain the performance database for the aircraft and a database of the route structure, and can perform the flight planning process at any time. The FMS can be used to verify the performance plan in the event of changes to the assumed route clearance, which may not become known until just before take-off, or even after the aircraft has commenced the flight.

The performance data used for flight planning are based on flight measured data, but these data will have been processed to account for the inevitable variability between aircraft of the same type. The manufacturing tolerance in the construction of the aircraft will mean that there will be a statistical variability in the aircraft weight and drag which, whilst small, will affect the performance of the individual aircraft with respect to the fleet mean. Similarly, the power or thrust output of the powerplant, and its specific fuel consumption, will vary statistically about a mean. In addition, the engine manufacturer rates the engine at its *guaranteed minimum power or thrust*, which is the lowest permissible output for that type of engine. The output of any engine in service, therefore, must not be less than the guaranteed minimum output. The statistical variation in individual engine performance is accounted for by reducing the fleet mean performance of the aircraft to give the performance that would be obtained by an aircraft with engines producing the guaranteed minimum output. This would be the aircraft *guaranteed minimum performance*, which is the lowest acceptable performance for any aircraft of that type. By this process, the performance data used for construction of the performance manual, which is the database for flight planning, are reduced to a datum that should always be lower than the actual performance of any individual aircraft.

The flight measured, or fleet mean, performance of the aircraft is referred to as the *gross performance*. It represents the actual performance the aircraft can achieve under specified WAT conditions. However, flight planning is based on expected WAT states, which may not occur at the time the flight commences, and optimum operating airspeeds, which may differ from the actual airspeeds used in flight; there are good reasons for these discrepancies. Since the flight planning takes place some time before the flight, it is based on forecast atmosphere states and wind strengths that may not correspond to the actual atmosphere states and wind strengths at the time of flight. Similarly, the estimated weight of the aircraft may differ from the actual weight. There are many reasons for this; for example, passengers are taken to be a mean weight, together with their hand baggage, and are not weighed individually. The OEW of the aircraft assumes a clean, dry aircraft but an aircraft that is wet on the outside or has condensation on the inside will be heavier. In addition, the airspeed indications are subject to permissible levels of instrument error which, whilst small, will mean that the aircraft may be flown at airspeeds that differ from those recommended in the flight manual. This will have a degrading influence on the performance. These deviations from the estimated WAT state and ideal operating speeds are accounted for statistically by reducing the gross performance of the aircraft by a factor or a margin to the *net performance*. The normal practice is to construct the aircraft performance manual in terms of net performance based on the *guaranteed minimum engine performance*. Consequently, all the statistical variations are accounted for and the level of performance achieved by the aircraft should never be less than that predicted by the performance manual. This principle was also noted in Chapter 9 in which the scheduled performance was considered as a regulatory issue. It is included again here because of its importance as an operational issue that influences the form of presentation of the performance data for flight planning.

In flight planning, the performance manual of the individual aircraft must be used since aircraft of the same type may differ in minor respects, which may have an effect on their individual performance. There are several reasons for this. During construction, an individual aircraft may have had modifications to its basic type-design which might affect its weight or drag. These may have been made to provide for an individual customer requirement or be the result of design changes required; for example, to accommodate modified equipment or to correct manufacturing defects. In addition, later in the life of the aircraft, it is likely that it will undergo modifications or that repairs to damage may have to be made. If they affect the weight or drag of the aircraft significantly, the performance manual will need to be amended to consider the changes. Each aircraft, therefore, has its own performance manual (identified by the manufacturer's aircraft construction number), which is amended from time to time as necessary. When the aircraft is certificated, its individual performance is measured and, following its correction to guaranteed minimum performance, is used to construct the performance manual which is unique to that aircraft. (In practice, a general performance manual, for the aircraft type will be issued, but it will be verified by the measured performance data for the individual aircraft and amended if necessary.) When the certificate of airworthiness of the aircraft is renewed, performance flight tests may be required to show that the measured performance is not less than the performance published in the performance manual. This proves that the performance of the aircraft has not deteriorated to a level at which it would fail to comply with the airworthiness performance criteria.

10.4.1 Fuel planning

The process of fuel planning is similar to that which was considered for the operational analysis. However, it will need to be carried out for an actual mission flown by a specific aircraft rather than for a proposed route and a 'fleet mean' aircraft. In the fuel planning, the aircraft OEW and payload weight will be known, as will the meteorological forecast for the en-route conditions. In addition, there will be 'local knowledge' of the route that may influence the decision to increase any contingency reserves if delays or holding are known to be likely to occur. The data are similar in form to those used for the operational analysis, but are usually specific to the individual aircraft.

As in the operational analysis, the company reserves and diversion fuel are calculated first, followed by the fuel for the primary mission and the percentage for en-route reserves and contingencies. Adding the fuel allowance for start-up and taxiing at the departure airfield gives the total fuel weight and hence the take-off weight of the aircraft.

10.4.2 Performance planning

Having now established the take-off weight of the aircraft, performance planning is required to show that the aircraft can comply with the minimum performance levels specified by the airworthiness requirements throughout its mission. Performance scheduling, considered in detail in Chapter 9, is concerned mainly with the take-off distances, climb gradients, one-engine inoperative en-route performance and landing distances so that the data required for performance planning are basically limited to these critical areas.

The data for performance planning need to be constructed to enable the maximum permitted take-off weight of the aircraft to be determined with respect to all critical points along a known route and in predicted meteorological conditions. In Chapter 9, it was seen that the maximum take-off weight was the least of the take-off weights that were determined by a number of criteria. These related to the distances available for take-off and landing, the minimum climb gradients, all engines operating and one (or more) engines inoperative, and the one (or more) engines inoperative ceiling. The performance datum used for the construction of the performance manual is *net performance*, and includes the factors and margins required to convert gross performance to net performance. These data can be used directly and no further factors or margins should need to be applied.

In the performance manual, the data are usually presented in a number of forms to enable the performance planning to be approached at different levels. Detailed performance data are required for all normal operations and consist of graphical data, which allow full interpolation of the flight conditions, leading to the most exact solution. The data for the terminal sectors of the flight, take-off, after-take-off climb and landing, need to enable the performance to be determined when the take-off weight is likely to be critical. In such cases, the take-off distances, decision speeds and rotation speeds will have to be found and shown to be acceptable. The data for the non-terminal sectors may be presented fully in graphical form or in tabular form for quicker reference. However, since this will be limited to discrete intervals of weight, altitude and temperature, the tabular solution generally will be less exact. In addition, a condensed version of the performance is normally included for *rapid flight planning*. Rapid flight planning can be used when it is known that the take-off weight of the aircraft will be less than the maximum permissible take-off weight and will not be limiting to the performance. It enables sector fuel, sector time and fuel reserves to be established quickly without having to go through the complete flight planning and fuel planning process in full detail.

The contents of the performance manual

The performance manual is normally divided into sections covering general information, simplified performance data and detailed performance data for each phase of the flight.

The general information includes charts for the conversion of wind speed and direction into head/tail wind components, temperatures as temperatures relative to ISA and pressure error corrections for the aircraft's air data system. It may also include some aircraft data relating to the critical speeds for performance.

Critical speeds. The performance of the aircraft will be governed or limited by a number of airspeeds, some of which are a function of aircraft weight. The first stage of the flight planning is to determine the weight dependent speeds in the appropriate airframe configuration at the take-off weight, these are the stalling speed, rotation speed, take-off safety speed and en-route climbing speed. En-route cruising speeds and approach and landing speeds, which are weight dependent, will be required for the later part of the mission.

The simplified performance data may combine a number of criteria into one chart so that the need to check each criterion to find the limiting case is avoided. The take-off and landing WAT limits are an example of a simplified performance data chart.

Take-off WAT limits. The critical take-off climb performance weight limit may be governed by a number of criteria. These include the first segment, second segment and fourth segment climb gradients with one engine inoperative, or by engine performance limited by atmosphere state, or a mechanical limit. By considering the limits set by each of the criteria a single chart can be constructed that gives the most limiting criteria, at any combination of altitude and temperature combination, with the maximum aircraft weight as the resultant variable. Figure 10.11 shows the form of a typical take-off WAT chart for a turbo-prop powered aircraft. In this case, the

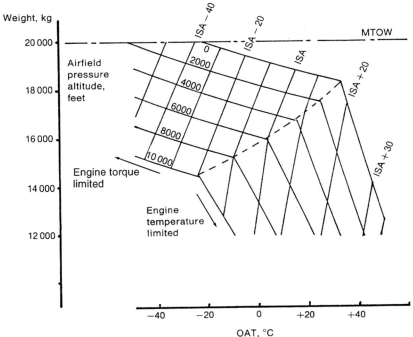

Fig. 10.11 Take-off WAT limit chart (turbo-prop).

WAT limits are set by maximum torque (a mechanical limit) and air temperature (a climb gradient limit set by power available).

By entering the chart with the airfield pressure altitude and atmosphere temperature, the maximum take-off weight at which the aircraft can comply with the scheduled climb performance criteria can be found. Since the after-take-off climb gradient with one engine inoperative is usually the most limiting criteria to the maximum take-off weight, the take-off WAT chart is one of the most essential pieces of information in the take-off performance analysis. A similar WAT chart can be constructed for the landing performance to determine the maximum landing weight; the limit in this case is based on the ability to climb away from a discontinued approach or baulked landing.

The detailed performance data enable each segment of the flight path to be fully analysed. The contents will depend on the aircraft performance class but will cover the following broad headings.

Take-off field performance. In Chapter 6, it was seen that the take-off distances were functions not only of weight, altitude and temperature but also of the component of the wind in the take-off direction and the mean slope of the runway. The take-off field performance charts need to enable all the variables to be taken into account in a logical sequence. There are two parts to a take-off field performance chart. One part, the 'runway correction' chart, deals with the influence of the airfield conditions of wind, runway slope and runway layout (clearway) on the take-off distances available. The other part, the 'aircraft datum performance' chart, deals with the influence of weight, altitude and temperature on the take-off performance required assuming a datum of a level runway, with no clearway, in zero wind. In the process of performance planning, the objective is to determine the maximum take-off weight at which the aircraft can be dispatched safely and the analysis of the take-off field performance is critical to that procedure. The process is best explained by following through part of the procedure related to take-off field performance, stage by stage.

The general procedure used to determine the maximum take-off weight, MTOW, of the aircraft can be broken down into a number of stages outlined below. (This is a simplified description, the full process may include other limitations and conditions which need to be met but have been omitted in the interests of clarity.)

Stage 1. Determination of the field length available for take-off, all engines operating (aeo) and one-engine inoperative (oei)

This stage involves the use of the 'runway correction' chart to account for the effect of clearway, slope and wind components on the take-off distances available. The distances required by the aircraft in the take off depend on the airfield conditions at the time; these will include the runway slope, the head (or tail) wind component and the length of clearway. The 'runway correction' chart is used to account for the airfield conditions and to convert the actual take-off distances available under those conditions to the equivalent distances in the case of a level runway, with no clearway, in zero wind. In this way, the airfield conditions are allowed for in the form of a 'corrected' distance available for take-off that is compatible with the statement of the 'aircraft datum performance'. In practice the take-off distance of interest is the shortest of:

(i) the TORA or TODA, all engines operating (D1), and
(ii) the balanced distance (D2), between,
      the Emergency distance available, ASDA, and
      the TORA or TODA, one engine inoperative.

A set of charts will be provided for correction of each of the 'actual' distances available to the 'equivalent' distances available if there was no clearway, no wind and no runway slope. A typical form of chart is shown in Fig. 10.12(a). The procedure is as follows.

(a) To apply the corrections to the all-engines operating case and to determine the corrected aeo take-off distance available, D1 (the clearway correction determines whether it is the TORA or the TODA that sets the limiting distance).

(b) To apply the corrections to the ASDA to give the corrected ASDA for the abandoned take-off. (In this case, by definition, a clearway correction is not required.)

(c) To apply the corrections to the TORA/TODA for the one-engine inoperative (oei) case to give the corrected oei take-off distance available for the continued take-off. (Again, the clearway correction determines whether it is the TORA or the TODA that sets the limiting distance.)

(d) The corrected oei TORA/TODA from the continued take-off (c), is then used with the corrected ASDA from the abandoned take-off (b), to give the balanced oei take-off distance, D2. Figure 10.12(b) shows the typical form of chart used for this purpose. This process balances the oei take-off distance available and emergency distance available to give the balanced field length, D2, and will also determine the decision speed ratio, V_1/V_R.

(e) The lesser of the distances D1 and D2 determines the minimum field length available and establishes the *field length limit*, 'D'.

Figures 10.12(a) and 10.12(b) have been separated here to illustrate the individual parts of the process. In practice, some of these charts may by combined to form a single chart, which could be more convenient for flight planning. There are many possible ways of constructing the flight planning airfield performance charts, but all are formed from these basic building blocks.

Stage 2. Determine the MTOW for the field length limit, WAT limit or other limits

This stage concerns the aircraft datum performance in the critical areas of the take-off and after-take-off climb and matches the space required by the aircraft to the space available determined in Stage 1. The after-take-off climb gradient and other possible limitations are also considered at this stage.

(a) The field length limited MTOW is found from the aircraft datum performance chart, Fig. 10.12(c). This chart uses the field length limit, 'D', determined from Stage 1 and the datum aircraft performance, at the airfield pressure altitude and temperature, to determine the MTOW at which the aircraft can achieve the take-off within the corrected distance available.

(b) The take-off WAT limit chart, of the form shown in Fig. 10.11 is used to determine the MTOW at which the aircraft can comply with the after-take-off climb gradient requirements.

(c) Other limitations may need to be considered, e.g. tyre limits and obstacle clearance. Tyres are structurally limited by their rotational speed and the wheel speed on take-off has to be kept below a safe maximum. If obstacles are present on the take-off flight path it may be necessary to reduce the MTOW to increase the gradient of climb available to ensure clearance of the obstacles, the MTOW is then an obstacle clearance limited weight. This is referred to in the 'take-off net flight path' below.

(d) The lowest of the MTOW's determined from (a), (b) or (c) is the scheduled MTOW set by the take-off requirements.

Stage 3. Determination of the critical take-off speeds

Following the determination of the MTOW in Stage 2, the weight-related take-off speeds can now be found, these include, the rotation speed, V_R, the decision speed, V_1, the lift-off speed, V_{LOF} and the take-off safety speeds, V_2 and V_3. In addition, it will be necessary to check that there are no further limitations on the take-off caused by these speeds, e.g. brake energy limits. If any limits are exceeded the MTOW may need to be reduced further.

Take-off net flight path

The take-off WAT chart should have confirmed that the aircraft is able to meet all the individual requirements for after-take-off climb in terms of the minimum gradients in each segment. However, the overall net flight path may need to be constructed to ensure that the aircraft clears not only the general obstacle limitation surface, but also any known obstacle at any point along the intended climb flight

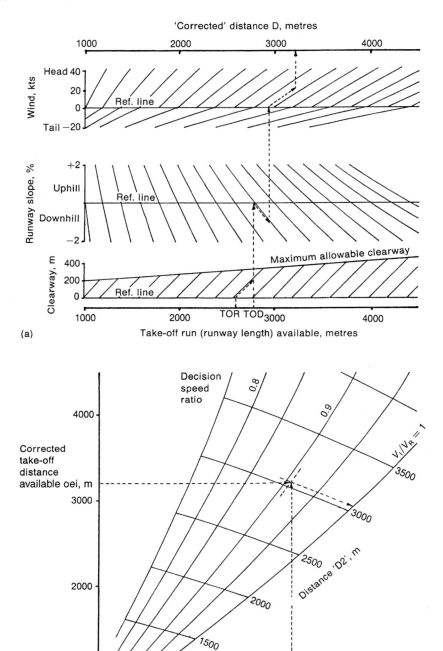

Fig. 10.12 (a) Runway length correction chart. (b) Runway length correction (oei) – balanced field length. (c) Maximum take-off weight – field length limits.

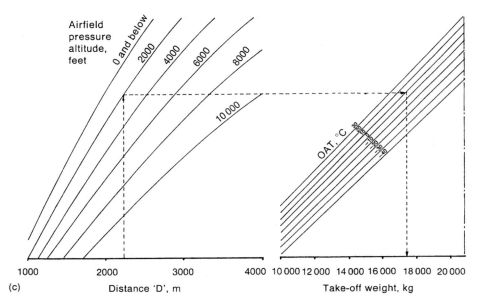

Fig. 10.12 Continued.

path. Figure 10.13 shows a typical take-off net flight path, extending to 1500 ft above the take-off surface, consisting of a number of segments, each representing a part of the overall after-take-off climb. The net flight path must be shown to clear the general obstacle limitation surface, or any isolated obstacle, by 35 ft.

Each segment of the after-take-off net flight path needs to be analysed to give the horizontal distance travelled and height gained. This enables the net gradient of the after-take-off flight path, measured from the brake release point, to be found and shown to comply with the obstacle clearance regulations. Each segment is assumed to

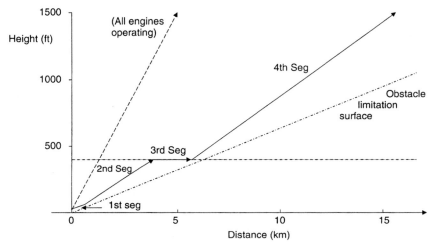

Fig. 10.13 Take-off net flight path.

consist either of a climb or of a level acceleration during which a change of airframe configuration is made. In the segments in which the aircraft is climbing, the net gradient of the flight path is used to give the height gain and horizontal distance travelled. This will require knowledge of the ground speed, which is the true airspeed corrected for the wind component. In addition, in the construction of the net take-off flight path, account needs to be taken of the time that elapses from the start of the take-off run. This is needed because the engines have a maximum operating time at take-off setting after which the power will have to be reduced to the *maximum continuous power (MCP)*. Following the take-off to the screen height, the net flight path segments in the climb are:

First segment. The first segment climb commences at the end of the take-off distance required, where the aircraft is 35 ft above the extended runway surface. At this point the landing gear retraction is initiated (and, if appropriate, the propeller of the failed engine is feathered) and continues until the landing gear is fully retracted. WAT charts, or tables, of the horizontal distance covered and height gained in the first segment can be formed from measured climb performance data with landing gear extended (and propeller windmilling) and included in the performance manual. These charts will be used to estimate the height gained and distance travelled in the first segment for the construction of the net take-off flight path. (Any headwind component will need to be taken into account in the calculation of the horizontal distance.)

Second segment. The second segment climb commences as soon as the landing gear retraction is completed. It continues at the take-off safety speed, with the critical engine inoperative, V_2, and the operating engines at take-off power, in the take-off configuration (with landing gear retracted) until the aircraft is at least 400 ft above the extended runway surface. A second segment climb WAT chart can be constructed from measured climb performance, and a typical form of chart is shown in Fig. 10.14.

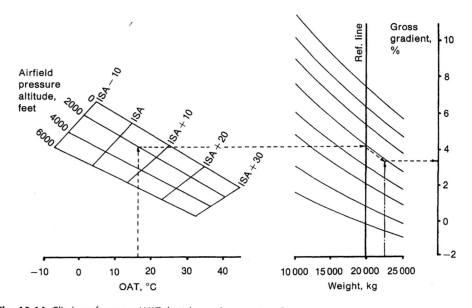

Fig. 10.14 Climb performance WAT chart (second segment, oei).

This chart is used to find the gradient of climb in the second segment to show compliance with the airworthiness requirements, which are based on the need to show that the aircraft will clear the general obstacle limitation surface.

Where it may be necessary to check for specific obstacle clearance in the second segment climb, the horizontal and vertical distances flown in the second segment can be calculated from the climb gradient and airspeed. This information may be presented as an obstacle clearance chart. This is a geometric construction of the net flight path in which the height of the aircraft is given in terms of both distance from the reference brake release point and the second-segment climb gradient. The height of the obstacle can then be compared with the predicted height of the aircraft as it passes the position of the obstacle. The minimum second-segment gradient to ensure obstacle clearance is read from the chart. The chart will include the height gained in the first-segment climb and allow any headwind component to be taken into account.

Third segment. Airframe configuration changes need to be made to improve the climb gradient; these are made in the third segment, which must not commence until the aircraft is at least 400 ft above the extended runway surface. The aircraft is accelerated to the best one-engine inoperative climb speed and the flaps are retracted from the take-off setting to the en-route setting (or an intermediate setting for best climb gradient). It is assumed that during the acceleration, the aircraft is in level flight and all the excess power is used for acceleration. At the end of the acceleration, the thrust or power may need to be reduced from the take-off setting to the maximum continuous power (MCP) setting for the continued climb. The horizontal distance flown during the acceleration can be calculated from the initial and final airspeeds, the excess thrust available and the weight of the aircraft. It can be formed into a chart, in terms of the WAT state, and will include a correction for headwind component. Since the gradient of climb in the second segment is proportional to the excess thrust, the third-segment distance flown is sometimes expressed as a function of the second-segment climb gradient.

Fourth segment. Following the acceleration in the third segment, the climb continues at the appropriate safety speed, with flaps retracted, until either the aircraft is 1500 ft above the extended runway surface or the transition from the take-off to the en-route configuration is complete. Charts for the gradient of climb in the fourth segment are generally similar in form to the second-segment climb charts.

Turning flight. If the net flight path needs to include a turn of more than 15° heading change to avoid obstacles then the reduction in performance due to the turn needs to be taken into account. In this case, there may be a need to calculate a plan position with respect to obstacles; this requires calculation of the radius of turn, time of turn and a correction for wind. Turns can only be accounted for in stabilized climbing conditions, therefore they should not be considered in the first or third segments. The increased drag due to the rate of turn will cause a reduction in the gradient of climb, which needs to be considered. A correction to the rate of climb based on the minimum rate of turn needed to avoid the obstruction can be applied to the climb gradient that would have been achieved in straight flight. Alternatively, it can be assumed that the turn is made at the maximum rate of turn that would result in level flight before resuming the climb in straight flight.

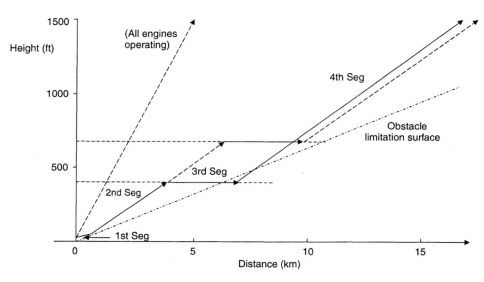

Fig. 10.15 Take-off net flight path – obstacle clearance.

In the construction of the net after-take-off flight path, the distance flown and height gained in each segment are found from the WAT charts and corrected to net performance if necessary. The data are then used to check that the net gradient at all points on the flight path are sufficient. Figure 10.15 shows that, although the aircraft would meet all the climb gradient requirements in each segment (this would have been shown by the take-off WAT chart), certain critical points may occur. The obstacle clearance surface may be penetrated or an obstacle not cleared if, for example, the third segment is initiated at the minimum height of 400 ft above the runway. In this example, the obstacle clearance can be achieved by continuing the second segment climb to a higher altitude before starting the third segment. It is to avoid this occurrence that the full net flight-path data are required and that the detailed information to calculate any obstacle clearance climb path is made available.

En-route climb performance

From the flight planning aspect, the principal information needed for the en-route phase of the flight concerns the cruising performance of the aircraft with one engine inoperative. In the case of aircraft with more than two engines, two-engines inoperative flight may need to be considered. With one engine inoperative and the remaining engines set at maximum continuous thrust, the aircraft may not be able to maintain its cruising altitude. It will be forced to reduce speed towards the minimum drag speed and to descend to an altitude at which the thrust available is equal to the drag of the aircraft and, again, height can be maintained.

The minimum information that will be needed will be the cruise ceiling of the aircraft with one engine inoperative. This will determine the altitude to which the aircraft will drift down following the failure of an engine in the cruising phase of the flight. This may be sufficient if the route does not take the aircraft over ground, or any

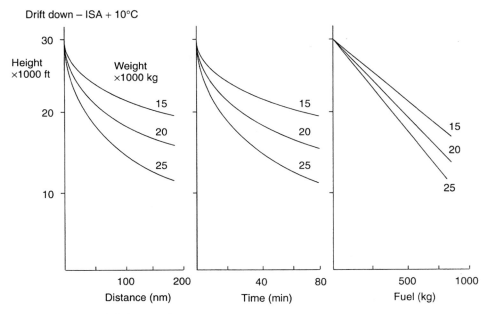

Fig. 10.16 En-route drift down performance.

obstruction, which is higher than the one-engine inoperative ceiling, taking into account the required 2000 ft en-route obstacle clearance margin. Further information concerning the drift-down distance and time may be needed if the obstruction level plus the clearance margin exceeds the one engine inoperative ceiling. In this case, there will be a critical distance on the route before which a diversion will be required; this was shown in Fig. 9.12.

The one-engine inoperative ceiling and drift-down distance will be dependent on the WAT state and will be affected by the head-wind or tail-wind component. Typical WAT charts of drift-down time, fuel and distance are shown in Fig. 10.16. Similarly, two-engine inoperative data will be required for aircraft with three or more engines.

Landing performance

The landing distances are required at the destination and at the alternate airfields. The landing distance chart, Fig. 10.17, differs from the take-off charts since there is no balanced field length to consider. The distance required can be reduced to an 'aircraft datum performance' chart based on aircraft weight and airfield altitude and a correction for wind component. The effects of temperature and of slopes less than 2% are often omitted and are considered to be covered by the landing distance factors, which range between 1.5 and 2. Landing distances at alternate airfields are proportional to the landing distances at the destination.

Since the landing is made at idle thrust setting, the one-engine inoperative landing is not significantly different from the all-engines operating landing from the performance aspect. However, reverse thrust for the speed reduction after touchdown may not be available, or may be limited, and handling considerations may require the approach and landing to be carried out at higher airspeeds.

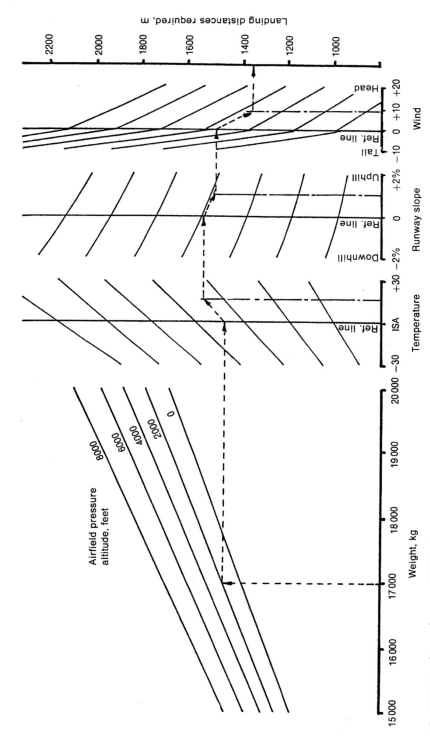

Fig. 10.17 Landing performance WAT chart.

10.5 Conclusions

The intention of this chapter is to illustrate the various practical applications and uses of performance data. It has covered a wide range of applications and shown, in general terms, how performance data are used and presented throughout the design, development and operation of the aircraft. Since it has been necessary to simplify some of the explanations of the performance terms, they may not have been defined with the rigour required for their application to the airworthiness codes of practice. They should be regarded as definitions for guidance only; full definitions should be sought before they are used in any analysis of the performance of an aircraft.

In the preamble to this chapter, it was stated that it was not intended to explain in detail how each user would apply the aircraft performance data to their particular requirements. Indeed it would be difficult to do so since there would be so many different applications to be considered and each aircraft manufacturer tends to have a different format for data presentation. To address the application of performance in practice, the approach adopted was to consider a number of topics ranging from the marketing of the aircraft to flight planning. Examples of performance data presentation that would be used for each purpose were then discussed. The forms of data presentation shown in this chapter are examples typical of those used throughout the aircraft industry, both civil and military. However, they may not be exactly as they will appear in the aircraft performance manual, or other documentation, in every case.

Bibliography

Airworthiness Authorities Steering Committee, *Joint Airworthiness Requirements, JAR-1, Definitions and Abbreviations*; *JAR-25, Large Aeroplanes* (CAA Printing and Publishing Services).

CAA (1996) *Specimen Performance Charts for Aeroplanes Certified in Performance Group A*. CAP 385.

Federal Aviation Administration, *Code of Federal Regulations*; *Title 14, Aeronautics and Space. Part 1, Definitions and Abbreviations*; *Part 25, Airworthiness Standards: Transport Category Airplanes* (Office of the Federal Register).

Grover, J. H. H. (1989) *Handbook of Aircraft Performance* (BSP Professional Books).

Grover, J. H. H. (1990) *Airline Route Planning* (BSP Professional Books).

ICAO *Annex 14, Aerodromes Vol 1, Aerodrome Design and Operations*.

Wagenmakers, J. (1991) *Aircraft Performance Engineering* (Prentice Hall).

Engineering sciences data unit items
Performance Series, Vols 1–14.
Variability of Standard Aircraft Performance Parameters. Vol 10, No 91020, 1991.

11

Performance examples

11.1 Introduction

To illustrate the performance processes discussed – and the application of the estimation methods developed – in the earlier chapters in context with the practical applications and regulatory considerations of the later chapters, a series of examples are offered. The examples are based on a medium range, subsonic, transport aircraft with two turbo-fan engines.

- The first examples illustrate the selection of the design points of the aircraft and the need to develop lift and drag characteristics that will enable the aircraft to achieve its performance targets. These must be attained within acceptable limits of, for example, pitch attitude and margins over the stalling speeds. (Other limitations may occur, e.g. handling qualities limitations and systems operation constraints, but these are beyond the scope of these examples, which are based on performance considerations only.)
- The second set of examples is a 'running calculation' of a typical transport aircraft mission from take-off to landing. This illustrates the means of estimating the time, distance and fuel required for the overall mission.
- In conjunction with the running calculation, and separately, performance summary information is developed.
- An example of a practical calculation for a specified mission is included.
- The regulatory considerations are addressed as they occur throughout the examples.

The examples are based on ISA conditions so that atmosphere data can be read directly from the tables of the International Standard Atmosphere (Appendix C). The reader will be invited to re-calculate the examples under alternative atmosphere states and operating conditions.

11.2 Aircraft characteristics

This is a simplified set of characteristics representative of a typical, medium range, 150 seat, transport aircraft. These data are supplied for the purposes of these examples only. They should not be used for any other purpose.

Weights

Max take-off weight (MTOW)	55 000 kg
Max zero fuel weight (MZFW)	45 000 kg
Operating empty weight (OEW)	30 000 kg
Max payload	15 000 kg
Max fuel	17 000 kg

Dimensions

Gross wing area   95 m^2
Aspect ratio   9

Powerplant (ISA conditions)

| Max take-off thrust (flat rated to 5000 ft ISA) | 70 kN/engine |
| Max continuous thrust, ISA datum | 63 kN/engine |

Altitude effect on climb/cruise thrust

$$F_N = F_{N0}\sigma^{0.8}$$

where F_{N0} is the ISA datum max continuous thrust.

Specific fuel consumption at max T-O	$10.2\theta^{\frac{1}{2}}$ mg/N s
SFC at max continuous thrust	$15.3\theta^{\frac{1}{2}}$ mg/N s
SFC at cruise/descent thrust	$20.8\theta^{\frac{1}{2}}$ mg/N s

Aerodynamic characteristics

Drag characteristic

	Cruise	Take-off/climb	Approach/landing
Lift-dependent drag factor, K	0.0440	0.0505	0.0544
Zero-lift drag coefficient, C_{Dz}	0.0222	0.0343	0.1248
C_{Dz} landing gear down, C_{Dz+lg}	–	0.0568	0.1473
C_{Dz+lg} with airbrakes			0.2200
$(L/D)_{max}$	16	12	6.07
$(L/D)_{max}$ landing gear down	–	9.34	5.59
$(L/D)_{max}$ landing gear + airbrakes			4.57
C_{Lmd}	0.710	0.824	1.515
C_{Lmd} landing gear down	–	1.061	1.646
C_{Lmd} landing gear + airbrakes			2.01

Lift characteristic

	Cruise	Take-off/climb	Approach/landing
Flap setting	Clean	LE + 20° TE	LE + 60° TE
$dC_L/d\alpha$ per rad.	5.73	5.73	5.73
α_0, deg	–2	–8	–20
C_{Lmax} clean	1.35		
C_{Lmax} with flap	–	2.4	3.4

Wing root datum is 2° LE up with respect to body axis datum.

The aerodynamic characteristics are shown in Figs 11.1 and 11.2.

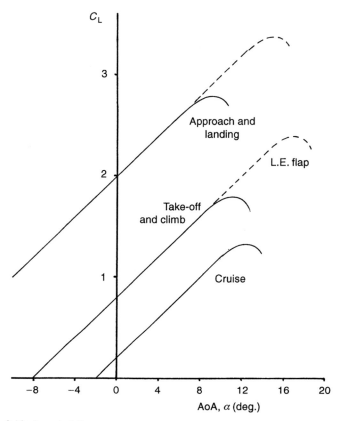

Fig. 11.1 Aircraft lift characteristic.

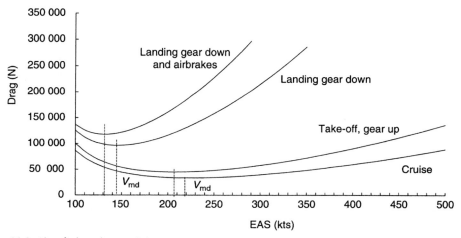

Fig. 11.2 Aircraft drag characteristic.

11.3 Basic operating points determined by aerodynamic characteristics

The purpose of these examples is to show the connection between the performance requirements of the aircraft and its aerodynamic design. When designing an aircraft, its aerodynamic characteristics must be developed to provide the performance required in each phase of flight. The set of data given above has been developed to illustrate that process.

From the lift and drag characteristics and operational limitations, the lift coefficients at the principal operating points can be determined. Operating speeds can then be determined from the weight of the aircraft. In the absence of any data on minimum control speeds, or other limitations on the operating speeds, it will be assumed that the stalling speed is the limiting criterion.

11.3.1 Cruise

The condition for optimization of cruise for range is that the aircraft is cruising at 1.316 times the minimum drag speed, assuming that the specific fuel consumption is independent of Mach number. From the aerodynamic characteristics of the aircraft in its cruising configuration, the C_L for minimum drag is 0.710. Thus the optimum C_L for cruise will be

$$0.710/(1.316)^2 = 0.410$$

which occurs at an angle of attack of about $+2°$. Since the aircraft body axis datum is $-2°$ with respect to the wing, this results in a level fuselage attitude during cruise. This is the ideal condition for cruise and will determine the design point for the $C_L-\alpha$ relationship for cruising flight.

The cruising speed will be limited by the drag rise Mach number; this is dealt with in Chapter 4, eqn (4.43) and Fig. 4.10. From eqn (4.43),

$$M_{\text{crit}} = 1.316 \left[\frac{2}{\gamma S}\right]^{\frac{1}{2}} \left[\frac{K}{C_{\text{Dz}}}\right]^{\frac{1}{4}} \left[\frac{W}{p}\right]^{\frac{1}{2}} \tag{4.43}$$

If the drag rise Mach number is taken to be 0.8, then the optimum pressure, and hence altitude, for cruise is a function of the aircraft weight only.

$$0.8 = 1.316 \left[\frac{2}{1.4 \times 95}\right]^{\frac{1}{2}} \left[\frac{0.044}{0.0222}\right]^{\frac{1}{4}} \left[\frac{W}{p}\right]^{\frac{1}{2}}$$

which is evaluated and shown in Fig. 11.3. This would be the ideal weight–height schedule in the cruise–climb.

11.3.2 Take-off and initial climb

The distance required for take-off needs to be minimized as far as possible. However, the aircraft also needs to be able to climb at a gradient exceeding a regulatory

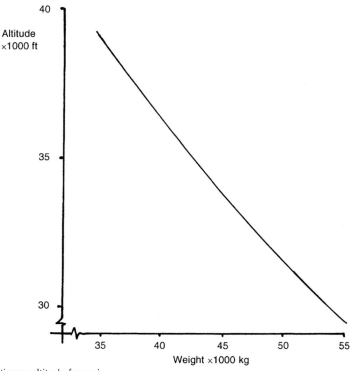

Fig. 11.3 Optimum altitude for cruise.

minimum value after the take-off. These requirements are conflicting since the flap setting required for a minimum ground run would probably produce a drag force that would limit the one-engine inoperative climb performance. A compromise is necessary in the aerodynamic characteristics of the aircraft in the take-off configuration, which will permit the aircraft to meet its climb gradient requirements and, at the same time, reduce the take-off distances to an acceptable length.

Take-off

In simplified terms, the distance, S, required to accelerate to the lift-off speed, V_{LOF}, is given by

$$S = \frac{V_{LOF}^2}{2a}$$

where a is the acceleration, which is governed by the equation of motion,

$$F = ma$$

Hence, for minimum distance, the acceleration must be maximized and the lift-off speed minimized.

- For maximum acceleration, the excess thrust, $F \; (= F_N - D)$, should be maximized. Since the available thrust is limited by the engine, the excess thrust can only be maximized by minimizing the drag. This is achieved by a low drag configuration, i.e. no flap deflection.

- For minimum lift-off speed, the flap should be deflected to increase the maximum lift coefficient and hence reduce the lift-off speed.
- The acceleration in the continued take-off following an engine failure after V_1 has been reached will need to be considered.

There is no simple rule to determine the optimum compromise and a series of calculations of the take-off distance required will need to be made using different flap settings until the best solution is found.

After-take-off climb

In the after-take-off climb, the aircraft must be able to achieve specified climb gradients, one-engine-inoperative, with landing gear up and down, and also with all engines operating. In addition, it must climb at a safe speed, defined as not less than $1.2V_s$ (or some other specified speed limitation). The drag increment resulting from the flap setting used to minimize the take-off distance may limit the gradient of climb in one or more of the critical climb segments. Therefore, a further compromise may be necessary to find a flap setting that is acceptable for all take-off and climb phases of the flight. Combinations of leading edge flap (which primarily extends the C_L–α characteristic to higher values of C_{Lmax}), trailing edge flap, (which primarily raises the C_L–α characteristic by an increment in C_L), and other high lift devices, will need to be investigated. In this example, the climb speed of $1.2V_s$ results in a C_L of 1.67, in the take-off configuration. The angle of attack will be 9° and hence, the pitch attitude will be 7° nose-up.

In these examples, a single take-off flap setting has been assumed. This is not unusual in the case of smaller aircraft but, particularly in the case of larger aircraft, the flap settings may be variable and their selection based on the aircraft weight and the take-off distances available.

11.3.3 Approach/landing

As in the case of the take-off, the speed of the aircraft in the landing manoeuvre needs to be as low as possible to minimize the stopping distance on the runway. However, the pitch attitude of the aircraft also needs to be considered. For safety reasons, the speed on the approach is required to be not less than $1.3V_s$ in the landing configuration; which sets the value of C_L on the approach. In this example, the value of C_{Lmax} in the approach configuration is 3.4. Hence, the C_L on approach will be 2.01 at an angle of attack of about 0° with respect to the wing root datum, and the aircraft attitude will be 2° nose down in level flight. On a typical 3° approach glide slope this would result in an aircraft attitude of 5° nose down, which would give the pilot a good view of the runway ahead. During the flare, the flight path gradient is reduced to zero and the speed allowed to decrease to the touchdown speed. The touchdown speed is not regulated but a speed of about $1.1V_s$ would produce a pitch attitude of about 5° nose up at touchdown. This would allow the nose wheel to be lowered onto the runway at a safe speed, reducing the angle of attack, and hence the lift, sufficiently to keep the aircraft firmly on the ground and under control.

A further consideration on the approach is the relationship between the approach speed and the minimum drag speed. In this case, the C_{Lmd} in the landing configuration, landing gear down, is 1.646, whereas the C_{Lapp} is 2.01. The aircraft would be operating at a speed below its minimum drag speed and therefore, it would be flight path unstable. It would be necessary to provide additional drag from airbrakes to increase the zero-lift drag coefficient, and hence C_{Lmd}, until the approach speed of the aircraft was above V_{md}. The airbrakes would need to increase the zero-lift drag coefficient to 0.2200 in order to increase the C_{Lmd} to the approach value of 2.01. On the approach, the thrust-to-weight ratio required would be

$$\frac{F_N}{W} = \frac{C_D}{C_L} - \sin 3°$$

Now C_L is 2.01 and C_D with landing gear down and airbrakes out is 0.4400, so that the net thrust required to maintain the flight path gradient is $F_N = 0.1666W$. Thus, the engines are producing about two thirds of their maximum continuous thrust on final approach. The high thrust setting on the approach assists in achieving maximum thrust quickly in the event of a discontinued approach. (In the event of a discontinued approach, if maximum take-off thrust was selected and the airbrakes retracted, but the flaps and landing gear left in the landing setting, the aircraft would achieve a climb gradient of about 6.75% with all engines operating.)

Comment

These are general conclusions that set the first approximation of the aerodynamic characteristics that will be required if the aircraft is to meet its performance specification and the regulatory requirements. By interpolation of the aerodynamic characteristics, the reader may look for a better optimization of the characteristics or flap settings.

11.4 Estimation of flight path performance

In this section, the aircraft will be assumed to be flying a mission from a departure point to a destination. The performance in each of the basic elements of the flight path will be estimated as a 'running example'. The atmosphere will be taken to be ISA and the aircraft takes off at its MTOW

The mission consists of a take-off at maximum take-off weight, 55 000 kg, a climb to 35 000 ft and a cruise segment of 2000 nm cruising at 0.8 M. From the cruise the aircraft descends to the terminal area and manoeuvres to a final approach for landing at its destination; this is referred to as the trip. A diversion from the decision height on the final approach is required. In the diversion, the aircraft climbs to 10 000 ft and cruises at 270 KEAS (knots, equivalent airspeed) to an alternate airfield 150 nm from the destination. Fuel reserves are calculated for the diversion and for regulatory requirements.

The time, distance and fuel required for each segment of the mission are calculated using the methods developed in the earlier chapters. The example progresses cumulatively with the weight reduction due to fuel consumption being taken into account.

11.4.1 Take-off distances

There are two parts to the estimation of the take-off distance required, the ground run and the airborne distance to clear the screen height of 35 ft (or 10.5 m).

Ground-run distance

The take-off ground run is made with the nose wheel on the ground in the take-off flap configuration. If the aircraft body datum is 2° nose down during the take-off run then the angle of attack of the wing will be 0°. The lift coefficient during the ground run will be 0.8 and, therefore, the drag coefficient will be 0.0891.

The take-off ground run required, TORR, can be estimated from the approximate expression (eqn (6.8)),

$$S_G = \frac{V_{LOF}^2}{2g(A - BV_{LOF}^2)_{0.7V_{LOF}}}$$

where

$$A = \left\{\frac{F_N}{W} - \mu_R\right\}$$

and

$$B = \frac{\frac{1}{2}\rho S C_L}{W}\left(\frac{C_D}{C_L} - \mu_R\right)$$

assuming a level runway.

Taking the rolling coefficient of friction, μ_R, to be 0.03 for a hard, dry surface and assuming ISA conditions at MTOW, A and B can be evaluated.

$$A = \frac{2 \times 70\,000}{55\,000 \times 9.81} - 0.03 = 0.229$$

In the take-off configuration

$$B = \frac{0.5 \times 1.225 \times 95 \times 0.8}{55\,000 \times 9.81}\left(\frac{0.0891}{0.8} - 0.03\right) = 0.00000702$$

Now, if V_R is taken to be $1.1V_s$ (a speed commensurate with V_2 being achieved by the screen height), and V_{LOF} occurs on rotation, then C_{LLOF} will be 1.98 and

$$V_{LOF} = \left[\frac{2W}{\rho S C_{LLOF}}\right]^{\frac{1}{2}} = 68.4\,\text{m/s} \quad \text{or} \quad 133\,\text{kts}$$

Thus $(A - BV_{LOF}^2)_{0.7V_{LOF}} = 0.245$ and

$$S_G = 973\,\text{m}$$

Airborne distance

The airborne distance to the screen height can be estimated from the expression

$$S_A = \frac{W}{(F_N - D)_{av}}\left\{\frac{V_2^2 - V_{LOF}^2}{2g} + 10.5\right\}$$

Now the mean speed in the airborne phase is at $1.15V_s$ and the lift coefficient is 1.815. Thus, the drag coefficient will be 0.2232.

$V_2 = 74.6\,\text{m/s}$ and $V_{LOF} = 68.4\,\text{m/s}$. The mean speed will be $71.5\,\text{m/s}$, and the drag force, $D = 66\,395N$. Thus, the airborne distance is given by

$$S_A = \frac{55\,000 \times 9.81}{140\,000 - 66\,395} \left\{ \frac{74.6^2 - 68.4^2}{2 \times 9.81} + 10.5 \right\} = 408\,\text{m}$$

Thus, the un-factored take-off distance required, TODR, is $973 + 408 = 1381\,\text{m}$.

Fuel used during take-off

The time, t, taken for the ground run is approximately,

$$t = \frac{2S_G}{V_{LOF}} = \frac{2 \times 973}{68.4} = 28.4\,\text{s}$$

and in the airborne distance is

$$t = \frac{2S_A}{V_2 + V_{LOF}} = \frac{2 \times 408}{74.6 + 68.4} = 5.7\,\text{s}$$

giving a total of $34.1\,\text{s}$.

Thus the fuel consumed will be

$$\text{Fuel} = F_N \times \text{sfc} \times t = 140\,000 \times 10.2 \times 10^{-6} \times 34.1 = 48.7\,\text{kg}$$

so that $50\,\text{kg}$ should be allowed for take-off to 35 ft.

11.4.2 After-take-off climb

In the after-take-off climb to 1500 ft above the runway, there are three critical, one-engine-inoperative climb conditions to be considered,

(i) Take-off, landing gear extended, in which the minimum gradient must be positive.
(ii) Take-off, landing gear retracted, in which the minimum gradient must not be less than 2.4%.
(iii) Final take-off, en-route configuration, maximum continuous power, in which the minimum gradient must not be less than 1.2%.

Take-off, landing gear extended

The climb gradient is given by

$$\frac{F_N}{W} - \frac{D}{W} = \frac{F_N}{W} - \frac{C_D}{C_L} = \sin\gamma$$

The speed for the climb is taken to be the speed at lift-off,

$$V = V_{LOF} = 1.1V_s$$

Hence

$$C_L = C_{Ls}/1.1^2 = 2.4/1.1^2 = 1.98$$

and

$$C_D = 0.568 + 0.505 \times 1.98^2 = 0.2548$$

$$C_D/C_L = 0.1287$$

Now, 50 kg fuel was allowed for the take-off, thus,

$$\frac{F_N}{W} = \frac{70\,000}{54\,950 \times 9.81} = 0.1299$$

The gradient is given by

$$0.1299 - 0.1287 = 0.0012 \quad \text{or} \quad 0.12\%$$

The gradient is positive and hence the aircraft complies with the requirement.

Take-off, landing gear retracted

The speed for climb is the take-off safety speed, $V_2 = 1.2V_s$, hence $C_L = 1.67$ and $C_D = 0.1751$. Following the process given above,

$$\frac{F_N}{W} = 0.1299$$

$$\frac{C_D}{C_L} = 0.1049$$

and the flight path gradient will be

$$0.1299 - 0.1049 = 0.025 = 2.5\%$$

The gradient is greater than 2.4%, and therefore the aircraft complies with the requirements.

Final take-off

This is made in the en-route configuration at a speed not less than $1.25V_s$ and with maximum continuous power set, assuming the take-off climb weight of 54 950 kg.

$$C_L = C_{Ls}/1.25^2 = 0.864$$

Hence $C_D = 0.0550$ and

$$\frac{C_D}{C_L} = 0.0637$$

$$\frac{F_N}{W} = \frac{63\,000}{54\,950 \times 9.81} = 0.1169$$

Therefore, $0.1169 - 0.0637 = 0.0532$, and the final take-off climb gradient is 5.33%. This is comfortably above the requirement of 1.2%.

Comment

This simplified set of examples illustrates the effect of the take-off flap deflection on the climb gradient. Considering the results of these calculations, it would be reasonable to re-estimate the take-off flap setting to reduce the drag and improve the climb performance at the expense of extending the take-off distances. As an extension to these examples the reader could re-calculate the take-off distances and climb performance using alternative flap settings. The changes to the drag characteristic can be deduced from the data given.

In practice, this procedure would need to take into consideration any limitations to the take-off distances available and the balanced field length required for engine failure accountability.

11.4.3 Climb performance

If the take-off has progressed normally, the aircraft will climb with all engines operating from the end of the take-off distance required to cruising altitude following an accepted climb schedule. The initial climb to 1500 ft will be made at V_3, the all-engines-operating take-off climb safety speed, at take-off thrust and in the take-off configuration, landing gear retracted. At 1500 ft, the transition to the en-route configuration will commence. In this phase the airspeed will be increased to the en-route climb speed, the flaps retracted and the thrust reduced to 'maximum continuous'. Since conditions are continually changing in this part of the climb, it is probably best analysed by an averaging process similar to that used for the airborne distance in the take-off flight path. Generally, in this phase, the aircraft will be following a standard departure route under air traffic control. Therefore, in practice, it may not be possible to include all or any of the distances calculated in this phase in the en-route flight path.

The en-route climb follows a climb-speed schedule that is both convenient for the control of the aircraft and gives a good rate and gradient of climb up to the cruise altitude. In the example, the schedule will call for a climb at 280 KEAS/0.7 M. An indicated airspeed of 280 KEAS will be used up to 25 000 ft, at which the Mach number is 0.7, above that, the climb will continue at 0.7 M. The climb performance will be estimated in segments of 10 000 ft.

Climb 0–1500 ft

Take-off configuration, landing gear retracted. Weight at end of take-off is 54 950 kg. Assume the take-off safety speed, $V_3 = 1.3V_s$.

$$C_L = 2.4/1.3^2 = 1.42 \qquad C_D = 0.1361 \qquad C_D/C_L = 0.0958$$

$$\frac{F_N}{W} = \frac{140\,000}{54\,950 \times 9.81} = 0.2597$$

$$\frac{F_N}{W} - \frac{C_D}{C_L} = 0.2597 - 0.0958 = 0.1639$$

This gives a gradient of climb of 16.6%.

The true airspeed is given by

$$V = \left[\frac{2W}{\rho S C_L}\right]^{\frac{1}{2}} = \left[\frac{2 \times 54\,950 \times 9.81}{1.172 \times 95 \times 1.42}\right]^{\frac{1}{2}} = 82.6\,\text{m/s} = 271\,\text{ft/s} = 160.5\,\text{kts}$$

Thus the rate of climb is

$$\text{TAS} \times \sin\gamma = 271 \times 0.1639 = 44.42\,\text{ft/s}$$

and the time taken to climb to 1500 ft is 33.8 s.

The distance covered to 1500 ft is given by

$$\text{TAS} \times \text{time} = 160.5 \times \frac{33.8}{3600} = 1.5\,\text{nm}$$

The fuel consumed in the climb is

$$F_N \times \text{sfc} \times \text{time} = 140\,000 \times \frac{10.2}{10^6} \times 33.8 = 48.3\,\text{kg}$$

Cumulative totals   From end of take-off distance.
Time   33.8 s/0.56 min
Distance   1.5 nm
Fuel   97 kg (including take-off)
Aircraft weight   54 903 kg (at end of the climb segment)

Climb 1500 ft–5000 ft

The thrust is reduced to maximum continuous thrust. The aircraft accelerates to the en-route climb speed, 280 KEAS, and flaps are retracted. Assume mean conditions at 3000 ft and a weight of 54 903 kg.

Since the speed and configuration are changing, the energy principle can be used to find the time taken to climb and accelerate from 1500 ft (a) to 5000 ft (b). This is given by

$$\left[V\frac{(F_N - D)}{W} \right]_{av} t = \left[(H_b - H_a) + \frac{V_b^2 - V_a^2}{2g} \right]$$

V_a is the TAS at 1500 ft, this was found to be 160.5 KTAS or 82.6 m/s.
V_b is the TAS at 5000 ft, 280 KEAS is equivalent to 301.6 KTAS or 155.2 m/s.
V_{av} is the mean of V_b and V_a; $V_{av} = 118.9$ m/s
Mean thrust at maximum continuous setting is found at 3000 ft,

$$F_{Nav} = F_{N0}\sigma^{0.8} = 2 \times 60\,000 \times 0.91512^{0.8} = 1\,117\,80\,\text{N}$$

D_a is the drag at 1500 ft $= 0.5 \times 1.172 \times 82.6^2 \times 95 \times 0.1361 = 51\,705\,\text{N}$
D_b is the drag at 5000 ft and 280 KEAS.

$$C_L = \frac{2W}{\rho_0 V_e^2 S} = \frac{2 \times 54\,903 \times 9.81}{1.225 \times 144^2 \times 95} = 0.4464$$

Hence

$$C_D = C_{Dz} + KC_L^2 = 0.0222 + 0.044 \times 0.4464^2 = 0.03097$$

and

$$D_b = 0.5 \times 1.225 \times 144^2 \times 95 \times 0.03097 = 37\,365\,\text{N}$$

The mean drag is

$$\frac{51\,705 + 37\,365}{2} = 44\,535\,\text{N}$$

The time taken to accelerate and climb can now be calculated.

$$\left[118.9\frac{(111\,780 - 44\,535)}{54\,903 \times 9.81}\right]t = \left[1066.8 + \frac{155.2^2 - 82.6^2}{2 \times 9.81}\right]$$

giving $t = 131.8\,\text{s}$.

The distance flown is found from the mean TAS as

$$\text{Dist} = \frac{118.9 \times 131.8}{0.51444 \times 3600} = 8.46\,\text{nm}$$

The fuel required is given by

$$\text{Fuel} = F_N \times \text{sfc} \times t = 111\,780 \times \frac{15.3}{10^6} \times 0.98963 \times 131.8 = 223\,\text{kg}$$

Cumulative totals   at 5000 ft
  Time   165.6 s/2.76 m
  Distance   10 nm
  Fuel   320 kg
  Weight   54 680 kg

Climb 5000 ft–15 000 ft (calculated at a mean height of 10 000 ft)

The climb is made at 280 KEAS in the en-route configuration.

$$10\,000\,\text{ft} \qquad \sigma = 0.73848 \qquad \sigma^{0.8} = 0.78464 \qquad F_N = 94\,157\,\text{N}$$

At 280 KEAS and 54 680 kg, $C_L = 0.4444$ and $C_D = 0.03089$

The gradient of climb is, therefore,

$$\frac{F_N}{W} - \frac{C_D}{C_L} = \frac{94\,157}{54\,680 \times 9.81} - \frac{0.03089}{0.4444} = 0.10602 = \sin\gamma$$

This is a gradient of 10.6%.

At 10 000 ft ISA, 280 KEAS is equivalent to 325.8 KTAS or 550.6 ft/s. The rate of climb is then

$$\text{TAS} \times \sin\gamma = 550.6 \times 0.10602 = 58.4\,\text{ft/s}$$

The time taken to climb is 171.3 s and the distance flown is

$$\text{TAS} \times t \times \cos\gamma = 325.8 \times \frac{171.3}{3600} \times 0.9927 = 15.4\,\text{nm}$$

and the fuel required is

$$F_N \times \text{sfc} \times t = 94\,157 \times \frac{15.3}{10^6} \times 0.96501 \times 171.3 = 238.1\,\text{kg}$$

Cumulative totals   at 15 000 ft
  Time   336.9 s/5.62 m
  Distance   25.4 nm
  Fuel   558.1 kg
  Weight   54 442 kg

Climb 15 000 ft–25 000 ft (calculated at a mean height of 20 000 ft)

The climb is made at 280 KEAS in the en-route configuration.

$$\text{At } 20\,000 \text{ ft} \qquad \sigma = 0.53281 \qquad \sigma^{0.8} = 0.60431 \qquad F_N = 72\,517 \text{ N}$$

At 280 KEAS and 54 442 kg, $C_L = 0.4425$ and $C_D = 0.03082$

$$\frac{F_N}{W} - \frac{C_D}{C_L} = \frac{72\,517}{54\,442 \times 9.81} - \frac{0.03082}{0.4425} = 0.0661 = \sin\gamma$$

This is a gradient of 6.63%.

At 20 000 ft ISA, 280 KEAS is equivalent to 383.6 KTAS or 648.3 ft/s. The rate of climb is then

$$\text{TAS} \times \sin\gamma = 648.3 \times 0.0661 = 42.9 \text{ ft/s}$$

The time taken to climb is 233.4 s and the distance flown is

$$\text{TAS} \times t \times \cos\gamma = 386.6 \times \frac{233.4}{3600} \times 0.9966 = 25 \text{ nm}$$

and the fuel required is

$$F_N \times \text{sfc} \times t = 72\,517 \times \frac{15.3}{10^6} \times 0.9287 \times 233.4 = 240.5 \text{ kg}$$

Cumulative totals	at 25 000 ft
Time	570.3 s/9.5 m
Distance	50.4 nm
Fuel	798.6 kg
Weight	54 201 kg

Climb 25 000 ft–35 000 ft (calculated at a mean of 30 000 ft)

At 25 000 ft, 280 KEAS is equivalent to 0.7 M, thus the climb continues at 0.7 M in the en-route configuration.

$$\text{At } 30\,000 \text{ ft} \qquad \sigma = 0.37413 \qquad \sigma^{0.8} = 0.45543 \qquad F_N = 54\,651 \text{ N}$$

At 0.7 M and 54 246 kg, $C_L = 0.5427$ and $C_D = 0.03516$

$$\frac{F_N}{W} - \frac{C_D}{C_L} = \frac{54\,651}{54\,201 \times 9.81} - \frac{0.03516}{0.5427} = 0.038 = \sin\gamma$$

This is a gradient of 3.8%.

At 30 000 ft ISA, 0.7 M is equivalent to 412.5 KTAS or 696.3 ft/s. The rate of climb is then

$$\text{TAS} \times \sin\gamma = 696.3 \times 0.038 = 26.5 \text{ ft/s}$$

The time taken to climb is 378 s and the distance flown is

$$\text{TAS} \times t \times \cos\gamma = 412.5 \times \frac{378}{3600} \times 0.9985 = 43.2 \text{ nm}$$

and the fuel required is

$$F_N \times sfc \times t = 54\,651 \times \frac{15.3}{10^6} \times 0.89092 \times 378 = 281.6\,kg$$

Cumulative totals at 35 000 ft
  Time   948.3 s/15.8 m
  Distance   93.6 nm
  Fuel   1080.2 kg
  Weight   53 920 kg

Climb performance summary

From the cumulative totals the climb summary can be drawn for MTOW, ISA conditions, at take-off. The fuel, distance and time required to climb to any altitude can be read from the summary. The summary is shown in Fig. 11.4.

This summary is based on a take-off weight of 55 000 kg. Since the climb performance is affected by the weight of the aircraft, the summary will normally include curves for a range of take-off weights so that the time, fuel and distance can be interpolated for the actual take-off weight.

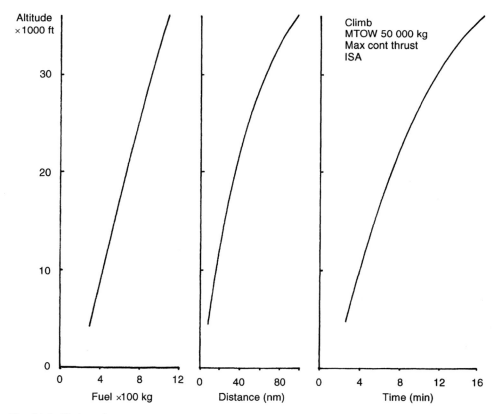

Fig. 11.4 Climb performance summary.

11.4.4 Cruise performance

The aircraft is required to cruise at 35 000 ft and 0.8 M for a cruise segment of 2000 nm. The initial weight at the start of cruise is 53 920 kg.

The cruise expression is given in Chapter 4, eqn (4.34) as,

$$R_3 = \left[\frac{V_{mdi}}{C} E_{max}\right] 2u_i \left\{ \tan^{-1}\left[\frac{1}{u_i^2}\right] - \tan^{-1}\left[\frac{1}{\omega u_i^2}\right] \right\}$$

The first stage is to evaluate the range factor. From the drag characteristic, the basic aerodynamic characteristics can be found,

$$C_{Lmd} = \left(\frac{C_{Dz}}{K}\right)^{\frac{1}{2}} = \left(\frac{0.0222}{0.044}\right)^{\frac{1}{2}} = 0.710$$

$$E_{max} = \frac{1}{2(C_{Dz}K)^{\frac{1}{2}}} = \frac{1}{2(0.0222 \times 0.044)^{\frac{1}{2}}} = 16$$

The minimum drag speed (TAS) at the start of cruise is

$$V_{mdi} = \left[\frac{2W}{\rho S C_{Lmd}}\right]^{\frac{1}{2}} = \left[\frac{2 \times 53\,920 \times 9.81}{0.3796 \times 95 \times 0.710}\right]^{\frac{1}{2}} = 203.3\,\text{m/s} \quad \text{or} \quad 395.1\,\text{KTAS}$$

The specific fuel consumption needs to be converted into units consistent with the other data. The most convenient form is in kg/Nhr.

$$C = 20.8\theta^{\frac{1}{2}}\,\text{mg/N s} = 0.07488\theta^{\frac{1}{2}}\,\text{kg/N hr}$$

It should be noted that, since the fuel is given in mass units and the thrust in force units, 'g' would be required in the range expression to maintain consistency of the units.

The range factor thus becomes

$$\left[\frac{V_{mdi}E_{max}}{Cg}\right] = \left[\frac{395.1 \times 16}{0.07488 \times 0.87141 \times 9.81}\right] = 9875.7\,\text{nm}$$

The range function now needs to be determined.

The TAS of the aircraft in cruise at 35 000 ft and 0.8 M is

$$0.8 \times 576.42 = 461.1\,\text{KTAS}$$

Hence the relative speed, u_i, at the start of cruise is

$$u_i = \frac{V}{V_{mdi}} = \frac{461.1}{395.1} = 1.167$$

The range function can now be partly evaluated

$$2u_i \left\{ \tan^{-1}\left[\frac{1}{u_i^2}\right] - \tan^{-1}\left[\frac{1}{\omega u_i^2}\right] \right\} = 2 \times 1.167 \left\{ 0.63336 - \tan^{-1}\left[\frac{1}{\omega u_i^2}\right] \right\}$$

$$2000 = 9875.7 \times 2 \times 1.167 \left\{ 0.63336 - \tan^{-1}\left[\frac{1}{\omega u_i^2}\right] \right\}$$

$$0.54659 = \tan^{-1}\left[\frac{1}{\omega u_i^2}\right]$$

leading to

$$\omega = 1.2068$$

and the final weight at the end of cruise is 44 678 kg.

Therefore, the fuel required for the cruise is 9241 kg.

The time required for the cruise is

$$t = \frac{\text{dist}}{\text{TAS}} = \frac{\text{dist}}{M \times a_0 \times \theta^{\frac{1}{2}}} = \frac{2000}{0.8 \times 661.48 \times 0.87141} = 4.337\,\text{hr} = 260.2\,\text{min}$$

Comment

Although the relative speed at the beginning of cruise, u_i, is 1.167 (which is less than the theoretical optimum of 1.316), the decrease in aircraft weight will reduce the minimum drag speed. Therefore, by the end of cruise the relative speed, u_f, will be 1.300, which is closer to the optimum. It would appear that the aircraft is not cruising at optimum altitude. From Fig. 11.3, the optimum cruise height for the aircraft would be about 30 000 ft at the beginning of cruise and 34 000 ft at the end of cruise. (This illustrates the principle of the 'cruise climb' in which the altitude is allowed to increase as fuel is burned to maintain the optimum cruise altitude.) The example shows the effect of cruising at non-optimum altitude, which is usually the case since the air traffic control will allocate a flight level for cruise based on the route and traffic separation. The reader could re-calculate the cruise for alternative cruise altitudes to look for the penalties incurred by not flying at the optimum altitude.

11.4.5 Descent performance

The descent from the cruise to the landing consists of three phases.

(i) The en-route descent from cruise down to the terminal area. This is flown in the en-route configuration and follows a speed schedule similar to that used for the climb.

(ii) The descent in the terminal area in which the aircraft is positioned for final approach; in this phase the speed will be regulated and the configuration changed.

(iii) The final approach, which is flown in the landing configuration.

The en-route descent follows a speed schedule that is both convenient for the control of the aircraft and gives a controlled rate and gradient of descent down to the terminal area. In the example, the schedule will call for a descent at 0.65 M/260 KEAS. An indicated Mach number of 0.65 M will be used down to 25 000 ft, at which the indicated airspeed will be 260 KEAS, below which the descent will continue at an airspeed of 260 KEAS. The descent performance will be estimated in segments of 10 000 ft down to 5000 ft at which height it will be assumed that the aircraft enters the terminal area.

To maintain a comfortable environment, the cabin of a transport aircraft is pressurized to the equivalent of 8000 ft pressure height. A constraint in the descent is that the aircraft must not descend at a rate that would require the rate of increase of cabin pressure to exceed 300 ft/min, otherwise the passengers may experience problems (the climb is not as limited in this way). This implies that there will be a minimum time required for the aircraft to descend determined by the cabin pressure height and the permissible rate of re-pressurization. If the cabin pressure is increased from 8000 ft in the cruise to 1000 ft at the terminal boundary, then the minimum time in the descent must be $7000/300 = 23.3$ min, or an average rate of descent of 1286 ft/min.

Descent 35 000 ft–25 000 ft (calculated at a mean height of 30 000 ft)

The descent is flown at 0.65 M in the en-route configuration. The initial weight is 44 678 kg and the maximum rate of descent is 1.286 ft/min or 21.4 ft/s
At 0.65 M the TAS is given by

$$V = M \times a_0 \times \theta^{\frac{1}{2}} = 0.65 \times 661.48 \times 0.89092 = 383.1 \, \text{KTAS} = 647.4 \, \text{ft/s}$$

The gradient of descent is

$$\sin\gamma = \frac{\mathrm{d}H/\mathrm{d}t}{V} = \frac{-21.4}{647.4} = -0.0331$$

$$C_{\text{L}} = \frac{2W}{\gamma p M^2 S} = \frac{2 \times 44\,678 \times 9.81}{1.4 \times 30\,090 \times 0.65^2 \times 95} = 0.5184$$

hence

$$C_{\text{D}} = 0.0222 + 0.044 \times 0.5184^2 = 0.03402$$

Thus, from the climb performance equation,

$$\frac{F_{\text{N}}}{W} = \frac{C_{\text{D}}}{C_{\text{L}}} + \sin\gamma = \frac{0.03402}{0.5184} - 0.0331 = 0.03252$$

Thus, the thrust required to maintain the rate of descent is

$$F_{\text{N}} = 0.03252 \times 44\,678 \times 9.81 = 14\,255 \, \text{N}$$

The time taken in the descent is set by the descent strategy. At 21.4 ft/s, it will be 467.3 s or 7.79 min.
The distance flown is

$$\text{TAS} \times t \times \cos\gamma = 383.1 \times \frac{467.3}{3600} \times 0.9995 = 49.7 \, \text{nm}$$

and the fuel required will be

$$F_{\text{N}} \times \text{sfc} \times t = 14\,255 \times \frac{20.8}{10^6} \times 0.89092 \times 467.3 = 123.4 \, \text{kg}$$

Cumulative totals   In descent only
   Time   467.3 s/7.79 m
   Distance   49.7 nm
   Fuel   123.4 kg
   Weight   44 555 kg (from take-off)

Descent 25 000 ft–15 000 ft (calculated at a mean height of 20 000 ft)

The descent is flown at 260 KEAS in the en-route configuration. The initial weight is 44 555 kg and the maximum rate of descent is 1.286 ft/min or 21.4 ft/s
   At 260 KEAS the TAS is given by

$$TAS = \frac{EAS}{\sigma^{\frac{1}{2}}} = \frac{260}{0.72994} = 356.2 \, KTAS = 602 \, ft/s = 183.2 \, m/s$$

The gradient of descent is

$$\sin\gamma = \frac{dH/dt}{V} = \frac{-21.4}{602} = -0.0355$$

$$C_L = \frac{2W}{\rho_0 V_e^2 S} = \frac{2 \times 44\,555 \times 9.81}{1225 \times 133.8^2 \times 95} = 0.4196$$

hence

$$C_D = 0.0222 + 0.044 \times 0.4196^2 = 0.02995$$

Thus, from the climb performance equation,

$$\frac{F_N}{W} = \frac{C_D}{C_L} + \sin\gamma = \frac{0.02995}{0.4196} - 0.0355 = 0.03587$$

Thus, the thrust required to maintain the rate of descent is

$$F_N = 0.03587 \times 44\,555 \times 9.81 = 15\,678 \, N$$

The time taken in the descent is set by the descent strategy. At 21.4 ft/s, it will be 467.3 sec or 7.79 min.
   The distance flown is

$$TAS \times t \times \cos\gamma = 356.2 \times \frac{467.3}{3600} \times 0.9994 = 46.2 \, nm$$

and the fuel required will be

$$F_N \times sfc \times t = 15\,678 \times \frac{20.8}{10^6} \times 0.9287 \times 467.3 = 141.5 \, kg$$

Cumulative totals	In descent only
Time	934.6 s/15.58 m
Distance	95.9 nm
Fuel	264.9 kg
Weight	44 414 kg (from take-off)

Descent 15 000 ft–5000 ft (calculated at a mean height of 10 000 ft)

The descent is flown at 260 KEAS in the en-route configuration. The initial weight is 44 414 kg and the maximum rate of descent is 1.286 ft/min or 21.4 ft/s.
   At 260 KEAS the TAS is given by

$$TAS = \frac{EAS}{\sigma^{\frac{1}{2}}} = \frac{260}{0.85935} = 302.6 \, KTAS = 511.3 \, ft/s = 155.6 \, m/s$$

The gradient of descent is

$$\sin \gamma = \frac{dH/dt}{V} = \frac{-21.4}{511.3} = -0.0419$$

$$C_L = \frac{2W}{\rho_0 V_e^2 S} = \frac{2 \times 44\,414 \times 9.81}{1225 \times 133.8^2 \times 95} = 0.4183$$

hence

$$C_D = 0.0222 + 0.044 \times 0.4183^2 = 0.02990$$

Thus, from the climb performance equation,

$$\frac{F_N}{W} = \frac{C_D}{C_L} + \sin \gamma = \frac{0.02990}{0.4183} - 0.0419 = 0.02958$$

Thus, the thrust required to maintain the rate of descent is

$$F_N = 0.02958 \times 44\,414 \times 9.81 = 12\,888\,N$$

The time taken in the descent is set by the descent strategy. At 21.4 ft/s, it will be 467.3 s or 7.79 min.

The distance flown is

$$\text{TAS} \times t \times \cos \gamma = 302.6 \times \frac{467.3}{3600} \times 0.9991 = 39.2\,nm$$

and the fuel required will be

$$F_N \times \text{sfc} \times t = 12\,888 \times \frac{20.8}{10^6} \times 0.96501 \times 467.3 = 118.3\,kg$$

Cumulative totals   In descent only to TMA boundary
 Time   1401.9 s/23.37 m
 Distance   135.1 nm
 Fuel   383.2 kg
 Weight   44 296 kg (from take-off)

Comment

This descent assumes that the aircraft is pressurized to 8000 ft cabin altitude in the cruise and that the descent is limited by the rate of cabin re-pressurization. If the cruise was at a lower altitude, the cabin could be pressurized to the same differential pressure resulting in a lower cabin pressure height. Therefore, the time taken in the descent could be reduced and the descent phase of the flight shortened. The descent performance summary assumes a pressurization level that allows the descent to be made at 1286 ft/min and so the fuel, distance and time required for a descent from any cruise altitude to 5000 ft can be read from the summary. The reader could calculate further descent profiles based on differing cabin pressurization levels and cruise altitudes. In the example, a simple cabin re-pressurization rate of 300 ft/min throughout the descent was assumed. In fact, the re-pressurization rate is 1095 Pa/min (which is equivalent to 300 ft/min at sea level), and gives a descent time of 20.5 min rather than 23.3 min. Recalculating for the higher rate of descent will show the benefit of the strategy.

Terminal manoeuvring area (TMA) descent

The operation of the aircraft in the TMA is under air traffic control and the descent flight path cannot always be flown at an optimum speed. Usually the airspeed is reduced to about 165 KEAS as the aircraft enters the TMA so that all traffic is operating at similar speeds to assist separation. This speed should not be less than the minimum drag speed of the aircraft. At the end of the TMA descent, the airspeed will be reduced further to the final approach speed. During the descent, in the TMA, the flaps will be selected down and the landing gear will be extended, changing the drag force. The engine thrust will have to be adjusted to give the airspeed and rate of descent required to maintain the flight path to the final approach.

The example assumes that the distance from the entry to the TMA to the touchdown point is 20 nm and that the final approach is made from a distance of 5 nm. If the gradient of the final approach is 3°, or 5.24%, then the height at the start of the final approach will be 1600 ft. The aircraft needs to descend from 5000 ft to 1600 ft in 15 nm, which is a mean gradient of 3.73%. The aircraft starts its deceleration from 260 KEAS as it enters the TMA by selecting take-off flap and landing gear down. The en-route descent thrust is maintained and the aircraft allowed to decelerate to 165 KEAS as it descends towards finals at the mean gradient of 3.73%. The weight at the TMA boundary is 44 296 kg.

The drag of the aircraft in the take-off configuration with landing gear extended is

$$68\,287\,\text{N at 260 KEAS} \qquad \text{and} \qquad 46\,558\,\text{N at 165 KEAS}$$

so that the mean drag during the deceleration is 57 432 N.

The mean decelerating force, F, is given by

$$F = F_N - D + W\sin\gamma = 12\,888 - 57\,432 + 44\,296 \times 9.81 \times 0.0373 = -28\,336\,\text{N}$$

and, using the simplified method from Chapter 6, the distance, S, to decelerate is given by

$$S = \int_{V_1}^{V_2} \frac{mV}{F}\,\mathrm{d}V = \frac{m}{2F}\left(\frac{V_{e2}^2 - V_{e1}^2}{\sigma}\right) = \frac{44\,296}{2 \times -28\,336}\left(\frac{84.9^2 - 133.8^2}{0.86167}\right)$$

$$= 9701\,\text{m} = 5.2\,\text{nm}$$

The time required will be

$$t = \frac{2S\sigma^{\frac{1}{2}}}{V_{e1} + V_{e2}} = \frac{2 \times 5.2 \times 0.92826 \times 60}{260 + 165} = 1.36\,\text{m} = 81.8\,\text{s}$$

and the fuel required will be

$$\text{Fuel} = F_N \times \text{sfc} \times t = 12\,888 \times \frac{20.8}{10^6} \times 0.98266 \times 81.8 = 21.5\,\text{kg}$$

The aircraft continues in a steady descent to the start of the final approach, a distance of 9.8 nm. The thrust in the steady descent at 165 KEAS is given by

$$F_N = D + W\sin\gamma = 46\,558 - 44\,296 \times 9.81 \times 0.0373 = 30\,350\,\text{N}$$

The TAS at the mean height of 2500 ft is

$$\text{TAS} = \frac{\text{EAS}}{\sigma^{\frac{1}{2}}} = \frac{165}{0.96378} = 171.2\,\text{KTAS}$$

and the time taken is

$$t = \frac{9.8 \times 60}{171.2} = 3.43\,\mathrm{m} = 206\,\mathrm{s}$$

and the fuel required is

$$\text{Fuel} = F_N \times \text{sfc} \times t = 30\,350 \times \frac{20.8}{10^6} \times 0.99136 \times 206 = 129\,\mathrm{kg}$$

The lift coefficient at minimum drag in the take-off configuration is 1.061 giving a minimum drag speed of 163 KEAS; using a speed of 165 KEAS will be acceptable for the approach.

Final approach

As the aircraft reaches the 3° glideslope on the final approach to touchdown the airspeed is further reduced and the descent gradient adjusted to 3°. The final approach is flown in the landing flap configuration, landing gear extended, at a speed not less than $1.3V_s$. This speed should not be less than the minimum drag speed so that the aircraft has flight path stability. To reduce the minimum drag speed the final approach will need to be flown with air brakes extended.

The stalling speed, based on the weight at the start of the TMA descent, will be

$$V_{es} = \left[\frac{2W}{\rho_0 S C_{L\,max}}\right]^{\frac{1}{2}} = \left[\frac{2 \times 44\,296 \times 9.81}{1.225 \times 95 \times 3.4}\right]^{\frac{1}{2}} = 46.9\,\mathrm{m/s} = 91.1\,\mathrm{KEAS}$$

This leads to a minimum approach speed of 118.5 KEAS.

The minimum drag speed occurs at $C_{Lmd} = 2.01 = C_{L\,max}/1.3^2$ so that the minimum approach speed based on the airworthiness requirements and the minimum drag speed are equal in this case. Therefore, an airspeed of 120 KEAS, giving a C_L of 1.96, would be an appropriate speed at which to fly the final approach. The drag coefficient at 120 KEAS is

$$C_D = C_{Dz} + KC_L^2 = 0.2200 + 0.0544 \times 1.96^2 = 0.4289$$

The thrust required at 120 KEAS in the approach configuration is

$$\frac{F_N}{W} = \frac{C_D}{C_L} - \sin 3° = \frac{0.4289}{1.96} - 0.0523 = 0.1665$$

giving

$$F_N = 44\,296 \times 9.81 \times 0.1665 = 72\,339\,\mathrm{N}$$

The time taken in the descent is given by

$$t = \frac{S}{\text{TAS}} = \frac{S\sigma^{\frac{1}{2}}}{\text{EAS}} = \frac{5 \times 3600 \times 0.98833}{120} = 151.8\,\mathrm{s}$$

and the fuel required is

$$\text{Fuel} = F_N \times \text{sfc} \times t = 72\,339 \times \frac{20.8}{10^6} \times 0.99725 \times 151.8 = 227.7\,\mathrm{kg}$$

Cumulative totals	From TMA boundary to touchdown
Time	439.6 s/7.33 m
Distance	20 nm
Fuel	378.2 kg
Weight	43 918 kg (from take-off)

and

Cumulative totals	For complete descent
Time	1841.5 s/30.69 m
Distance	155.1 nm
Fuel	761.4 kg
Weight	43 918 kg (from take-off)

Comment

A summary of the descent performance can be drawn in a similar fashion to the climb performance, and is shown in Fig. 11.5. Since the weight of the aircraft in the descent and landing phase will not be very variable, a single summary chart will usually be sufficient. However, since there may be cases in which an alternative descent strategy might be used, e.g. long-range descent or high-speed descent, summaries for the different descent strategies may be required. The reader may like to calculate further descent strategies and compare them with that given above.

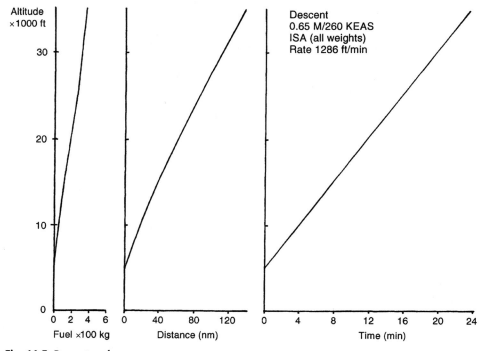

Fig. 11.5 Descent performance summary.

11.4.6 Landing

From the screen height of 50 ft (15 m) the thrust is reduced to idle and the rate of descent reduced to zero. The touchdown speed should be above the stalling speed which is 90.7 KEAS at the landing weight, and a touchdown speed of 100 KEAS will be assumed. Idle thrust will be taken to be 1000 N per engine and the drag of the aircraft can be estimated at the average airspeed in the flare, 110 KEAS, to be 99 555 N.

The airborne distance can be estimated from eqn (6.13),

$$S_a = \frac{-W}{(F_N - D)_{av}} \left\{ \frac{V_{app}^2 - V_{td}^2}{2g} + 50\,ft \right\} = \frac{-43\,918 \times 9.81}{(2000 - 99\,555)} \left\{ \frac{61.7^2 - 51.4^2}{2 \times 9.81} + 15 \right\}$$

$$= 328\,m$$

The nose wheel is lowered onto the runway as soon as possible after the touchdown and braking is applied. The pitch attitude of the aircraft in the ground run is $-2°$ so that the angle of attack of the wing is $0°$ and the lift coefficient is 2.0. Taking the braking coefficient of friction to be 0.4 and the runway to be level, the ground run distance can be calculated. From Chapter 6, eqns (6.16) and (6.17), the estimation of the ground run distance is given by

$$S_g = \frac{1}{2gB} \log_e(BV^2 + A)$$

where

$$BV^2 + A = \left[\frac{\rho SC_L}{2W} \left(\frac{C_D}{C_L} - \mu \right) V^2 + \left\{ \mu + \sin\gamma_R - \frac{F_N}{W} \right\} \right]$$

In the ground run at $C_L = 2.0$, $C_D = 0.4376$ so that

$$B = \frac{\rho SC_L}{2W} \left(\frac{C_D}{C_L} - \mu \right) = \frac{1.225 \times 95 \times 2.0}{2 \times 43\,918 \times 9.81} \left(\frac{0.4376}{2.0} - 0.4 \right) = -0.000049$$

and

$$A = \mu - \frac{F_N}{W} = 0.4 - \frac{2000}{43\,918 \times 9.81} = 0.395$$

and, when evaluated at $0.7V_{td}$,

$$BV^2 + A = -0.000049 \times 36^2 + 0.395 = 0.3315$$

Thus, the ground run is given by

$$S_g = \frac{1}{2gB} \log_e(BV^2 + A) = \frac{-1}{2 \times 9.81 \times 0.000049} \log_e(0.3315) = 1148\,m$$

Thus the ground run distance is 1148 m and the landing distance from the screen height is 1476 m.

11.4.7 Summary of trip time, distance and fuel required

The overall totals for the trip are as follows.

	Time (mins)	Distance (nm)	Fuel (kg)
Take-off	0.6	–	50
Climb	15.8	94	1080
Cruise	260.2	2000	9241
Descent/landing	30.7	155	761
Total	307.3	2249	11 133

11.4.8 Diversion and reserves

The flight plan must declare a diversion in the event of the destination airfield being unavailable for landing. The diversion is calculated on the assumption that it is flown from the decision height on the final approach (this is usually about 200 ft above the runway). It consists of a climb to a cruising height, cruise to the diversion airfield, a descent and a landing. The diversion fuel and time are calculated in the same way as the trip.

Fuel reserves must also be carried. There are two forms of fuel reserves.

(i) Regulatory reserves, usually fuel for 45 minutes cruise at the landing weight calculated at the best endurance speed. This is a 'final' reserve and should remain in the tanks after a diversion to the alternate airfield.

(ii) Contingency reserves, these are usually determined by the operator but, in some cases, may be subject to regulation. The reserve is normally a percentage of the trip fuel calculated for the flight between the departure and destination airfields. The purpose of the contingency reserve is to allow for deviations of the actual flight from the assumed flight plan. These deviations might include, cruise at a different flight level, differences between forecast and actual headwinds and temperatures, route changes and holding at the destination. The percentage may be at the discretion of the operator, and depends on experience of the route, but is often 10% of the trip fuel.

Regulatory fuel reserve
Maximum endurance is achieved by flying at the minimum drag speed. The reserve will be calculated for a typical landing weight of 45 000 kg and assume that the aircraft is flying at 5000 ft. The thrust required is given by

$$F_N = \frac{W}{(C_L/C_D)_{max}} = \frac{45\,000 \times 9.81}{16} = 27\,591\,N$$

and, therefore, the fuel required for 45 min flying at 5000 ft is given by

$$\text{Fuel} = F_N \times \text{sfc} \times t = 27\,591 \times \frac{20.8}{10^6} \times 0.98266 \times 60 \times 45 = 1523\,kg$$

Diversion

For the purposes of the exercise it will be assumed that the diversion airfield is 150 nm from the destination and that the diversion will be flown at 10 000 ft and cruising at 270 KEAS with standard climb and descent schedules. As an example of the use of the performance summary, the climb and descent segments of the diversion will be read from the performance summaries calculated above, Fig. 11.4. (The weight of the aircraft, however, will be the weight at landing, not at take-off.) The descent in the TMA to the diversion airfield will be assumed similar to that at the destination. The fuel reserves will be taken to be 45 minutes cruise at the landing weight plus 10% trip fuel.

	Time (min)	Distance (nm)	Fuel (kg)
Climb to 10 000 ft from decision height	4.2	17	420
Descent from 10 000 ft to 5000 ft at TMA	3.9	18	70
Descent in TMA to landing	7.3	20	378
Totals	15.4	55	868

Thus, the cruise segment of the diversion is 95 nm and the weight at beginning of cruise will be

$$43\,918 - 420 = 43\,498\,\text{kg}$$

and the time required for the diversion cruise is

$$t = \frac{\text{dist.}}{\text{TAS}} = \frac{95 \times 0.85935}{270} = 0.302\,\text{hr}$$

Since the cruise segment of the diversion is so short, it would be reasonable to approximate the diversion cruise fuel by taking the thrust at the start of cruise and assuming it is constant throughout the cruise segment. The lift coefficient at the start of cruise is given by,

$$C_\text{L} = \frac{2W}{\rho_0 V_\text{e}^2 S} = \frac{2 \times 43\,498 \times 9.81}{1.225 \times 138.9^2 \times 95} = 0.3801$$

and the drag coefficient by

$$C_\text{D} = 0.0222 + (0.044 \times 0.3801^2) = 0.02856$$

so that the thrust required will be

$$F_\text{N} = D = \tfrac{1}{2}\rho_0 V_\text{e}^2 S C_\text{D} = \tfrac{1}{2} \times 1.225 \times 138.9^2 \times 95 \times 0.02856 = 32\,059\,\text{N}$$

and the fuel required is

$$\text{Fuel} = F_\text{N} \times \text{sfc} \times t = 32\,059 \times \frac{20.8}{10^6} \times 0.96501 \times 3600 \times 0.302 = 700\,\text{kg}$$

(If the full range expression had been used to calculate the fuel required it would have given 697 kg, thus justifying the simplified approach used here for the diversion.)

Cumulative totals   For the diversion segment
  Time   33.5 mins
  Distance   150 nm
  Fuel   1568 kg
  Weight   42 798 kg (from take-off)

Fuel reserves and total fuel required

The fuel reserves required for the mission will be

  Regulatory   1523 kg
  Diversion   1568 kg
  Contingency (10% trip fuel)   1113 kg
  Total reserves   4204 kg

and the total fuel required will be

$$4204 + 11\,133 = 15\,337\,\text{kg}$$

11.4.9 Weight breakdown

The take-off weight of the aircraft was the maximum take-off weight of 55 000 kg which, when reduced by the operating empty weight of 30 000 kg, leaves a maximum disposable load of 25 000 kg. The fuel required for the mission was 15 337 kg so that the payload available was 9663 kg. Therefore, the aircraft was payload limited in this mission.

11.5 Payload–range diagram

The critical points on the payload–range diagram are as follows.

(i) The maximum payload range, which is the range over which the maximum structural payload can be transported.
(ii) The maximum fuel range, which is the maximum range over which the maximum disposable load can be carried, the payload being limited by fuel load.
(iii) The ferry range, which is the range that can be achieved with full fuel and no payload.

The conditions under which the payload–range diagram is calculated need to be specified. The example will be based on a standard climb and descent procedure and cruise at 35 000 ft, 0.8 M. Reserves for 45 min cruise at maximum endurance and a 150 nm diversion are assumed, based on the running example above, together with 10% trip fuel as contingency reserve.

(a) *Maximum payload range*

Fuel weight

	MTOW	55 000 kg
Less	Max payload	15 000 kg
Less	OEW	30 000 kg
Gives	Total fuel available	10 000 kg
	Total fuel available	10 000 kg
Less	45 min reserve	1523 kg
Less	Diversion	1568 kg
Gives	Total trip fuel	6909 kg
Less	10% contingency	691 kg
Gives	Trip fuel available	6218 kg
	Trip fuel available	6218 kg
Less	Take off and climb fuel	1130 kg
Less	Descent and landing	761 kg
Gives	Cruise fuel available	4327 kg

Aircraft weights

Initial cruise weight $55\,000 - 1130 = 53\,870$ kg

Final cruise weight $53\,870 - 4327 = 49\,543$ kg

Hence, fuel ratio, $\omega = \dfrac{W_i}{W_f} = \dfrac{53\,870}{49\,543} = 1.08734$

Initial minimum drag speed, V_{mdi},

$$V_{mdi} = \left[\frac{2W}{\rho S C_{Lmd}}\right]^{\frac{1}{2}} = \left[\frac{2 \times 53\,870 \times 9.81}{0.3796 \times 95 \times 0.71}\right]^{\frac{1}{2}} = 203.2\,\text{m/s} = 394.9\,\text{KTAS}$$

Cruise TAS, V,

$$V = M a_0 \theta^{\frac{1}{2}} = 0.8 \times 661.48 \times 0.87141 = 461.1\,\text{KTAS}$$

Hence

$$u_i = \frac{V}{V_{mdi}} = \frac{461.1}{394.9} = 1.1677$$

The range in constant altitude and Mach number cruise is given by the expression,

$$R = \left[\frac{V_{mdi}}{C} E_{max}\right] 2u_i \left\{ \tan^{-1}\left[\frac{1}{u_i^2}\right] - \tan^{-1}\left[\frac{1}{\omega u_i^2}\right]\right\}$$

Thus,

$$R = \left[\frac{394.9 \times 10^6 \times 16}{20.8 \times 3600 \times 0.87141 \times 9.81}\right] \times 2 \times 1.1677 \times \{0.63276 - 0.59337\} = 908\,\text{nm}$$

Now, including the distance flown in the climb, 93.6 nm, and in the descent, 155.1 nm. The maximum payload range is 1156.7 nm.

(b) *The maximum fuel range*

Payload weight

	MTOW	55 000 kg
Less	Max fuel	17 000 kg
Less	OEW	30 000 kg
Gives	Payload available	8000 kg
	Total fuel available	17 000 kg
Less	45 min reserve	1523 kg
Less	Diversion	1568 kg
Gives	Total trip fuel	13 909 kg
Less	10% contingency	1391 kg
Gives	Trip fuel available	12 518 kg
	Trip fuel available	12 518 kg
Less	Take off and climb fuel	1130 kg
Less	Descent and landing	761 kg
Gives	Cruise fuel available	10 627 kg

Aircraft weights

Initial cruise weight $\qquad 55\,000 - 1130 = 53\,870\,\text{kg}$

Final cruise weight $\qquad 53\,870 - 10\,627 = 43\,243\,\text{kg}$

Hence, fuel ratio, $\omega = \dfrac{W_i}{W_f} = \dfrac{53\,870}{43\,243} = 1.2457$

Initial minimum drag speed, V_{mdi},

$$V_{mdi} = \left[\frac{2W}{\rho S C_{Lmd}}\right]^{\frac{1}{2}} = \left[\frac{2 \times 53\,870 \times 9.81}{0.3796 \times 95 \times 0.71}\right]^{\frac{1}{2}} = 203.2\,\text{m/s} = 394.9\,\text{KTAS}$$

Cruise TAS, V,

$$V = Ma_0\theta^{\frac{1}{2}} = 0.8 \times 661.48 \times 0.87141 = 461.1\,\text{KTAS}$$

Hence

$$u_i = \frac{V}{V_{mdi}} = \frac{461.1}{394.9} = 1.1677$$

The range in constant altitude and Mach number cruise is given by the expression,

$$R = \left[\frac{V_{mdi}}{C}E_{max}\right]2u_i\left\{\tan^{-1}\left[\frac{1}{u_i^2}\right] - \tan^{-1}\left[\frac{1}{\omega u_i^2}\right]\right\}$$

Thus,

$$R = \left[\frac{394.9 \times 10^6 \times 16}{20.8 \times 3600 \times 0.87141 \times 9.81}\right] \times 2 \times 1.1677 \times \{0.63276 - 0.53210\} = 2320\,\text{nm}$$

Now, including the distance flown in the climb, 93.6 nm, and in the descent, 155.1 nm. The maximum fuel range is 2568.7 nm with a payload of 8000 kg.

(c) *Ferry range*

Aircraft take-off weight

	OEW	30 000 kg
Plus	Max fuel	17 000 kg
Gives	Take-off weight	47 000 kg
	Total fuel available	17 000 kg
Less	45 min reserve	1523 kg
Less	Diversion	1568 kg
Gives	Total trip fuel	13 909 kg
Less	10% contingency	1391 kg
Gives	Trip fuel available	12 518 kg
	Trip fuel available	12 518 kg
Less	Take-off and climb fuel*	1130 kg
Less	Descent and landing	761 kg
Gives	Cruise fuel available	10 627 kg

Aircraft weights

Initial cruise weight	$47\,000 - 1130 = 45\,870$ kg
Final cruise weight	$45\,870 - 10\,627 = 35\,243$ kg

Hence, fuel ratio, $\omega = \dfrac{W_i}{W_f} = \dfrac{45\,870}{35\,243} = 1.3015$

Initial minimum drag speed, V_{mdi},

$$V_{mdi} = \left[\frac{2W}{\rho S C_{Lmd}}\right]^{\frac{1}{2}} = \left[\frac{2 \times 45\,870 \times 9.81}{0.3796 \times 95 \times 0.71}\right]^{\frac{1}{2}} = 187.5\,\text{m/s} = 364.5\,\text{KTAS}$$

Cruise TAS, V,

$$V = Ma_0\theta^{\frac{1}{2}} = 0.8 \times 661.48 \times 0.87141 = 461.1\,\text{KTAS}$$

Hence

$$u_i = \frac{V}{V_{mdi}} = \frac{461.1}{364.5} = 1.265$$

Using the expression for range in constant altitude and Mach number cruise gives:

$$R = \left[\frac{364.5 \times 10^6 \times 16}{20.8 \times 3600 \times 0.87141 \times 9.81}\right] \times 2 \times 1.265 \times \{0.55854 - 0.44764\} = 2559\,\text{nm}$$

Now, including the distance flown in the climb, 93.6 nm, and in the descent, 155.1 nm, the maximum ferry range is 2808 nm.

The payload range diagram is shown in Fig. 11.6.

---

* The take-off and climb fuel has been taken from the running calculation for convenience. At the lower take-off weight of 47 000 kg, the fuel required to climb will be less than 1130 kg.

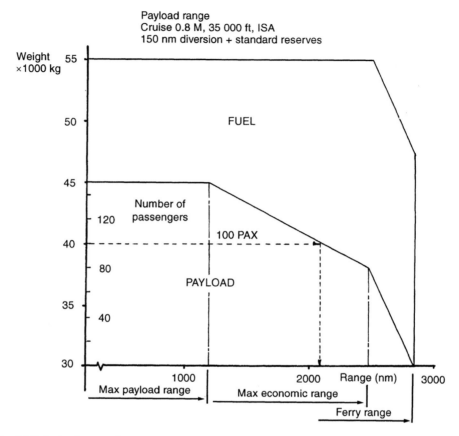

Fig. 11.6 Payload–range diagram.

11.6 Fuel required for a specified mission

In the examples above, the aircraft is assumed to take-off at its MTOW. In practice, the aircraft may be dispatched with a partial payload and on a mission that does not require all the fuel capacity available; the dispatch weight will, therefore be less than MTOW. Since carrying unnecessary fuel incurs a fuel penalty, the minimum fuel for the mission needs to be calculated. The process of calculating the fuel required for a specified mission progresses in reverse, from the final weight at the diversion to the take-off weight at the point of departure. In the following example, the aircraft is required to carry a payload of 9500 kg from its point of departure to a destination 1650 nm away. The mission is to be flown at 35 000 ft cruising at 0.8 M in ISA conditions. The alternate airfield is 150 nm from the destination.

Aircraft Weight

	OEW	30 000 kg
Plus	Payload	9500 kg
Gives	Zero fuel weight	39 500 kg
Plus	45 min reserve	1523 kg
Plus	Diversion	1568 kg
Gives	Min. weight at destination	42 591 kg
Plus	10% contingency (Est)[†]	500 kg
Gives	Weight at destination	43 091 kg
Plus	Descent and landing	761 kg
Gives	Weight at end of cruise	43 852 kg

The cruise distance will be the distance between the departure and destination airfields less the climb and descent distances,

$$1650 - (93.6 + 155.1) = 1401.3 \, \text{nm}$$

To calculate the fuel required, the range expression is written in terms of the final weight rather than the initial weight, this leads to a modified form of eqn (4.34),

$$R = \left[\frac{V_{mdf}}{C} E_{max} \right] 2u_f \left\{ \tan^{-1} \left[\frac{\omega}{u_f^2} \right] - \tan^{-1} \left[\frac{1}{u_f^2} \right] \right\}$$

thus the minimum drag speed and speed ratio are calculated for the aircraft weight at the end of the cruise rather than at the start of cruise.

The final minimum drag speed, V_{mdf}, is given by,

$$V_{mdf} = \left[\frac{2W_f}{\rho S C_{Lmd}} \right]^{\frac{1}{2}} = \left[\frac{2 \times 43\,852 \times 9.81}{0.3796 \times 95 \times 0.71} \right]^{\frac{1}{2}} = 183.3 \, \text{m/s} = 356.3 \, \text{KTAS}$$

Cruise TAS, V,

$$V = Ma_0 \theta^{\frac{1}{2}} = 0.8 \times 661.48 \times 0.87141 = 461.1 \, \text{KTAS}$$

Hence

$$u_f = \frac{V}{V_{mdf}} = \frac{461.1}{356.3} = 1.294$$

$$1401.3 = \left[\frac{2 \times 356.3 \times 10^6 \times 16}{20.8 \times 0.87141 \times 3600 \times 9.81} \right] \times 2 \times 1.294 \times \left\{ \tan^{-1} \left[\frac{\omega}{u_f^2} \right] - 0.53836 \right\}$$

giving,

$$\tan^{-1} \left[\frac{\omega}{u_f^2} \right] = 0.53836 + 0.03040$$

and

$$\omega = 1.0703$$

Thus the initial cruise weight is 46 936 kg and the cruise fuel required is 3084 kg.

[†] Since the cruise fuel is not known yet, an estimate has to be used here.

The trip fuel is, therefore,

$$1130 + 3084 + 761 = 4975 \, \text{kg}$$

(This figure justifies the estimated contingency fuel reserve of 500 kg.)

Adding the take-off and climb fuel to the weight at start of cruise gives the take-off weight of the aircraft to be 48 066 kg, which is well below the maximum take-off weight.

The fuel required for the mission, including reserves, is the take-off weight less the zero fuel weight,

$$\text{TOW} - \text{ZFW} = 48\,066 - 39\,500 = 8566 \, \text{kg}$$

From the payload–range diagram the fuel required for a mission of 1650 nm would be about 12 700 kg; however, this would assume a take-off at 55 000 kg and a payload of 12 300 kg. This shows the importance of calculating the fuel required for each individual case.

A number of simplifying assumptions have been made in this example. In particular, the take-off and climb fuel and distances have been taken from the fuel requirement at MTOW; this will cause an overestimate of the fuel required in this example. In practice, the take-off and climb performance summary will usually be presented for a range of take-off weights. The descent and landing will usually be performed at a weight that is somewhat less variable since it is the weight of the aircraft with normal payload and fuel reserves. Because of the small weight variation it may only be necessary to provide a descent and landing performance summary at one landing weight, in this case about 45 000 kg would be appropriate.

11.7 Extension of the performance analysis

These sample performance calculations are typical of those used in the design and development of the aircraft and in the construction of the operating data. They have been simplified by omitting some of the criteria that may limit the performance of the aircraft. These criteria include, minimum control speeds and EAS limits, and by assuming a simple parabolic drag polar and constant specific fuel consumption so that the basic performance algorithms can be applied. Nevertheless, the results produced are typical of a medium-range transport aircraft and illustrate the practical performance estimation process.

The reader is invited to extend the examples given in this chapter. This can be done in several ways.

- By using atmosphere states other than ISA and to consider alternative operating strategies for climb, cruise or descent.
- The effect of alternative flap settings can be seen by interpolation of the aerodynamic characteristics to optimize the trade-off between take-off and climb performance.
- Modifications to the basic aircraft can be proposed that will affect the datum weight and the aerodynamic characteristics or the powerplant characteristics. This enables cost-benefit analyses to be made on a proposed modification programme. An example of such a programme might be an increased capacity, short-range version of the aircraft – for example, by the extension of the fuselage to accommodate 180 seats, using the same wing and with no increase in MTOW.

Appendix A

Reference axis systems

The analysis of the flight path of an aircraft requires the definition of a framework of axes to which the motions can be referred. Several axis systems are used, since the forces acting on an aircraft arise from different sources. The equations of motion derived from these axes can be written in terms of the flight path in Earth-related axes (performance equations of motion) or as perturbation equations in aircraft related axes (stability equations of motion). All axis systems are right-handed with their origin, O, at the centre of gravity, CG, of the aircraft so that only the angular relationships between the axis systems need be considered. Figure A1 summarizes the most common axis systems and provides the definition of some of the important parameters in flight path performance and the relationships between them.

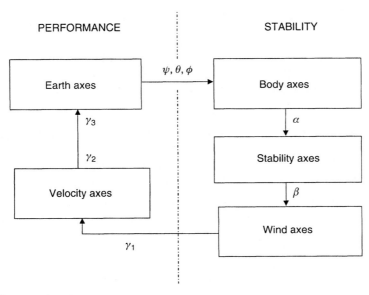

Fig. A1 Reference axis systems.

Earth axes

The datum for aircraft motions is taken to be the local Earth axes defined by the horizontal plane at the geographical location of the aircraft. Earth axes are oriented OX_e Northerly and OY_e Easterly in the horizontal plane with OZ_e vertically downward, see Fig. A2. In some special circumstances, for example, inertial navigation systems, the curvature of the Earth must be taken into account and the consequent rotation of the axes with global position considered. These special cases will not be considered here and a 'flat Earth' system will be assumed to exist.

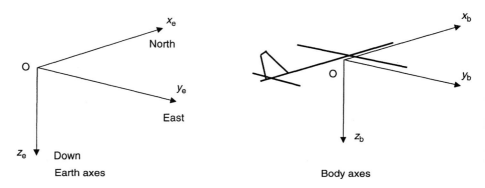

Fig. A2 Earth and body axis systems.

Body axes

Usually the aircraft has a plane of symmetry that is taken as the basis for the body axis system. The OX_b axis is taken to be forwards parallel to a convenient aircraft datum, such as the fuselage construction datum line, and OZ_b downwards in the plane of symmetry. The OY_b axis is normal to the plane of symmetry, positive to starboard. Aircraft instrumentation for the measurement of attitudes, rates of rotation and accelerations is usually aligned to the body axis system.

Body-to-earth axis relationship

In flight, the aircraft will change its orientation with respect to the Earth. The body axes are related to the Earth axes by the aircraft attitude angles; these are as follows.

(i) The yaw attitude, or azimuth angle, ψ, provides the aircraft heading with respect to North. It is taken to be positive in rotation from North towards East. In most cases the aircraft heading does not affect the analysis and it is assumed that the datum value of ψ is zero; ψ can then be taken to be the yaw attitude of the aircraft with respect to an arbitrary heading.

(ii) The pitch attitude, θ, is the angle between the aircraft longitudinal body axis, OX_b, and the horizontal plane; θ is positive, aircraft nose up.

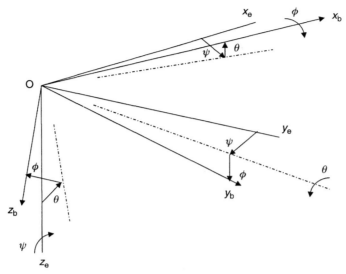

Fig. A3 Earth to body axis transfer.

(iii) The bank attitude, ϕ, is the angle between the aircraft lateral body axis, OY_b, and the horizontal plane; ϕ is positive, starboard wing down.

The transfer of force vectors between Earth axes and body axes is effected by rotation through heading, pitch attitude and bank attitude respectively. It is shown in Fig. A3. The order in which the rotations are performed is important in the development of the equations of motion.

$$[\mathbf{F}]_b = \begin{bmatrix} 1 & 0 & 0 \\ 0 & \cos\phi & \sin\phi \\ 0 & -\sin\phi & \cos\phi \end{bmatrix} \begin{bmatrix} \cos\theta & 0 & -\sin\theta \\ 0 & 1 & 0 \\ \sin\theta & 0 & \cos\theta \end{bmatrix} \begin{bmatrix} \cos\psi & \sin\psi & 0 \\ -\sin\psi & \cos\psi & 0 \\ 0 & 0 & 1 \end{bmatrix} [\mathbf{F}]_e$$

Stability axes and wind axes

An aircraft derives its lift from the angle of attack, α, of the wing relative to the airflow, thus the body axis system is inclined to the flight path to provide the necessary angle of attack. It should be noted that generally, the longitudinal body axis is not aligned with the zero-lift angle of attack of the wing and so the angle of attack may be measured from an arbitrary body axis datum. Since the analysis of the stability of an aircraft involves forces generated by a disturbance from the steady state, or trim condition, the reference axes for stability are relative to the undisturbed flow direction. Therefore, the body axis system is rotated through an angle α about the OY_b axis; this is now the stability axis system, OX_s, OY_s, OZ_s. Conventionally, the angle of attack is taken to be positive when the aircraft is nose-up with respect to the wind, thus the transfer from body to stability axes involves a 'negative' rotation about OY_b.

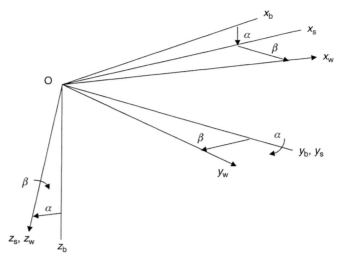

Fig. A4 Body to stability and stability to wind axis transfer.

Whilst the stability axis system provides the basis for analysis of the forces resulting from a disturbance from the symmetric flight state, cases exist in which the datum condition of flight is not symmetric. In asymmetric flight, the aircraft has an angle of sideslip, β, about the OZ_s axis and the flight axes now become the wind axis system in which the principal direction, OX_w, is the direction of flight. Stability axes are, therefore, a special case of wind axes in which $\beta = 0$. The sideslip angle is taken to be positive to starboard so that the transfer from stability to wind axes is made by a 'positive' rotation about OZ_s.

The transfer from body axes to stability axes and stability axes to wind axes is effected by rotation through the angle of attack, α, and the angle of sideslip, β, respectively and is shown in Fig. A4.

$$[\mathbf{F}]_s = \begin{bmatrix} \cos\alpha & 0 & \sin\alpha \\ 0 & 1 & 0 \\ -\sin\alpha & 0 & \cos\alpha \end{bmatrix} [\mathbf{F}]_b$$

$$[\mathbf{F}]_w = \begin{bmatrix} \cos\beta & \sin\beta & 0 \\ -\sin\beta & \cos\beta & 0 \\ 0 & 0 & 1 \end{bmatrix} [\mathbf{F}]_s$$

Velocity axes

The velocity axis system describes the path of the aircraft relative to the Earth by defining the track, γ_3, (relative to North in the horizontal plane), and the flight path gradient, γ_2, (the flight path angle relative to the horizontal plane). The reference direction, OX_v, is along the flight path and is co-incident with the wind axis OX_w.

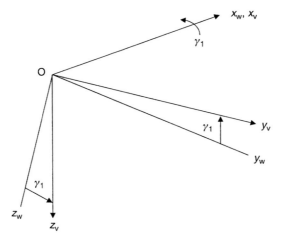

Fig. A5 Wind to velocity axis transfer.

To define the flight path in terms of its gradient the OX_vZ_v plane must be vertical. Therefore, the velocity axes are produced by rotating the lateral wind axis, OY_w, back into the horizontal plane through the bank angle, γ_1, in the 'negative' direction. The transfer of the force vector between wind axes and velocity axes is shown in Fig. A5.

$$[\mathbf{F}]_v = \begin{bmatrix} 1 & 0 & 0 \\ 0 & \cos\gamma_1 & -\sin\gamma_1 \\ 0 & \sin\gamma_1 & \cos\gamma_1 \end{bmatrix} [\mathbf{F}]_w$$

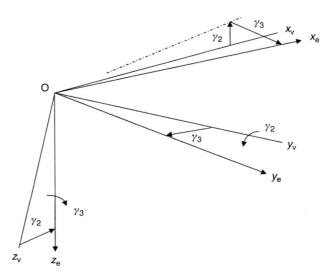

Fig. A6 Velocity to earth axis transfer.

The velocity axes are related to the Earth axes by the track, γ_3, and the flight path gradient, γ_2. Figure A6 shows the transfer of the force vector from velocity to Earth axes, both of which will be 'negative' rotations in that direction.

$$[\mathbf{F}]_e = \begin{bmatrix} \cos\gamma_3 & -\sin\gamma_3 & 0 \\ \sin\gamma_3 & \cos\gamma_3 & 0 \\ 0 & 0 & 1 \end{bmatrix} \begin{bmatrix} \cos\gamma_2 & 0 & \sin\gamma_2 \\ 0 & 1 & 0 \\ -\sin\gamma_2 & 0 & \cos\gamma_2 \end{bmatrix} [\mathbf{F}]_v$$

Angular relationships

The axis transfers described above are summarized in Table A1.

Table A1. Axis transfers and angular relationships

Axis transfer	Earth to body	Body to stability	Stability to wind	Wind to velocity	Velocity to earth
Pitch	θ	α	–	–	γ_2
Roll	ϕ	–	–	γ_1	–
Yaw	ψ	–	β	–	γ_3

(i) *Pitch related angles*

$$\theta = \alpha + \gamma_2$$

Pitch attitude = angle of attack + flight path gradient

The pitch related angles are shown in Fig. A7. The pitch attitude, θ, is the angle between the aircraft longitudinal body axis and the horizontal plane and is the angle of the aircraft relative to the horizon observed from the aircraft (positive nose up). The flight path gradient, γ_2, is the angle of the velocity vector, or principal velocity axis, relative to the horizontal. It is taken to be positive in the climb since it is calculated from the rate of climb, dH/dt, and true airspeed. It is not usually possible to observe the flight path gradient unless there is a physical reference near the aircraft, for example a cloud layer, and then any observation will only be qualitative. The angle of attack, α, (positive nose up) is the difference between the pitch attitude and flight path gradient.

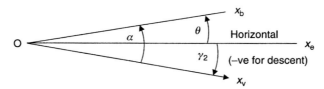

Fig. A7 Angular relationships in pitch.

(ii) *Roll related angles*

$$\gamma_1 \quad \text{and} \quad \phi$$

The bank attitude, ϕ, is the angle between the horizontal plane and the aircraft lateral body axis and can be regarded as the attitude of the aircraft as seen from the Earth. The bank angle, γ_1, is the angle between the plane of the flight path and the aircraft lateral axis and can be regarded as the horizon as seen from the aircraft. In horizontal flight the bank angle and bank attitude are equal but in climbing or descending flight the bank angle will be modified by a component of the flight path gradient.

(iii) *Yaw related angles*

$$\psi = \gamma_3 - \beta$$

$$\text{Heading} = \text{track} - \text{sideslip}$$

The heading, ψ, is the direction in which the aircraft body axis is pointing; this is the 'compass' direction. The track, γ_3, is the path of the aircraft projected onto the horizontal plane, assuming still air conditions. The difference, in horizontal flight, between track and heading is sideslip, β, which is due to asymmetry of flight; in symmetric flight the sideslip angle is zero. Conventionally, track and heading are measured from North towards East on the compass rose. Figure A8 shows the yaw related angles.

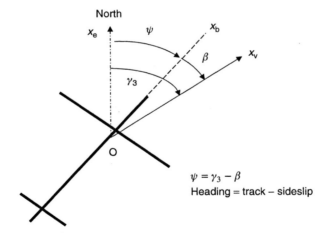

Fig. A8 Angular relationships in azimuth.

The performance equations of motion

In performance, the motion of the aircraft is considered relative to the Earth, or more exactly, relative to the general air mass assuming still air conditions. The motion of the aircraft is determined by the forces acting on it, and the changes that occur in those forces as the flight conditions vary produce the performance characteristics of the aircraft. The first stage of the development of the performance characteristics must be the establishment of the forces acting on the aircraft that arise from the gravitational, aerodynamic and powerplant sources. These determine the external forces acting on the aircraft which, under Newton's law, cause the aircraft to accelerate and change its state of motion. The general equation of motion can thus be expressed as,

$$[\mathbf{F_a}] + [\mathbf{F_p}] + [\mathbf{F_g}] = [\mathbf{F_I}] \tag{B1}$$

where $[\mathbf{F_a}]$, $[\mathbf{F_p}]$ and $[\mathbf{F_g}]$ are the aerodynamic, propulsive and gravitational force vectors respectively and $[\mathbf{F_I}]$ is the resultant inertial force vector.

The aircraft is considered to be in trim (that is, all the forces and moments are in balance), and therefore its trajectory can be considered to be affected only by the forces acting at its centre of gravity (CG). The moments acting about the axes through the CG have no direct contribution to the trajectory and so have no relevance to the performance equations of motion. The trajectory, or flight path, of the aircraft is related to the Earth by its velocity vector. Therefore, it is necessary to derive the equations of motion of the aircraft in terms of the velocity axis system as defined in Appendix A.

The performance equations of motion will be developed using simple statements of the aerodynamic and propulsive forces, the characteristics of these forces and their variation with the flight variables, weight, altitude, temperature and Mach number, are considered in Chapter 3.

The inertial forces

The external forces acting on the aircraft produce the accelerations that change its flight path; they can be written in terms of the velocity axes as,

$$[\mathbf{F_I}]_\upsilon = \begin{bmatrix} F_x \\ F_y \\ F_z \end{bmatrix}_\upsilon = \frac{\mathrm{d}(mV)_\upsilon}{\mathrm{d}t} \tag{B2}$$

leading to,

$$[\mathbf{F_I}]_\upsilon = m(\dot{\mathbf{V}} + \boldsymbol{\omega} \times \mathbf{V})_\upsilon + \dot{m}(\mathbf{V})_\upsilon \tag{B3}$$

where ($\boldsymbol{\omega}$) is the angular velocity vector of the aircraft defined by the rate of pitch, $\dot{\gamma}_2$, and rate of turn, $\dot{\gamma}_3$. These can be resolved into the velocity axis system giving,

$$(\boldsymbol{\omega})_\upsilon = \begin{bmatrix} 0 \\ \dot{\gamma}_2 \\ 0 \end{bmatrix} + \begin{bmatrix} \cos\gamma_2 & 0 & -\sin\gamma_2 \\ 0 & 1 & 0 \\ \sin\gamma_2 & 0 & \cos\gamma_2 \end{bmatrix} \begin{bmatrix} 0 \\ 0 \\ \dot{\gamma}_3 \end{bmatrix} = \begin{bmatrix} -\dot{\gamma}_3\sin\gamma_2 \\ \dot{\gamma}_2 \\ \dot{\gamma}_3\cos\gamma_2 \end{bmatrix}_\upsilon \tag{B4}$$

Equation (B3) thus becomes,

$$[\mathbf{F_I}]_\upsilon = m\begin{bmatrix} \dot{V} \\ 0 \\ 0 \end{bmatrix}_\upsilon + m\begin{bmatrix} -\dot{\gamma}_3\sin\gamma_2 \\ \dot{\gamma}_2 \\ \dot{\gamma}_3\cos\gamma_2 \end{bmatrix} \times \begin{bmatrix} V \\ 0 \\ 0 \end{bmatrix}_\upsilon + \dot{m}\begin{bmatrix} V \\ 0 \\ 0 \end{bmatrix}_\upsilon \tag{B5}$$

and,

$$\begin{bmatrix} F_x \\ F_y \\ F_z \end{bmatrix}_\upsilon = m\begin{bmatrix} \dot{V} \\ V\dot{\gamma}_3\cos\gamma_2 \\ -V\dot{\gamma}_2 \end{bmatrix}_\upsilon + \dot{m}\begin{bmatrix} V \\ 0 \\ 0 \end{bmatrix}_\upsilon \tag{B6}$$

Usually the rate of change of mass, which represents the fuel flow, is small enough to be negligible and it will not be included in the further development of the equations of motion. In some cases, however, it may be large; for example, a missile with a rocket engine, and the rate of change of mass will need to be considered in the inertial force term.

The gravitational force

The gravitational force acts, by definition, in the Earth axis system, thus,

$$[\mathbf{F_g}]_e = \begin{bmatrix} F_{gx} \\ F_{gy} \\ F_{gz} \end{bmatrix}_e = \begin{bmatrix} 0 \\ 0 \\ W \end{bmatrix}_e \tag{B7}$$

This can be transferred from Earth to velocity axes by the transfer matrix, eqn (A2), giving

$$[\mathbf{F_g}]_\upsilon = \begin{bmatrix} -W\sin\gamma_2 \\ 0 \\ W\cos\gamma_2 \end{bmatrix}_\upsilon \tag{B8}$$

The aerodynamic forces

The general aerodynamic force acts in the wind axis system and can be expressed as,

$$[\mathbf{F_a}]_w = \begin{bmatrix} F_{ax} \\ F_{ay} \\ F_{az} \end{bmatrix}_w = \begin{bmatrix} -D \\ Y \\ -L \end{bmatrix}_w \tag{B9}$$

This can be transferred from wind axes to velocity axes through the bank angle, γ_1, using eqn (A3), to give,

$$[\mathbf{F}_a]_v = \begin{bmatrix} -D \\ Y\cos\gamma_1 + L\sin\gamma_1 \\ Y\sin\gamma_1 - L\cos\gamma_1 \end{bmatrix}_v \tag{B10}$$

The propulsive forces

The force vector of a thrust-producing engine will be used in this development of the equations of motion. The force vector of any other propulsive system – for example, propellers or rockets – can be written in a similar form and the resulting equations of motion will differ only in detail.

Figure B1 shows the thrust force of an aircraft with thrust-producing engines. The gross thrust, F_G, of each engine is produced by the gas flow from the exhaust, and may have both vertical, τ_1, and lateral, τ_2, deflection with respect to the aircraft body axis system. The reason for these deflections may be to reduce the asymmetric thrust due to engine failure or to reduce the effects of exhaust temperature or acoustic pressure levels on the local structure. The gross thrust must first be resolved from engine axes into aircraft body axes

$$[\mathbf{F}_G]_b = \begin{bmatrix} \cos\tau_1 & 0 & \sin\tau_1 \\ 0 & 1 & 0 \\ -\sin\tau_1 & 0 & \cos\tau_1 \end{bmatrix} \begin{bmatrix} \cos\tau_2 & \sin\tau_2 & 0 \\ -\sin\tau_2 & \cos\tau_2 & 0 \\ 0 & 0 & 1 \end{bmatrix} [\mathbf{F}_G]_E \tag{B11}$$

where subscript E refers to engine axes.

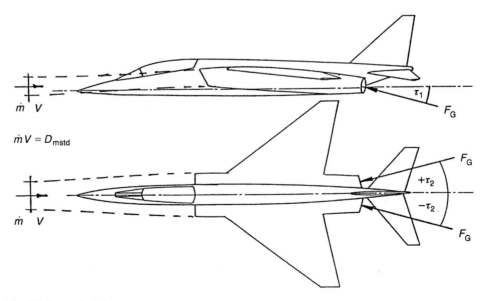

Fig. B1 Powerplant thrust components.

This gives,

$$[\mathbf{F_G}]_b = \begin{bmatrix} \cos\tau_1 \cos\tau_2 \\ -\sin\tau_2 \\ -\sin\tau_1 \cos\tau_2 \end{bmatrix} [\mathbf{F_G}]_E \qquad (B12)$$

The gross thrust components can then be transferred from body to velocity axes by the transfer matrices given in Appendix A. This involves transfer through stability and wind axes and results in the following expression,

$$[\mathbf{F_G}]_v = \begin{bmatrix} (\cos\tau_1 \cos\tau_2 \cos\alpha\cos\beta - \sin\tau_1 \cos\tau_2 \sin\alpha\cos\beta - \sin\tau_2 \sin\beta) \\ (-\cos\tau_1 \cos\tau_2 \cos\alpha\sin\beta\cos\gamma_1 + \sin\tau_1 \cos\tau_2 \sin\alpha\sin\beta\cos\gamma_1 + \\ \cos\tau_1 \cos\tau_2 \sin\alpha\sin\gamma_1 + \sin\tau_1 \cos\tau_2 \cos\alpha\sin\gamma_1 - \sin\tau_2 \cos\beta\cos\gamma_1) \\ (-\cos\tau_1 \cos\tau_2 \cos\alpha\sin\beta\sin\gamma_1 + \sin\tau_1 \cos\tau_2 \sin\alpha\sin\beta\sin\gamma_1 - \\ \cos\tau_1 \cos\tau_2 \sin\alpha\cos\gamma_1 - \sin\tau_1 \cos\tau_2 \cos\alpha\cos\gamma_1 - \sin\tau_2 \cos\beta\sin\gamma_1) \end{bmatrix} [\mathbf{F_G}]_E$$

$$(B13)$$

This expression is completely general and enables the components of the gross thrust of each engine to be determined under all flight conditions.

Usually, certain simplifying assumptions can be applied. If pairs of engines are disposed symmetrically about the longitudinal body axis of the aircraft and are producing equal thrust then, since the lateral angle τ_2 is positive to starboard and negative to port, the sideforce components involving $\sin\tau_2$ will cancel. This enables the gross thrust to be taken to be T, where,

$$T = n\cos\tau_2 \mathbf{F_G} \qquad (B14)$$

The powerplant gross thrust vector can now be simplified into the form,

$$[\mathbf{F_G}]_v = \begin{bmatrix} \cos(\alpha+\tau_1)\cos\beta \\ -\cos(\alpha+\tau_1)\sin\beta\cos\gamma_1 + \sin(\alpha+\tau_1)\sin\gamma_1 \\ -\cos(\alpha+\tau_1)\sin\beta\sin\gamma_1 + \sin(\alpha+\tau_1)\cos\gamma_1 \end{bmatrix} T \qquad (B15)$$

Here it should be noted that τ_1 and τ_2 are usually constants but in the case of a vectored thrust engine they may be variables in which case the combination with angle of attack, α, may not be convenient. In the case of aircraft with an odd number of engines, or operating with one engine inoperative, the odd engine will have to be considered separately if the value of either τ_1 or τ_2 differs from the other engines.

The powerplant net thrust vector, $\mathbf{F_N}$, is found by including the momentum drag with the gross thrust. The momentum drag represents the loss of momentum of the airflow entering the engine intake. The air entering the engine is then processed and ejected from the exhaust nozzle to produce the gross thrust from the exhaust momentum. The net thrust is the difference between the momentum of the intake and exhaust flows,

$$[\mathbf{F_N}]_v = [\mathbf{F_G}]_v + [\mathbf{D_m}]_v \qquad (B16)$$

where the momentum drag, which acts in the velocity axis system, is given by,

$$[\mathbf{D}_m]_\upsilon = \begin{bmatrix} -D_m \\ 0 \\ 0 \end{bmatrix}_\upsilon \tag{B17}$$

Usually the net thrust is assumed to refer to the net propulsive force in the direction of flight, the Ox_υ direction, and the components of the gross thrust in the lateral and normal directions are assumed negligible. This assumption is reasonable for most aircraft but should be justified before being applied generally.

When the momentum drag is included in eqn (B15), the net propulsive thrust vector is given as,

$$[\mathbf{F}_p]_\upsilon = \begin{bmatrix} T\cos(\alpha + \tau_1)\cos\beta - D_m \\ T(-\cos(\alpha + \tau_1)\sin\beta\cos\gamma_1 + \sin(\alpha + \tau_1)\sin\gamma_1) \\ T(-\cos(\alpha + \tau_1)\sin\beta\sin\gamma_1 + \sin(\alpha + \tau_1)\cos\gamma_1) \end{bmatrix} \tag{B18}$$

Although the intake momentum drag appears as a drag force it is a part of the engine thrust and not part of the airframe drag since it depends on the air mass flow that passes through the engine. It is important that it should be included in the thrust and not in the airframe drag otherwise the estimation of the minimum drag speed will be in error.

Equation (B18) represents the powerplant forces in their most convenient form for development of the equations of motion of the aircraft.

The performance equations of motion

Equation (B1) can now be completed from eqns (B6), (B8), (B10) and (B18) to give the general performance equations of motion,

$$\left.\begin{aligned} -D + \{T\cos(\alpha + \tau_1)\cos\beta - D_m\} - W\sin\gamma_2 &= m\dot{V} + \dot{m}V \\ Y\cos\gamma_1 + L\sin\gamma_1 + T\{-\cos(\alpha + \tau_1)\sin\beta\cos\gamma_1 + \sin(\alpha + \tau_1)\sin\gamma_1\} &= mV\dot{\gamma}_3\cos\gamma_2 \\ Y\sin\gamma_1 - L\cos\gamma_1 + T\{-\cos(\alpha + \tau_1)\sin\beta\sin\gamma_1 - \sin(\alpha + \tau_1)\cos\gamma_1\} + W\cos\gamma_2 &= -mV\dot{\gamma}_2 \end{aligned}\right\} \tag{B19}$$

These are now the generalized equations of motion for an aircraft with thrust producing engines in their most convenient form. However, they do contain the assumptions made in eqn (B14) and they should only be used when those assumptions apply.

Special flight conditions

In many cases, one or more of the terms in the performance equations of motion can be taken to be zero, simplifying their application to a specific problem. These cases are,

(a) Non-accelerating flight $\dot{V} = 0$ steady TAS.
(b) Non-manoeuvring flight $\dot{\gamma}_2 = 0$ steady rate of climb or descent.
(c) Straight flight $\dot{\gamma}_3 = 0$ no rate of turn.
(d) Symmetric flight $\beta = 0$ no sideslip.
(e) Wings level flight $\gamma_1 = 0$ no bank angle.
(f) Level flight $\gamma_2 = 0$ no rate of climb or descent.
(g) Constant mass $\dot{m} = 0$ negligible fuel mass flow.

Steady flight is usually satisfied by conditions (a), (b) and (g).

The performance equations for steady, level, straight, symmetric flight with wings level will reduce to the form,

$$\left. \begin{aligned} -D + F_{\text{N}} &= 0 \\ Y &= 0 \\ -L - T\sin(\alpha + \tau_1) + W &= 0 \end{aligned} \right\} \tag{B20}$$

A large proportion of performance analysis can be dealt with in terms of these simplified equations of motion.

The International Standard Atmosphere model

The development of the International Standard Atmosphere (ISA) model was discussed fully in Chapter 2. Here, the principal characteristics are summarized as a background to the tables of properties of the ISA model. Although the International Standard Atmosphere is defined in terms of a metric height scale, aircraft operations are normally conducted with the height indicaions in terms of the Imperial unit of the foot. For this reason, the tables of the International Standard Atmosphere given in this Appendix will be expressed in metric units and in imperial units.

The International Standard Atmosphere model datum

The ISA model datum is taken to be the height in the atmosphere at which the static pressure is $101\,325\,\text{N/m}^2$. (The datum is often referred to, incorrectly, as 'sea level' and it should be noted that the actual pressure in the real atmosphere at sea level generally would not be the same as the ISA datum pressure.)

The temperature in the ISA model at the datum pressure height is taken to be $288.15\,\text{K}$ ($15°\text{C}$).

Since the atmosphere obeys the equation of state, $p = \rho T R$, (eqn (2.1)), in which the gas constant, $R = 287.05287\,\text{Nm/kg K}$, the ISA datum properties can be summarized as,

$$\left. \begin{array}{l} p_0 = \text{datum pressure} = 101\,325\,\text{N/m}^2 \\ T_0 = \text{datum temperature} = 288.15\,\text{K} \\ \rho_0 = \text{datum density} = 1.225\,\text{kg/m}^3 \end{array} \right\} \tag{C1}$$

The ISA model consists of three segments, shown in Fig. 2.6, these are as follows.

- The *troposphere*, which extends from datum to the *tropopause* at a geopotential height of 11 km (36 089 ft).
- The *lower stratosphere* which extends from the tropopause to a geopotential height of 20 km (65 617 ft).
- The *middle stratosphere* which extends from 20 km to a geopotential height of 32 km (104 987 ft).

Temperature–height profile

The temperature–height profile in the ISA model is a defined series of three linear segments that determine the profiles of the characteristic properties of the atmosphere,

$$T = T_i + L_i(H - H_i) \tag{2.2}$$

where subscript i refers to the datum level of each segment of the atmosphere, see Fig. 2.6.

The temperature lapse rates, L_i, on which the ISA model is defined, are,

$$L_0 = -0.00650 \, \text{K/m}$$
$$L_{11} = 0 \, \text{K/m} \tag{C2}$$
$$L_{20} = +0.001 \, \text{K/m}$$

Pressure–height profile

The pressure–height relationship in the ISA model was given in eqns (2.7) and (2.8) as,

$$p = p_0 \left[1 + \frac{L_i}{T_i} (H - H_i) \right]^{\frac{-g_i}{RL_i}} \tag{2.7}$$

where $L_i \neq 0$ and,

$$p = p_i \exp \left[-\frac{g_0}{RT_i} (H - H_i) \right] \tag{2.8}$$

where $L_i = 0$.

Numerically, these become,

$$\left. \begin{aligned} &p = 101\,325[1 - 0.000022558H]^{5.25588} \text{ in the troposphere,} \\ &p = 22\,632 \exp[-0.000157688(H - 11\,000)] \text{ in the lower stratosphere, and} \\ &p = 5474.9[1 + 0.000004616(H - 20\,000)]^{-34.1632} \text{ in the middle stratosphere} \end{aligned} \right\} \tag{C3}$$

where the heights are in metres.

The pressure–height relationships are independent of temperature and so they are valid for all atmosphere states. They form the calibration equations for the altimeter, Section 2.5.1, so that, when the altimeter is referenced to standard datum pressure, 1013 mb, heights indicated by the altimeter can be converted into atmosphere static pressures. They are known as pressure heights since they are derived from the atmosphere pressure. The ISA tables, therefore, can be used to convert pressure heights into atmosphere static pressures in any atmosphere state.

The density–height profile

The density–height profile is derived from the equation of state using the pressure and temperature determined from eqns (C2) and (C3).

Derived relationships

The relative properties of the ISA model in eqn (2.10) are

$$\left. \begin{array}{l} \delta = p/p_0, \text{ the relative pressure} \\ \theta = T/T_0, \text{ the relative temperature, and} \\ \sigma = \rho/\rho_0, \text{ the relative density} \end{array} \right\} \tag{2.10}$$

The speed of sound, a, is a function of temperature, T, such that,

$$a = \sqrt{\gamma R T_0 \theta} = 340.294\sqrt{\theta} \text{ m/s} \tag{2.17}$$

as a true speed of sound, or

$$a_e = \sqrt{\gamma R T_0 \theta \sigma} = 340.294\sqrt{\delta}\, \text{m/s}$$

as an equivalent speed of sound, thus,

$$V = \text{TAS} = Ma$$

and

$$V_e = \text{EAS} = Ma_e.$$

Table C1 The International Standard Atmosphere model. Metric height

Height H, m	All atmospheres		International Standard Atmosphere only				Speed of sound, a	
	Pressure p, N/m^2	Relative pressure, δ	Tempera-ture T, K	Relative tempera-ture, θ	Density ρ, kg/m^3	Relative density, σ	m/s (TAS)	m/s (EAS)
0	101 325	1.00000	288.15	1.00000	1.22500	1.00000	340.29	340.29
200	98 945	0.97651	286.85	0.99549	1.20165	0.98094	339.53	336.27
400	96 611	0.95348	285.55	0.99098	1.17864	0.96216	338.76	332.28
600	94 322	0.93088	284.25	0.98647	1.15598	0.94365	337.98	328.32
800	92 076	0.90872	282.95	0.98195	1.13364	0.92542	337.21	324.39
1000	89 874	0.88699	281.65	0.97744	1.11164	0.90746	336.43	320.49
1200	87 715	0.86568	280.35	0.97293	1.08997	0.88977	335.66	316.62
1400	85 599	0.84479	279.05	0.96842	1.06862	0.87234	334.88	312.77
1600	83 523	0.82431	277.75	0.96391	1.04759	0.85518	334.10	308.96
1800	81 489	0.80423	276.45	0.95940	1.02688	0.83827	333.31	305.17
2000	79 495	0.78455	275.15	0.95488	1.00649	0.82162	332.53	301.42
2200	77 541	0.76527	273.85	0.95037	0.98640	0.80523	331.74	297.69
2400	75 625	0.74636	272.55	0.94586	0.96663	0.78908	330.95	293.99
2600	73 749	0.72784	271.25	0.94135	0.94716	0.77319	330.16	290.32
2800	71 910	0.70969	269.95	0.93684	0.92799	0.75754	329.37	286.68
3000	70 108	0.69191	268.65	0.93233	0.90912	0.74214	328.58	283.06
3200	68 343	0.67450	267.35	0.92782	0.89054	0.72697	327.78	279.48
3400	66 615	0.65744	266.05	0.92330	0.87226	0.71205	326.98	275.92
3600	64 922	0.64073	264.75	0.91879	0.85426	0.69736	326.18	272.39
3800	63 263	0.62436	263.45	0.91428	0.83655	0.68290	325.38	268.89
4000	61 640	0.60834	262.15	0.90977	0.81912	0.66867	324.58	265.42
4200	60 050	0.59265	260.85	0.90526	0.80197	0.65467	323.77	261.97
4400	58 494	0.57729	259.55	0.90075	0.78510	0.64090	322.97	258.55
4600	56 970	0.56225	258.25	0.89623	0.76850	0.62735	322.16	255.16
4800	55 479	0.54753	256.95	0.89172	0.75217	0.61402	321.34	251.80
5000	54 019	0.53313	255.65	0.88721	0.73611	0.60091	320.53	248.47
5200	52 591	0.51903	254.35	0.88270	0.72031	0.58801	319.71	245.16
5400	51 194	0.50524	253.05	0.87819	0.70477	0.57532	318.90	241.88
5600	49 826	0.49175	251.75	0.87368	0.68949	0.56285	318.08	238.63
5800	48 489	0.47855	250.45	0.86917	0.67446	0.55058	317.25	235.41
6000	47 181	0.46564	249.15	0.86465	0.65969	0.53852	316.43	232.21
6200	45 901	0.45301	247.85	0.86014	0.64516	0.52666	315.60	229.04
6400	44 650	0.44066	246.55	0.85563	0.63088	0.51501	314.77	225.89
6600	43 426	0.42858	245.25	0.85112	0.61685	0.50355	313.94	222.78
6800	42 230	0.41677	243.95	0.84661	0.60305	0.49229	313.11	219.69
7000	41 060	0.40523	242.65	0.84210	0.58949	0.48122	312.27	216.62
7200	39 917	0.39395	241.35	0.83758	0.57617	0.47034	311.44	213.59
7400	38 800	0.38293	240.05	0.83307	0.56308	0.45965	310.60	210.58
7600	37 708	0.37215	238.75	0.82856	0.55021	0.44915	309.75	207.59
7800	36 641	0.36162	237.45	0.82405	0.53757	0.43884	308.91	204.64
8000	35 599	0.35134	236.15	0.81954	0.52516	0.42870	308.06	201.70
8200	34 581	0.34129	234.85	0.81503	0.51296	0.41875	307.21	198.80
8400	33 587	0.33148	233.55	0.81052	0.50099	0.40897	306.36	195.92
8600	32 616	0.32189	232.25	0.80600	0.48923	0.39937	305.51	193.07
8800	31 668	0.31254	230.95	0.80149	0.47768	0.38994	304.65	190.24
9000	30 742	0.30340	229.65	0.79698	0.46634	0.38069	303.79	187.44
9200	29 838	0.29448	228.35	0.79247	0.45521	0.37160	302.93	184.66
9400	28 956	0.28577	227.05	0.78796	0.44428	0.36268	302.07	181.91
9600	28 095	0.27728	225.75	0.78345	0.43355	0.35392	301.20	179.19
9800	27 255	0.26899	224.45	0.77893	0.42303	0.34533	300.33	176.49
10 000	26 436	0.26090	223.15	0.77442	0.41270	0.33690	299.46	173.82
10 200	25 636	0.25301	221.85	0.76991	0.40256	0.32862	298.59	171.17
10 400	24 856	0.24531	220.55	0.76540	0.39262	0.32050	297.71	168.54
10 600	24 096	0.23781	219.25	0.76089	0.38286	0.31254	296.83	165.95

Table C1 The International Standard Atmosphere model. Metric height (contd)

Height H, m	All atmospheres		International Standard Atmosphere only					
	Pressure p, N/m^2	Relative pressure, δ	Tempera- ture T, K	Relative tempera- ture, θ	Density ρ, kg/m^3	Relative density, σ	Speed of sound, a m/s (TAS)	m/s (EAS)
10 800	23 354	0.23049	217.95	0.75638	0.37329	0.30473	295.95	163.37
11 000	**22 632**	**0.22336**	**216.65**	**0.75187**	**0.36391**	**0.29707**	**295.07**	**160.82**
11 200	21 929	0.21643	216.65	0.75187	0.35262	0.28785	295.07	158.31
11 400	21 249	0.20971	216.65	0.75187	0.34167	0.27892	295.07	155.83
11 600	20 589	0.20320	216.65	0.75187	0.33106	0.27026	295.07	153.40
11 800	19 950	0.19689	216.65	0.75187	0.32079	0.26187	295.07	151.00
12 000	19 330	0.19078	216.65	0.75187	0.31083	0.25374	295.07	148.63
12 200	18 730	0.18485	216.65	0.75187	0.30118	0.24586	295.07	146.31
12 400	18 149	0.17911	216.65	0.75187	0.29183	0.23823	295.07	144.02
12 600	17 585	0.17355	216.65	0.75187	0.28277	0.23083	295.07	141.77
12 800	17 039	0.16817	216.65	0.75187	0.27399	0.22366	295.07	139.55
13 000	16 510	0.16294	216.65	0.75187	0.26548	0.21672	295.07	137.36
13 200	15 998	0.15789	216.65	0.75187	0.25724	0.20999	295.07	135.22
13 400	15 501	0.15298	216.65	0.75187	0.24925	0.20347	295.07	133.10
13 600	15 020	0.14823	216.65	0.75187	0.24152	0.19716	295.07	131.02
13 800	14 554	0.14363	216.65	0.75187	0.23402	0.19104	295.07	128.97
14 000	14 102	0.13917	216.65	0.75187	0.22675	0.18510	295.07	126.95
14 200	13 664	0.13485	216.65	0.75187	0.21971	0.17936	295.07	124.96
14 400	13 240	0.13067	216.65	0.75187	0.21289	0.17379	295.07	123.01
14 600	12 829	0.12661	216.65	0.75187	0.20628	0.16839	295.07	121.08
14 800	12 430	0.12268	216.65	0.75187	0.19988	0.16317	295.07	119.19
15 000	12 045	0.11887	216.65	0.75187	0.19367	0.15810	295.07	117.33
15 200	11 671	0.11518	216.65	0.75187	0.18766	0.15319	295.07	115.49
15 400	11 308	0.11160	216.65	0.75187	0.18183	0.14844	295.07	113.68
15 600	10 957	0.10814	216.65	0.75187	0.17619	0.14383	295.07	111.90
15 800	10 617	0.10478	216.65	0.75187	0.17072	0.13936	295.07	110.15
16 000	10 287	0.10153	216.65	0.75187	0.16542	0.13504	295.07	108.43
16 200	9968	0.09838	216.65	0.75187	0.16028	0.13084	295.07	106.73
16 400	9659	0.09532	216.65	0.75187	0.15531	0.12678	295.07	105.06
16 600	9359	0.09236	216.65	0.75187	0.15049	0.12285	295.07	103.42
16 800	9068	0.08950	216.65	0.75187	0.14581	0.11903	295.07	101.80
17 000	8787	0.08672	216.65	0.75187	0.14129	0.11534	295.07	100.21
17 200	8514	0.08403	216.65	0.75187	0.13690	0.11176	295.07	98.64
17 400	8250	0.08142	216.65	0.75187	0.13265	0.10829	295.07	97.10
17 600	7993	0.07889	216.65	0.75187	0.12853	0.10492	295.07	95.58
17 800	7745	0.07644	216.65	0.75187	0.12454	0.10167	295.07	94.08
18 000	7505	0.07407	216.65	0.75187	0.12068	0.09851	295.07	92.61
18 200	7272	0.07177	216.65	0.75187	0.11693	0.09545	295.07	91.16
18 400	7046	0.06954	216.65	0.75187	0.11330	0.09249	295.07	89.74
18 600	6827	0.06738	216.65	0.75187	0.10978	0.08962	295.07	88.33
18 800	6615	0.06529	216.65	0.75187	0.10637	0.08684	295.07	86.95
19 000	6410	0.06326	216.65	0.75187	0.10307	0.08414	295.07	85.59
19 200	6211	0.06130	216.65	0.75187	0.09987	0.08153	295.07	84.25
19 400	6018	0.05939	216.65	0.75187	0.09677	0.07900	295.07	82.93
19 600	5831	0.05755	216.65	0.75187	0.09377	0.07654	295.07	81.64
19 800	5650	0.05576	216.65	0.75187	0.09086	0.07417	295.07	80.36
20 000	**5475**	**0.05403**	**216.65**	**0.75187**	**0.08803**	**0.07187**	**295.07**	**79.10**
20 200	5305	0.05236	216.85	0.75256	0.08522	0.06957	295.21	77.86
20 400	5140	0.05073	217.05	0.75325	0.08250	0.06735	295.34	76.65
20 600	4981	0.04916	217.25	0.75395	0.07987	0.06520	295.48	75.45
20 800	4827	0.04764	217.45	0.75464	0.07733	0.06313	295.61	74.27
21 000	4678	0.04617	217.65	0.75534	0.07487	0.06112	295.75	73.12
21 200	4533	0.04474	217.85	0.75603	0.07249	0.05918	295.89	71.98
21 400	4393	0.04336	218.05	0.75672	0.07019	0.05730	296.02	70.86

Table C1 The International Standard Atmosphere model. Metric height (contd)

	All atmospheres		International Standard Atmosphere only					
Height H, m	Pressure p, N/m^2	Relative pressure, δ	Temperature T, K	Relative temperature, θ	Density ρ, kg/m^3	Relative density, σ	Speed of sound, a m/s (TAS)	m/s (EAS)
21 600	4258	0.04202	218.25	0.75742	0.06796	0.05548	296.16	69.76
21 800	4127	0.04073	218.45	0.75811	0.06581	0.05372	296.29	68.67
22 000	4000	0.03947	218.65	0.75881	0.06373	0.05202	296.43	67.61
22 200	3877	0.03826	218.85	0.75950	0.06171	0.05037	296.56	66.56
22 400	3758	0.03708	219.05	0.76019	0.05976	0.04878	296.70	65.53
22 600	3642	0.03595	219.25	0.76089	0.05787	0.04724	296.83	64.52
22 800	3530	0.03484	219.45	0.76158	0.05604	0.04575	296.97	63.52
23 000	3422	0.03378	219.65	0.76228	0.05428	0.04431	297.11	62.54
23 200	3318	0.03274	219.85	0.76297	0.05257	0.04291	297.24	61.57
23 400	3216	0.03174	220.05	0.76366	0.05091	0.04156	297.38	60.63
23 600	3118	0.03077	220.25	0.76436	0.04931	0.04026	297.51	59.69
23 800	3023	0.02983	220.45	0.76505	0.04776	0.03899	297.65	58.77
24 000	2930	0.02892	220.65	0.76575	0.04626	0.03777	297.78	57.87
24 200	2841	0.02804	220.85	0.76644	0.04481	0.03658	297.92	56.98
24 400	2755	0.02718	221.05	0.76714	0.04341	0.03544	298.05	56.11
24 600	2671	0.02636	221.25	0.76783	0.04205	0.03433	298.19	55.25
24 800	2590	0.02556	221.45	0.76852	0.04074	0.03325	298.32	54.40
25 000	2511	0.02478	221.65	0.76922	0.03946	0.03222	298.45	53.57
25 200	2435	0.02403	221.85	0.76991	0.03823	0.03121	298.59	52.75
25 400	2361	0.02330	222.05	0.77061	0.03704	0.03024	298.72	51.94
25 600	2289	0.02259	222.25	0.77130	0.03588	0.02929	298.86	51.15
25 800	2220	0.02191	222.45	0.77199	0.03477	0.02838	298.99	50.37
26 000	2153	0.02125	222.65	0.77269	0.03369	0.02750	299.13	49.60
26 200	2088	0.02061	222.85	0.77338	0.03264	0.02664	299.26	48.85
26 400	2025	0.01998	223.05	0.77408	0.03163	0.02582	299.40	48.11
26 600	1964	0.01938	223.25	0.77477	0.03064	0.02502	299.53	47.37
26 800	1905	0.01880	223.45	0.77546	0.02969	0.02424	299.66	46.66
27 000	1847	0.01823	223.65	0.77616	0.02877	0.02349	299.80	45.95
27 200	1792	0.01768	223.85	0.77685	0.02788	0.02276	299.93	45.25
27 400	1738	0.01715	224.05	0.77755	0.02702	0.02206	300.07	44.57
27 600	1686	0.01664	224.25	0.77824	0.02619	0.02138	300.20	43.89
27 800	1635	0.01614	224.45	0.77893	0.02538	0.02072	300.33	43.23
28 000	1586	0.01565	224.65	0.77963	0.02460	0.02008	300.47	42.58
28 200	1539	0.01519	224.85	0.78032	0.02384	0.01946	300.60	41.93
28 400	1493	0.01473	225.05	0.78102	0.02311	0.01886	300.74	41.30
28 600	1448	0.01429	225.25	0.78171	0.02239	0.01828	300.87	40.68
28 800	1405	0.01386	225.45	0.78240	0.02171	0.01772	301.00	40.07
29 000	1363	0.01345	225.65	0.78310	0.02104	0.01718	301.14	39.47
29 200	1322	0.01305	225.85	0.78379	0.02039	0.01665	301.27	38.87
29 400	1283	0.01266	226.05	0.78449	0.01977	0.01614	301.40	38.29
29 600	1245	0.01228	226.25	0.78518	0.01916	0.01564	301.54	37.72
29 800	1208	0.01192	226.45	0.78588	0.01858	0.01517	301.67	37.15
30 000	1172	0.01156	226.65	0.78657	0.01801	0.01470	301.80	36.59
30 200	1137	0.01122	226.85	0.78726	0.01746	0.01425	301.94	36.05
30 400	1103	0.01089	227.05	0.78796	0.01693	0.01382	302.07	35.51
30 600	1071	0.01057	227.25	0.78865	0.01641	0.01340	302.20	34.98
30 800	1039	0.01025	227.45	0.78935	0.01591	0.01299	302.33	34.46
31 000	1008	0.00995	227.65	0.79004	0.01543	0.01259	302.47	33.94
31 200	978	0.00966	227.85	0.79073	0.01496	0.01221	302.60	33.44
31 400	949	0.00937	228.05	0.79143	0.01450	0.01184	302.73	32.94
31 600	921	0.00909	228.25	0.79212	0.01406	0.01148	302.87	32.45
31 800	894	0.00883	228.45	0.79282	0.01364	0.01113	303.00	31.97
32 000	**868**	**0.00857**	**228.65**	**0.79351**	**0.01322**	**0.01079**	**303.13**	**31.49**

Table C2 The International Standard Atmosphere model. Imperial height

Height H, ft	Pressure p, N/m^2	Relative pressure, δ	Tempera- ture T, K	Relative tempera- ture, θ	Density ρ, kg/m^3	Relative density, σ	Speed of sound, a m/s (TAS)	m/s (EAS)
	All atmospheres		International Standard Atmosphere only					
0	101 325	1.00000	288.15	1.00000	1.22500	1.00000	340.29	340.29
1000	97 717	0.96439	286.17	0.99312	1.18955	0.97106	339.12	334.18
2000	94 213	0.92981	284.19	0.98625	1.15490	0.94277	337.95	328.13
3000	90 812	0.89624	282.21	0.97937	1.12102	0.91512	336.77	322.16
4000	87 510	0.86366	280.23	0.97250	1.08790	0.88808	335.58	316.25
5000	84 307	0.83205	278.24	0.96562	1.05554	0.86167	334.39	310.40
6000	81 199	0.80138	276.26	0.95875	1.02392	0.83586	333.20	304.63
7000	78 185	0.77163	274.28	0.95187	0.99304	0.81064	332.00	298.92
8000	75 262	0.74278	272.30	0.94500	0.96287	0.78601	330.80	293.28
9000	72 428	0.71481	270.32	0.93812	0.93340	0.76196	329.60	287.71
10 000	69 681	0.68770	268.34	0.93124	0.90463	0.73847	328.39	282.20
11 000	67 019	0.66143	266.36	0.92437	0.87655	0.71555	327.17	276.76
12 000	64 440	0.63598	264.38	0.91749	0.84913	0.69317	325.95	271.38
13 000	61 942	0.61132	262.39	0.91062	0.82238	0.67133	324.73	266.07
14 000	59 523	0.58745	260.41	0.90374	0.79627	0.65002	323.50	260.82
15 000	57 181	0.56434	258.43	0.89687	0.77081	0.62923	322.27	255.64
16 000	54 915	0.54197	256.45	0.88999	0.74597	0.60896	321.03	250.52
17 000	52 721	0.52032	254.47	0.88312	0.72175	0.58919	319.79	245.46
18 000	50 599	0.49938	252.49	0.87624	0.69814	0.56991	318.54	240.47
19 000	48 547	0.47912	250.51	0.86936	0.67512	0.55112	317.29	235.55
20 000	46 563	0.45954	248.53	0.86249	0.65269	0.53281	316.03	230.68
21 000	44 645	0.44061	246.54	0.85561	0.63083	0.51496	314.77	225.88
22 000	42 791	0.42231	244.56	0.84874	0.60953	0.49758	313.50	221.14
23 000	41 000	0.40464	242.58	0.84186	0.58879	0.48065	312.23	216.47
24 000	39 270	0.38757	240.60	0.83499	0.56860	0.46416	310.95	211.85
25 000	37 600	0.37109	238.62	0.82811	0.54894	0.44811	309.67	207.30
26 000	35 988	0.35518	236.64	0.82123	0.52980	0.43249	308.38	202.80
27 000	34 433	0.33982	234.66	0.81436	0.51118	0.41729	307.09	198.37
28 000	32 932	0.32501	232.68	0.80748	0.49306	0.40250	305.79	194.00
29 000	31 484	0.31073	230.70	0.80061	0.47544	0.38811	304.48	189.69
30 000	30 089	0.29696	228.71	0.79373	0.45830	0.37413	303.17	185.44
31 000	28 744	0.28368	226.73	0.78686	0.44164	0.36053	301.86	181.25
32 000	27 448	0.27089	224.75	0.77998	0.42545	0.34731	300.54	177.11
33 000	26 200	0.25858	222.77	0.77311	0.40972	0.33446	299.21	173.04
34 000	24 998	0.24672	220.79	0.76623	0.39443	0.32199	297.87	169.03
35 000	23 842	0.23530	218.81	0.75935	0.37959	0.30987	296.54	165.07
36 000	22 729	0.22432	216.83	0.75248	0.36517	0.29810	295.19	161.17
36 089	**22 632**	**0.22336**	**216.65**	**0.75187**	**0.36391**	**0.29707**	**295.07**	**160.83**
37 000	21 663	0.21379	216.65	0.75187	0.34833	0.28435	295.07	157.34
38 000	20 646	0.20376	216.65	0.75187	0.33198	0.27101	295.07	153.61
39 000	19 677	0.19420	216.65	0.75187	0.31641	0.25829	295.07	149.96
40 000	18 754	0.18509	216.65	0.75187	0.30156	0.24617	295.07	146.40
41 000	17 874	0.17640	216.65	0.75187	0.28741	0.23462	295.07	142.92
42 000	17 035	0.16812	216.65	0.75187	0.27392	0.22361	295.07	139.53
43 000	16 236	0.16023	216.65	0.75187	0.26107	0.21311	295.07	136.22
44 000	15 474	0.15271	216.65	0.75187	0.24881	0.20311	295.07	132.98
45 000	14 748	0.14555	216.65	0.75187	0.23714	0.19358	295.07	129.82
46 000	14 056	0.13872	216.65	0.75187	0.22601	0.18450	295.07	126.74
47 000	13 396	0.13221	216.65	0.75187	0.21540	0.17584	295.07	123.73
48 000	12 767	0.12600	216.65	0.75187	0.20530	0.16759	295.07	120.79
49 000	12 168	0.12009	216.65	0.75187	0.19566	0.15972	295.07	117.93
50 000	11 597	0.11446	216.65	0.75187	0.18648	0.15223	295.07	115.13
51 000	11 053	0.10908	216.65	0.75187	0.17773	0.14509	295.07	112.39
52 000	10 534	0.10397	216.65	0.75187	0.16939	0.13828	295.07	109.72

Table C2 The International Standard Atmosphere model. Imperial height (contd)

Height H, ft	All atmospheres		International Standard Atmosphere only					
	Pressure p, N/m²	Relative pressure, δ	Temperature T, K	Relative temperature, θ	Density ρ, kg/m³	Relative density, σ	Speed of sound, a m/s (TAS)	m/s (EAS)
53 000	10 040	0.09909	216.65	0.75187	0.16144	0.13179	295.07	107.12
54 000	9568.9	0.09444	216.65	0.75187	0.15386	0.12560	295.07	104.57
55 000	9119.8	0.09001	216.65	0.75187	0.14664	0.11971	295.07	102.09
56 000	8691.9	0.08578	216.65	0.75187	0.13976	0.11409	295.07	99.67
57 000	8284.0	0.08176	216.65	0.75187	0.13320	0.10874	295.07	97.30
58 000	7895.2	0.07792	216.65	0.75187	0.12695	0.10364	295.07	94.99
59 000	7524.8	0.07426	216.65	0.75187	0.12100	0.09877	295.07	92.73
60 000	7171.6	0.07078	216.65	0.75187	0.11532	0.09414	295.07	90.53
61 000	6835.1	0.06746	216.65	0.75187	0.10991	0.08972	295.07	88.38
62 000	6514.4	0.06429	216.65	0.75187	0.10475	0.08551	295.07	86.28
63 000	6208.7	0.06127	216.65	0.75187	0.09983	0.08150	295.07	84.24
64 000	5917.3	0.05840	216.65	0.75187	0.09515	0.07767	295.07	82.24
65 000	5639.6	0.05566	216.65	0.75187	0.09068	0.07403	295.07	80.28
65 617	**5474.8**	**0.05403**	**216.65**	**0.75187**	**0.08803**	**0.07186**	**295.07**	**79.10**
66 000	5375.0	0.05305	216.77	0.75227	0.08638	0.07052	295.15	78.38
67 000	5123.1	0.05056	217.07	0.75333	0.08222	0.06712	295.36	76.52
68 000	4883.3	0.04819	217.38	0.75439	0.07826	0.06389	295.56	74.71
69 000	4655.0	0.04594	217.68	0.75544	0.07450	0.06081	295.77	72.94
70 000	4437.7	0.04380	217.99	0.75650	0.07092	0.05789	295.98	71.22
71 000	4230.8	0.04176	218.29	0.75756	0.06752	0.05512	296.18	69.54
72 000	4033.9	0.03981	218.60	0.75862	0.06429	0.05248	296.39	67.90
73 000	3846.3	0.03796	218.90	0.75968	0.06121	0.04997	296.60	66.30
74 000	3667.8	0.03620	219.21	0.76073	0.05829	0.04758	296.80	64.74
75 000	3497.7	0.03452	219.51	0.76179	0.05551	0.04531	297.01	63.22
76 000	3335.8	0.03292	219.81	0.76285	0.05287	0.04316	297.22	61.74
77 000	3181.5	0.03140	220.12	0.76391	0.05035	0.04110	297.42	60.30
78 000	3034.6	0.02995	220.42	0.76496	0.04796	0.03915	297.63	58.89
79 000	2894.7	0.02857	220.73	0.76602	0.04569	0.03729	297.83	57.52
80 000	2761.4	0.02725	221.03	0.76708	0.04352	0.03553	298.04	56.18
81 000	2634.4	0.02600	221.34	0.76814	0.04146	0.03385	298.25	54.87
82 000	2513.4	0.02481	221.64	0.76920	0.03950	0.03225	298.45	53.60
83 000	2398.1	0.02367	221.95	0.77025	0.03764	0.03073	298.66	52.35
84 000	2288.3	0.02258	222.25	0.77131	0.03587	0.02928	298.86	51.14
85 000	2183.6	0.02155	222.56	0.77237	0.03418	0.02790	299.07	49.96
86 000	2083.9	0.02057	222.86	0.77343	0.03257	0.02659	299.27	48.80
87 000	1988.8	0.01963	223.17	0.77448	0.03105	0.02534	299.47	47.67
88 000	1898.2	0.01873	223.47	0.77554	0.02959	0.02416	299.68	46.58
89 000	1811.8	0.01788	223.78	0.77660	0.02821	0.02302	299.88	45.50
90 000	1729.5	0.01707	224.08	0.77766	0.02689	0.02195	300.09	44.46
91 000	1651.0	0.01629	224.39	0.77872	0.02563	0.02092	300.29	43.44
92 000	1576.2	0.01556	224.69	0.77977	0.02444	0.01995	300.50	42.44
93 000	1504.8	0.01485	225.00	0.78083	0.02330	0.01902	300.70	41.47
94 000	1436.8	0.01418	225.30	0.78189	0.02222	0.01814	300.90	40.52
95 000	1372.0	0.01354	225.61	0.78295	0.02119	0.01729	301.11	39.60
96 000	1310.1	0.01293	225.91	0.78400	0.02020	0.01649	301.31	38.69
97 000	1251.1	0.01235	226.22	0.78506	0.01927	0.01573	301.51	37.81
98 000	1194.9	0.01179	226.52	0.78612	0.01838	0.01500	301.72	36.95
99 000	1141.2	0.01126	226.83	0.78718	0.01753	0.01431	301.92	36.11
100 000	1090.1	0.01076	227.13	0.78824	0.01672	0.01365	302.12	35.30
101 000	1041.2	0.01028	227.43	0.78929	0.01595	0.01302	302.32	34.50
102 000	994.7	0.00982	227.74	0.79035	0.01522	0.01242	302.53	33.72
103 000	950.2	0.00938	228.04	0.79141	0.01452	0.01185	302.73	32.95
104 000	907.9	0.00896	228.35	0.79247	0.01385	0.01131	302.93	32.21
104 987	**867.9**	**0.00857**	**228.65**	**0.79351**	**0.01322**	**0.01079**	**303.13**	**31.49**

Index

Page numbers in **bold** are definitions, page numbers in *italic* are figures

Acceleration 85, 87, 96, 105
    lateral **125**
    linear **125**
    normal **126**, 127
Accelerate–stop distance **174**
Accelerate–stop distance available,
    ASDA **172**, **218**
Advance ratio **58**, 95
Aerodynamic efficiency, *see* Lift-drag ratio
Airborne re-fuelling 3, 193
Aircraft force system 39, **40–2**, **268–72**
    aerodynamic 39, 41, 42–53, **269**
    gravitational 39, 41, **269**
    inertial 39, 41, **268**
    propulsive 39, **270–2**
Aircraft mission profile **2–5**
Air data **24–36**
    measurement of 24–35
    systems 24, 216
    system pressure error 35, 86
Air data computer 25, 35–6
Airflow direction detector ADD 24, 35
Airspeed indicator 30–2
Airspeed 28
    calibrated **30**, 31, 86
    equivalent **30**, 32, 37, 86
    indicated **31**
    measurement of **29–32**
    true **32**, 37, 86
Air superiority 1, 138, 198
Air temperature sensor 24, 33
    recovery factor 34
Air traffic control 137

Airworthiness
    requirements and regulations 1, 2, 201,
        228
    certification, *see* Certification
All engines operating flight 172
Alternate airfield, *see* Diversion
Altimeter 17, 26
    full law calibration 30
    simple law calibration 32
    scale altitude error **30**, 31, 86, 116
Altitude **26**
Angle of attack 42, 45, 263
    measurement of 35
    post-stall 136
    stalling **43**
    zero-lift 42
Aspect ratio 42
Asymmetric flight 46, 110
Atmosphere 5, 141
    datum 16
    density 11, 274
    design 6, **19**, *20*, 141
    international standard 6, 14–24, **16**, 141,
        274–81
    off-standard 19
    pressure 11, 274
    relative properties of **21–3**, 276
    state of 40, 141
    structure of 16
    temperature 11, 274
Attitude 85, 99, 231, 233, 262
    pitch 101, 113, **266**
    bank 125, 131, **267**

Axis systems 39, **261–7**
  body 54, **262**
  earth 41, **262**
  stability **263**
  thrust 54
  velocity 39, 41, **264**
  wind 41, **263**

Balanced field length **178**
Baulked landing 190
Block performance 3, 201–2
  block fuel 199, 202
  block time 199, 202
Boundary layer control 108, 109
Breguet range, *see* Cruise–Climb
By-pass ratio 56

Cabin pressure 85, 99, 245
Ceiling height 60, 170
  absolute 187
  en-route cruise 188, 225
Certification 6, 141, 198, **199**
Certificate of airworthiness 168, 170, 199,
    215
  type certificate 6
  type record 7
Clearway **171**, 218
Climb 85
  after take-off 137, 181, 233, 236
  first segment **184**, 222
  fourth segment **187**, 223
  gradient 37, **87**, 88, 149, 181
  gross gradient 183, 214
  minimum fuel 97–8
  net gradient 183, 187, 214
  percentage gradient **87**
  rate 87, 125, 149, 181
  second segment **185**, 222
  third segment **186**, 223
  total energy 87
Climb performance 88–97
  combat aircraft 104
  en-route 224, 238
  measurement of 95–6, 149–52, 158, 164–5
  mixed powerplant 94–5
  operational 96–8
  power producing engines 91–4
  thrust producing engines 88–91
  wind, effect of 102–4
Compressibility 45
Continued take-off distance 175

Critical engine **183**
Critical field length **178**
Critical speeds 216, 219
Cruise–Climb **67**, 81, 231
Cruising performance 64–83
  optimum altitude 78–9
  power producing engines 79–81
  thrust producing engines 65–79
  mixed powerplants 81–2
  measurement of 147–9, 156–7, 160–4
Cruise methods
  constant AoA, mach 67–71
  constant AoA, altitude 71–2
  constant mach, altitude 72–4

Decision speed 111, 173, **177**
Decision distance 178
Decision height 191, 212
Density
  ISA reference 16
  relative **21**,
Density altitude **23–4**
Descent 85, 98–102
  approach 99–101
  final approach 101, 249
  gradient 98–9
  rate 100, 125
Descent performance 244
  emergency 99, 101
  operational 98–102
  wind, effect of 102–4
Design speeds
  cruise, V_C/M_C **129**
  diving, V_D **129**
  manoeuvring, V_A **129**, 133
  maximum, V_H **129**
Dimensional analysis 143
Dimensionless
  thrust, power **61**
  climb rate 90
Discontinued approach **191**, 234
Disposable load 193, 203
Diversion 3, 192, 212, 253
Drag 46–53
  area 47
  characteristic 74
  intake momentum **54**, 272
  lift-dependent 46, 48–9, *49*
  lift-independent 46, 47–8, *48*, 101, 117
  parabolic polar 46, *50,* 72
  pressure 47

profile 47
  standard momentum **54**, 272
  surface friction 47
  volume dependent wave 46, **49**, *50*
  vortex, *see* lift dependent
  zero-lift, *see* Lift-independent
Drag coefficient **46**
  lift dependent 46
  zero-lift, lift independent 46, *47*, 101
Drag-minimum drag ratio **61**
Drift-down **188**, 225

Endurance 2, 64
  factor **69**
  function **69**
Energy height 104, 130
Energy, kinetic, potential 104
Engine failure 107
  accountability **108**, 109
  speed **174**
En-route performance 169
  climb **187**
  descent 100, **187**
  terrain clearance 170
  turns 137
Emergency distance, *see* Accelerate–stop
    distance
Equation of state **11,** 16, 274
Equations of motion for performance **39**,
    88, **268–73**
  dimensionless **61**
  general 272
  simplified 27
Equivalent shaft horsepower 82
Equivalent-weight, e-W
  climb-e-W 156
  power-e-W 155
  speed-e-W 155
  see also Performance, measurement
Excess thrust 4, **40**, 57, 85, 88, 96
Excess power **59**, 96

Ferry range 20
Field length
  balanced **178**
  critical **178**
  limit 219
  unbalanced **178**
Flaps, flap systems 100, 101, 233
  leading edge 43, 44
  slots, slats 44

trailing edge 43, 44
Flare 101, 116, 191
Fleet mean performance 200
Flight level **27**
Flight management system, FMS 100, 213
Flight manual, *see* Performance manual
Flight path 37, 85, 102, 140, 142–3
  reconstruction 153
Flight path stability 74, 76, 77, **98**, 101, 117,
    234
Flight plan 3, 140, 167, 192
Flight planning 3, 169–70, 198, **201**,
    213–26
Flight safety 96, 137, 141, 168
Flight variables, *see* Performance variables
Fuel
  contingency 64, 194, 204, 252
  diversion 193, **195**
  flow laws 65–6, 147
  mass flow **64**
  maximum load **204**
  reserves 193, 194, 195, 204, 234, 252
  trip **193**, 204
Fuel planning 7, 168, 192, 195, 215
Fuel tankering 195, 206
Fuel ratio **67**

Gas constant **11**, 274
Generalized performance
  climb 61, 92, 151
  characteristic 145
  specific air range 148
Go-around 101
Gliding flight 93
Gravitational acceleration 17
  standard value of **18**
Gross performance 170

Height **26**
  geopotential **18**, 28
  measurement of 26–8
  pressure 18, **19–21**, 27, 275
  scales 17–18
  true 17
High-lift devices 108, 117
Holding pattern 137, 194

International Standard Atmosphere,
    see Atmosphere

Kinetheodolite 153

Landing 116–22
  approach 117, 233
  baulked 190, 191
  discontinued approach 191
  flare 116, 117
  nose-down speed 118
  reference speed 119
  touchdown speed 116, 118
Landing distance 107, 117, 119
  airborne 107, 118, 119
  available, LDA **190**
  braking 118
  free-roll 118
  ground run 107, 118, 119
  required, LDR **191**
Landing performance 116–22, 169, 189–92, 225
  measurement of 152–5, 158–9, 165
Level acceleration 96
Lift 42–5
  characteristic 42, *43*, *44*
  curve slope 42
  effect of Mach number 45
  force 42
  power induced 109
Lift coefficient 42
  maximum 43
  minimum drag **60**
Lift–drag ratio **60**, 98
Lift-off speed 110, 114, 115
Load factor **126**, 127

Mach meter 33
Mach number 28, 37, 42, 86
  critical 49, 231
  effect on lift 45
  measurement of 32–3
Manoeuvring flight 125–7
  'cobra' 136
  co-ordinated 126
  load limits 132
  longitudinal 130
  pull-up or loop 134–7
  speed 133
  turning 131–4
Manoeuvre envelope 127–9, *128*, 132
  airspeed boundaries 129
  load limits 128, 132
  structural boundaries 128
Maximum endurance speed 137
Maximum refusal speed 173, **174**

Maximum take-off weight, **MTOW** 8, 154, 170, 177
  authorized, MTWA 168, 186
Maximum landing weight 170, 189, 195, 204
Mesosphere 13
Minimum continue speed 173, **175**
Minimum control speed 110, 111, 178, 181
Minimum drag speed **51**, *51*, 60, 66, 69, 80, 91, 134, 150, 231
Minimum unstick speed 111
Minimum power speed **52**, *52*, 60, 80, 91, 92
Minimum sink rate 91
Missed approach 190
Mission
  civil *3*, 63
  military *4*, 63
  profile 2–5, 85
Modification 5, 199

n–V diagram 127
Net performance 170, 183
Net propulsive force 53, 271
Noise limitations 98, 108

Obstacle limitation surface 96, 101, **181**, 190
Obstacle clearance range **188**
One engine inoperative flight 97, 134, 137, 172, 181, 185–6
Operating data manual, *see* Performance manual
Operational analysis 198, **200**, 207–13
Operational performance 2, 198

Parametric performance analysis 143–55
Parametric performance variables 70, 144, 146
  form of 144
  groups 144
Partial climb 95
Path performance 67
Payload 2, 193
  maximum structural 204
Payload–range 201, **203–6**, *203*, 206, 254–8
Percentage gradient **87**
Performance
  fleet mean 200
  gross 170, 183, 214
  guaranteed minimum 214
  measured 170
  net 170, 183, 214, 216

Performance classification 168–9
Performance manual 6, 7, 103, 140, 149,
    201, 216
Performance measurement methods
    data reduction 159–65
    equivalent-weight 155–9
    parametric form 143–55
Performance planning 7, 167, 195
Performance scheduling 201
Performance summary 198, **201–6**, 242
    block fuel 199, *202*
    block time 199, *202*
    cruise *209*
    en-route climb *208*
    en route descent *209*
    payload–range 203–6
Performance variables, WAT 7, 40, 114,
    121, 140, 142, 153
Pitot-static system 24
Pitch
    rate of 126
Point performance 65, 159
Powerplant
    power producing 52, 58–60, 88
    thrust producing 51, 53–8, *53*, 88
    turbo-prop/mixed 94
Power
    coefficient **58**
    excess 92
    guaranteed minimum 170
    shaft 79, 81
    equivalent shaft 82
Prandtl–Glauert factor 45
Pressure
    dynamic **41**, 127
    impact **29**
    global distribution 12
    Pitot **29**
    relative **21**, 28
    ISA reference 16
    static 28, 29
    total **29**
Propeller
    efficiency 58, 80, 95
Pull-up manoeuvre 134

Range 2, 65
    factor **68**
    function **68**
    radius of action 2
Rapid flight planning 216

Recovery factor, (air thermometer)
Reference zero **182**, 184
Relative airspeed **61**, 68
Reverse thrust 118, 120
Reynolds number 41
Rotation speed 110, 113, **173**, 174, 178
Runway **171**
    contaminated **180**
    friction 115, 116, 122
    slope 116, 122, 218
    wet **180**

Safety factors 21, **178**
Scale–altitude error, *see* Altimeter
Sea level; mean sea-level 16, 27
Scheduled performance 107
Screen height 111, 113, 117, 119, 169, 182
Side force 45
Sideslip; sideslip angle 45
Simplified performance data 216
Space available, required 7, 167, **171–2**,
    190–1
Span efficiency factor 48
Specific air range **64–5**, 148
Specific climb **97**
Specific energy **130**
Specific endurance **64–5**
Specific excess power **104**, 130
Specific fuel consumption **65**, 80
Specific total energy **104**
Speed of sound **29**, 32
Speed ratio, *see* Relative speed
Stall
    angle of attack **43**
    buffet boundary 43, 129
    speed 93, 115, 129, 181
    post-stall manoeuvre 136
    warning 43
Step-climb **69**
Stratosphere **13**, 70, 274
Stopway **171**
System pressure error, *see* Air data

Take-off
    airborne distance 107, 111, **113**, 115, 154,
        235
    ground-run distance 107, 111, **112**, 115,
        154, 235
    distance required/available 110, 111
    flight path 169
    net flight path 170, **181**, 183, 221

Take-off (*cont.*)
  safety speed 111, 113, 114, 181, 186
  ski-jump 109
Take-off distance
  available, TODA **172**, 217–19
  required, TODR **172–80**
  see also Field strength
Take-off performance 110–16, 169, 171–80
  measurement of 152–5, 158–9, 165
Take-off run
  available, TORA **172**, 217–19
  required, TORR **172–80**
Temperature
  global distribution 11
  ISA reference 16
  lapse rate **13**
  measurement of 33–4
  relative **21**
  static 33
  total 33
Temperature–height profile 13, *15*, 17, *20*
Terminal flight phase **107**
  CTOL, conventional take-off and
    landing **108**
  RTOL, reduced take-off and landing **108**
  STOL, short take-off and landing **109**
  STOVL, short take-off vertical
    landing **109**
  VTOL, vertical take-off and landing **109**
Thrust
  excess 4, 104
  gross **53**, 270
  guaranteed minimum **170**, 213
  net **54**, 271
  static 58
Thrust-drag accounting 54

Thrust-power **58**, 89, 157
Thrust–weight ratio 4, 88, 104, 108
Transition 107, 111
Trim 64, 70, 263, 268
Tropopause **13**, 274
Troposphere **13**, 17, 70, 274
Turning performance 223
  rate of turn 125, 132
Type certificate/record, *see* Certificate of
  airworthiness

Unbalanced field length **178**

Vectored thrust 54, 109, 122, 126, 136
Vertical take-off/landing 122
Viffing 136

WAT limits 170, 216, *217*
  charts 182, 192
Weight
  aircraft prepared for service 193, 195, 203
  fuel 193
  maximum take-off, *see* Maximum take-off
    weight
  maximum landing, *see* Maximum landing
    weight
  operating empty 193, 203
  payload 195
  ramp 195
  zero-fuel 195, 204
Wind
  headwind 103, 115, 116, 121
  tailwind 103
  gradient 103
Wind-shear 103
Wing area 144

LaVergne, TN USA
02 December 2009

165603LV00004B/1/A

9 780340 758977